Innovations in Drug Resistance: A Commemorative Issue in Honor of Prof. Salvador Luria

Innovations in Drug Resistance: A Commemorative Issue in Honor of Prof. Salvador Luria

Guest Editors

Maria A. Bonifacio
Antonio Mazzocca
Anna Volpe

MDPI

Basel • Beijing • Wuhan • Barcelona • Belgrade • Novi Sad • Cluj • Manchester

Guest Editors
Maria A. Bonifacio
Department of Precision and Regenerative Medicine and Ionian Area
University of Bari Aldo Moro
Bari
Italy

Antonio Mazzocca
Interdisciplinary Department of Medicine
University of Bari
Bari
Italy

Anna Volpe
Department of Biomedical Sciences and Human Oncology
University of Bari Aldo Moro
Bari
Italy

Editorial Office
MDPI AG
Grosspeteranlage 5
4052 Basel, Switzerland

This is a reprint of the Special Issue, published open access by the journal *Biomedicines* (ISSN 2227-9059), freely accessible at: https://www.mdpi.com/journal/biomedicines/special_issues/Drug_Resistance_.

For citation purposes, cite each article independently as indicated on the article page online and as indicated below:

Lastname, A.A.; Lastname, B.B. Article Title. *Journal Name* **Year**, *Volume Number*, Page Range.

ISBN 978-3-7258-5249-9 (Hbk)
ISBN 978-3-7258-5250-5 (PDF)
https://doi.org/10.3390/books978-3-7258-5250-5

Contents

About the Editors

Maria A. Bonifacio

Maria A. Bonifacio works for the Department of Precision and Regenerative Medicine, University of Bari Aldo Moro, Italy. Her main research interests include the study of drug resistance in HIV infections and the development of smart drug delivery systems. She is dedicated to translational research, bridging fundamental biology and clinical practice. Her work spans from molecular diagnostics of immune-mediated diseases to biological characterization of composite biomaterials for tissue engineering applications. Maria completed the Specialization School in Clinical Pathology and Biochemistry in 2023 at the University of Bari Aldo Moro, Italy. In 2020 she became a collaborator of the National Interuniversity Consortium of Materials Science and Technology. In 2018 she received her European PhD in Chemical and Molecular Sciences at the same University. She was a visiting PhD at the Medical School and the School of System Engineering at Newcastle University, UK and was awarded the "Julia Polak European Doctorate Award" from the European Society for Biomaterials.

Antonio Mazzocca

Antonio Mazzocca earned his MD degree from the University of Bari School of Medicine, Italy, and his PhD in Pathophysiology from the University of Florence, Italy. He conducted postdoctoral research in the laboratory of Professor Alex Toker in the Department of Pathology at Beth Israel Deaconess Medical Center, Harvard Medical School, Boston, MA, under the supervision of Harold Dvorak. Subsequently, he joined the Department of Pathology at Vanderbilt University, Nashville, TN. Upon returning to Italy, he was appointed Assistant Professor of Pathology and Laboratory Medicine at the University of Bari Medical School and was later promoted to Full Professor of General Pathology. His scientific interests focus on tumor pathobiology, specifically the metabolic mechanisms that contribute to the origin and development of cancer from an evolutionary perspective. In this context, he has formulated and proposed a theory of cancer origin known as the Systemic–Evolutionary Theory of Cancer Origin (SETOC).

Anna Volpe

Anna Volpe is a researcher at the Clinic of Infectious Diseases, Department of Precision and Regenerative Medicine and Ionian Area, University of Bari Aldo Moro, Italy. For more than 30 years, she has been leading the Molecular Virology Unit of the Clinical Pathology Section, Azienda Ospedaliero-Universitaria Consorziale Policlinico di Bari, Italy. Her main research interests involve immunology, molecular biology and virology, with a special focus on HIV and HCV infections. She is dedicated to the training of young researchers and clinicians, having three decades of expertise in a multitude of molecular methods for the diagnosis and follow up of infectious diseases. She is responsible for hospital surveillance for the development of antiretroviral resistance, monitored through next-generation sequencing techniques. Her research interests also include the detection of inflammatory biomarkers by cytofluorimetric methods, on which she holds an extensive experience. Over the years, she established research collaborations with the main Italian research institutes. She is actively involved in national and international multicenter studies and serves as a member of the Italian HIV NGS network and the ARCA Cohort.

Preface

After more than eight decades from the description of their molecular mechanism, drug resistance is still a hot topic of scientific literature. A quick search on Scopus shows more than 45,000 papers, related to drug resistance, published only in 2024. Therefore, the scientific interest in this research challenge is far from extinguished.

Several reasons make drug resistance a still buzzing research field, among which there is the alarming spread of resistant microrganisms (i.e., bacteria, fungi) and the development of chemoresistant cancer cells. Therefore, biotechnologists and pharmacologists are developing computational approaches to speed-up the discovery of new drugs or to boost the effectiveness of those already available. Smart devices are assembled to answer the challenging demands of clinicians, exploiting innovative, stimuli-responsive delivery systems to achieve controlled release kinetics or to clarify ex vivo biological mechanisms causing drug resistance.

In such a research field in continuous development, this reprint aims at providing new insights into the management of drug resistance. We collected eleven original manuscripts reporting on the latest strategies to contrast drug resistance development, with a focus on resistant pathogenic microrganisms and chemoresistant forms of cancer.

We wish to thank all the Authors who contributed to this Reprint, as well as the Editor-in-Chief and the Managing Editors of Biomedicines, for their kind and timely assistance.

Maria A. Bonifacio, Antonio Mazzocca, and Anna Volpe
Guest Editors

biomedicines

MDPI

Article

Mode of the Interaction of Efflux Inhibitor Phenylalanyl-arginyl-β-naphtylamide with Bacterial Cells

Sandra Sakalauskaitė [1,2,*], Valeryia Mikalayeva [1,3], Simona Sutkuvienė [1,4] and Rimantas Daugelavičius [1,5,*]

1 Department of Biochemistry, Vytautas Magnus University, 44248 Kaunas, Lithuania; valeryia.mikalayeva@lsmu.lt (V.M.); simona.sutkuviene@vdu.lt (S.S.)
2 Laboratory of Immunology, Department of Immunology and Allergology, Lithuanian University of Health Sciences, 50161 Kaunas, Lithuania
3 Institute of Cardiology, Lithuanian University of Health Sciences, 50162 Kaunas, Lithuania
4 Department of Biochemistry, Lithuanian University of Health Sciences, 50161 Kaunas, Lithuania
5 Research Institute of Natural and Technological Sciences, Vytautas Magnus University, 44248 Kaunas, Lithuania
* Correspondence: sandra.sakalauskaite@lsmu.lt (S.S.); rimantas.daugelavicius@vdu.lt (R.D.)

Abstract: An increased efflux activity is one of the major reasons for bacterial antibiotic resistance. The usage of efflux pump inhibitors could be a promising approach to restoring the activity of inefficient antibiotics. The interaction of the RND family efflux pump inhibitor phenylalanyl-arginyl-β-naphthylamide (PAβN) with *Salmonella enterica* ser. Typhimurium cells was assayed using traditional microbiological techniques and a novel PAβN-selective electrode. Monitoring the PAβN concentration in the medium using the electrode enabled the real-time measurements of this compound's interaction with bacterial cells. We showed that *S.* Typhimurium cells accumulate a high amount of PAβN because of its high affinity to lipopolysaccharides (LPSs), the major constituent of the outer layer of the outer membrane, and does not affect the functioning of the plasma membrane. EDTA enhanced the binding of PAβN to *S.* Typhimurium cells and the purified *E. coli* LPSs, but the energization of the cells by glucose does not affect the cell-bound amount of this inhibitor. Polycationic antibiotic Polymyxin B released both the cells accumulated and the suspended LPS-bound PAβN.

Keywords: efflux pump inhibitor; ion-selective electrode; LPS; phenylalanyl-arginyl-β-naphthylamide; Polymyxin B; *S.* Typhimurium

1. Introduction

The resistance to antibiotics is one of the top global public health and development threats. In addition to a low outer membrane (OM) permeability to lipophilic compounds and a high activity of periplasmic β-lactamases, highly expressed efflux pumps are one of the major causes of antibiotic resistance in *Salmonella enterica* ser. Typhimurium [1,2]. It is known that efflux pumps also contribute to bacterial virulence [3] and biofilm formation [4]. Therefore, an inhibition of the efflux could be a promising strategy to restore the potency of antibacterial agents [5]. The main and clinically most important efflux pumps in Gram-negative bacteria belong to the resistance–nodulation–division (RND) superfamily [6] and phenylalanyl-arginyl-β-naphthylamide (PAβN) is one of the most popular inhibitors of these devices [7,8]. Although PAβN is the best thoroughly studied efflux inhibitor, the mechanism of its action still needs to be fully understood. Two opinions prevail: PAβN is the OM permeabilizer [9] and/or it changes the conformation of the PM-located lipophilic pocket of the AcrB pump, decreasing its affinity to antibiotics [10].

The most common method for monitoring the efflux activity in bacteria is the fluorimetric estimation of the cell-bound ethidium, which is a substrate of many efflux pumps [11]. The higher levels of the accumulated ethidium inside the cells lead to increased fluorescence, indicating that the efflux is reduced and the indicator compound is not removed from the cytosol [12]. However, ethidium binds to the cell DNA and it is difficult to determine the exact concentration of this indicator accumulated in the cytosol. Therefore, the risk of an underestimated efflux level remains [11].

The efficiency of efflux pumps can be monitored using lipophilic cation-selective electrodes [13,14]. These cations are indicators of membrane voltage (transmembrane difference of electrical potential, $\Delta\psi$) in mitochondria and bacteria. The most popular representative of these ions is tetraphenylphosphonium (TPP^+). The $\Delta\psi$ (negative inside) bacteria accumulate TPP^+ and the concentration of this cation inside the cells can be more than 10^3 times higher than in the incubation medium [15]. However, despite a high $\Delta\psi$ on the PM, this lipophilic cation is not able to equilibrate across the envelope of Gram-negative bacteria because of the OM barrier. By measuring the interaction of TPP^+ with Gram-negative bacteria, it is also possible to detect the changes in the permeability of the OM barrier [15,16]. In 1986, Midgley [17] showed that the level of TPP^+ accumulation in bacteria depends also on the efficiency of pumps extruding this indicator cation back to the incubation medium. Further studies revealed that lipophilic cations are the most universal substrates of the efflux pumps in prokaryotic [13,14,18,19] and eukaryotic [19,20] cells. The monitoring of TPP^+ accumulation in bacterial cells in the presence and absence of PAβN allows us to evaluate the totalefflux activity [21].

It is known [7] that PAβN interacts with a hydrophobic pocket of AcrB or analogous pumps in the bacterial PM, but it is not clear why this compound is so effective as an inhibitor, and how PAβN gets into the place of action. Based on our experience in the development of TPP^+, phenyldicarbaundecaborane [15], and ethidium-selective [21] electrodes, we have constructed an electrode that is selective to PAβN. This device allowed us to assess the affinity of the *S.* Typhimurium cells to PAβN. In the current study, we demonstrated that *S.* Typhimurium cells bind a high amount of PAβN because of the affinity of this inhibitor to LPSs in the OM barrier. The high affinity of PAβN to bacterial LPSs and the accumulation of this inhibitor in the pump neighborhood could be an important factor in strengthening efflux inhibition in Gram-negative bacteria.

2. Materials and Methods

2.1. Bacterial Strains and Chemicals

Salmonella enterica ser. Typhimurium strain SL1344 and ΔacrB were obtained from Prof. Séamus Fanning (Institute of Food and Health, University College Dublin, Ireland). Phenylalanyl-arginyl-β-naphthylamide (PAβN) dihydrochloride, lipopolysaccharide (LPS, serotype *Escherichia coli* O11: B4, phenol extracted), polymyxin B (PMB) sulfate, Luria–Bertani broth (LB), $Na_2S_2O_4$, chloramphenicol (Cm), tungstosilicic acid hydrate, poly(vinyl chloride) (PVC, average Mw ~80,000), and dioctyl phthalate were obtained from Sigma-Aldrich (Seelze, Germany). Ethylenediaminetetraacetic acid (EDTA), glucose, and HCl were obtained from Sharlau (Barcelona, Spain). Tris(hydroxymethyl)-aminomethane (Tris) and tetracycline (Tet) were obtained from Roth (Karlsruhe, Germany).

2.2. Conditions of Bacterial Growth and Determination of Minimal Inhibitory Concentration (MIC) of Antimicrobial Compounds

Overnight culture of *S.* Typhimurium cells was diluted 1:50 in fresh LB medium and grown with aeration at 37 °C to OD_{600} of 0.8. The cells were collected with centrifugation at 3000× *g* (Allegra TM64R; Beckman Coultier Inc., Fullerton, CA, USA) at 4 °C for 10 min. Pelleted cells were resuspended in 100 mM Tris/HCl (pH 8.0) buffer to obtain 1 mL of concentrated suspension (~2×10^{11} cells/mL), kept on ice, and used within 4 h. To permeabilize the OM, the pelleted cells were resuspended in 10 mL 100 mM Tris/HCl containing 10 mM EDTA, pH 8.0, then pelleted and re-suspended in 100 mM Tris/HCl, pH 8.0, to obtain 1 mL of concentrated suspension (~2×10^{11} cells/mL).

Heat-inactivated cells were prepared by incubating the concentrated suspension at 95 °C for 10 min. To evaluate the antimicrobial activity of the compounds, we applied the broth dilution method. The procedure involves serial two-fold dilutions of the antimicrobial compounds in the LB growth medium in 96-well microtitration plates. Each well was inoculated with a microbial inoculum where the concentration of bacteria cells was 5×10^5 cfu/mL [22]. Microplates were incubated without agitation at 37 °C. The turbidity (λ = 612 nm) of the cell suspensions was measured using a "TECAN GENios Pro™" (Männedorf, Switzerland) plate reader after 16–20 h of incubation.

2.3. Electrochemical Measurements

A PAβN sensor was developed using electroactive material–ion–association complexes of PAβN cation with tungstosilicate anion dispersed in a PVC matrix. The membrane of the sensor was prepared following a well-known procedure [23] by dissolving poly(vinyl) chloride (PVC), dioctyl phthalate plasticizer, and tungstosilicic acid hydrate under stirring in freshly distilled tetrahydrofuran. The membrane components were taken in the following proportion (by weight): tungstosilicic acid hydrate, 1%; dioctyl phthalate, 66%; and PVC, 33%. The membrane was fixed to the sensor body, soaked overnight in 1 mM PAβN solution and washed in deionized water before the experiments. The sensors were stored dry at room temperature. Assembling the PAβN-selective electrode, the sensor was filled with 0.1 mM PAβN solution in 100 mM NaCl and connected to an internal Ag/AgCl reference electrode.

PAβN$^+$ and TPP$^+$-selective electrodes were connected to the potential-amplifying system based on an ultralow input bias current operational amplifier AD549JH (Analog Devices, Norwood, MA, USA) [21]. The amplifying system was connected to a computer through PowerLab 4/35 logger (ADInstruments Pty Ltd., Bella Vista, Australia). Ag/AgCl reference electrodes (model Orion 9001, Thermo Inc., Cambridge, MA, USA) were indirectly connected to the measuring vessels by agar salt bridges. Potentiometric measurements of the PAβN$^+$ and TPP$^+$ concentration in the incubation medium were performed simultaneously in two to three thermostated reaction vessels. The experiments were performed at 37 °C in 5 mL of magnetically stirred 100 mM Tris/HCl, pH 8.0, and TPP$^+$ ions to 3 μM concentration were added to the cuvette. The concentrated cell suspension was added to obtain OD_{600} of 1. After adding the cells, the change in TPP$^+$ concentration in the medium was registered due to the accumulation of these cations in the cells. Efflux pumps are non-selective, and they extrude different lipophilic compounds—antibiotics, dyes, etc. TPP$^+$ is known as a universal efflux pump substrate and according to its accumulation in the cells, the efflux pump activity can be determined using a potentiometric system and TPP$^+$-selective electrode. A representative set of curves from three independent experiments is presented.

3. Results

3.1. Inhibitory Efficiency of PAβN Depends on the Concentration of Treated Cells

In determining the minimal inhibitory concentration (MIC) of antimicrobials, it is important to select a proper initial concentration of the tested bacteria [22]. The MICs of chloramphenicol and tetracycline on *S.* Typhimurium cells grown in the LB medium were 4 and 2 mg/L, correspondingly, when the initial cell concentration was 10^5 cfu/mL. Considerably lower, at 0.25 mg/L, concentrations of chloramphenicol and tetracycline affected the growth of *S.* Typhimurium cells when 60 μM of PAβN was present in the medium at the initial cell concentration of 10^5 cfu/mL (Figure 1a,b). However, PAβN did not stimulate the inhibitory action of chloramphenicol when the initial concentration of *S.* Typhimurium cells was higher—10^6 or 10^7 cfu/mL (Figure 1a). PAβN had some stimulatory effect on the action of tetracycline when the concentration of *S.* Typhimurium was 10^6 cfu/mL but had no effect on the efficiency of this antibiotic when the initial concentration of bacteria was 10^7 cfu/mL (Figure 1b). The decreased efficiency of PAβN at higher concentrations of cells can be expected if PAβN bind to bacteria and this inhibitor is distributed to a 10- or 100-times higher number of cells.

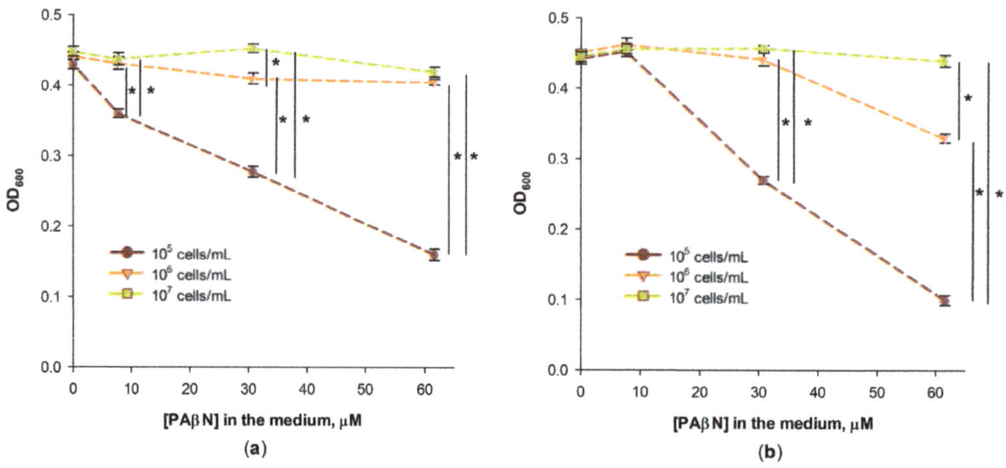

Figure 1. Effect of the initial concentration of *S.* Typhimurium cells on their growth in chloramphenicol (**a**) or tetracycline (**b**) containing LB medium in the presence of PAβN. The antibiotics were added to the final concentration of 0.25 mg/L, and the initial concentrations of cells are indicated in the figure. The cells were grown overnight in 96-well plates at 37 °C. Three biological replicates were performed for each set of experiments; *—$p < 0.05$.

3.2. At Low PAβN Concentrations TPP^+ Is not Able to Penetrate the Cells

The results of the TPP^+ microdilution test showed that PAβN can be characterized by some threshold values on TPP^+ action on the cells. A considerable bacteriostatic effect of TPP^+ on the growth of *S.* Typhimurium wt cells was achieved in the presence of 60 μM and higher concentrations of PAβN (Figure A1). Determining the effects of PAβN on the efficiency of antimicrobials using the microdilution method, we registered the efficiency of this efflux inhibitor on the growth of cells after 16–24 h of incubation in the LB medium.

In the subsequent series of experiments, we delved into the impact of varying PAβN concentrations on the accumulation of TPP^+ within *S.* Typhimurium cells using an ion-selective electrode, enabling the monitoring of the concentration of this lipophilic cation in the incubation medium (Figure 2). Cells with an intact OM exhibited a discernible TPP^+ accumulation when exposed to PAβN concentrations exceeding 24 μM (Figure 2a). In contrast, a distinct and more pronounced accumulation of TPP^+ by the cells was observed when Tris/EDTA-permeabilized cells were studied (Figure 2b). Even at a considerably lower PAβN concentration of 3 μM, these cells exhibited an augmented TPP^+ accumulation (Figure 2b). The supplementation of EDTA to the medium induced an additional uptake of TPP^+ by the cells with an intact OM (Figure 2a). In both cases, the increasing concentration of the efflux inhibitor in the medium resulted in a proportional increase in TPP^+ uptake.

However, the effect of the increasing PAβN concentrations was different when ΔacrB mutant cells were analyzed (Figure A2). The initial uptake of TPP^+ was the highest at 3 μM PAβN. When the OM of ΔacrB cells was permeabilized by adding EDTA, the maximal TPP^+ uptake was achieved in the presence of 24 μM PAβN. The following increase in PAβN concentrations in the medium resulted in a lower accumulation of TPP^+ by the cells and in the presence of 96 μM PaβN, the TPP^+ accumulation in ΔacrB cells was suppressed. An assessment of the PAβN effect on the medium-dissolved oxygen concentration showed that this efflux inhibitor did not affect the respiration rate and, most probably, the $\Delta\psi$ of *S.* Typhimurium cells at concentrations ranging from 3 to 120 μM (see Figure A3).

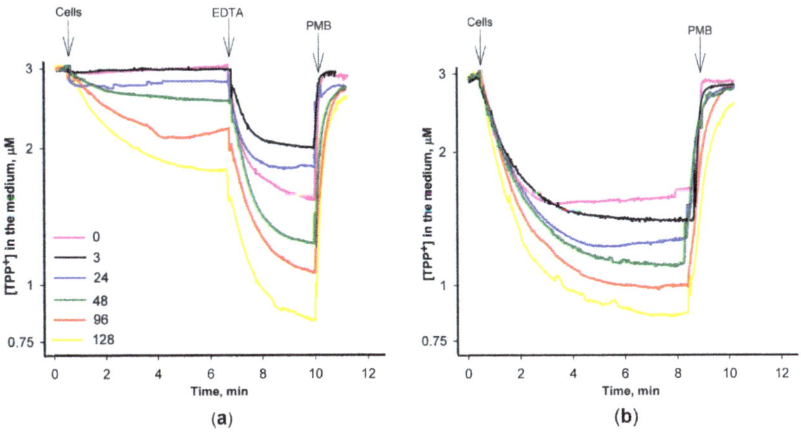

Figure 2. Influence of PAβN concentration on TPP^+ accumulation in *S.* Typhimurium wt cells with (**a**) intact and (**b**) Tris/EDTA-permeabilized OM. The experiments were performed at 37 °C in 100 mM Tris/HCl buffer, pH 8.0, containing 0.1% of glucose. The cells were added to obtain the suspension OD_{600} of 1. EDTA (**a**)—to the final concentration of 1 mM, and Polymyxin B (PMB)—to 100 mg/L. PAβN was added to the medium before the cells to concentrations indicated in the figure (µM).

3.3. Most of PAβN Is Bound to the Cell Surface

To learn more about the mode of interaction between *S.* Typhimurium cells and the efflux inhibitor, a PAβN-selective electrode was developed. It allowed us to monitor 0.2 µM and higher concentrations of PAβN (see Figure 3a, insert). The electrode showed a stable PAβN concentration-dependent potential and near Nernstian response, approximately 56.5 +/− 0.5 mV/decade at 37 °C. The tested components of microbiological media and buffers did not affect the potential of the electrode and its response time was 10–15 s.

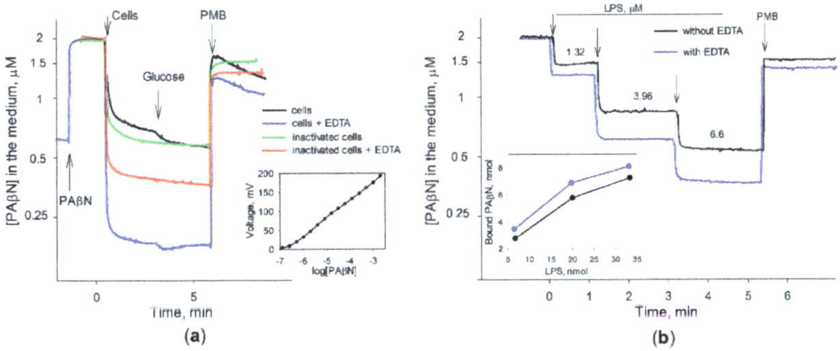

Figure 3. Binding of PAβN to *S.* Typhimurium cells (**a**) and LPS (**b**). The experiments were performed in 5 mL of 100 mM Tris-HCl, pH 8.0, at 37 °C. In (**a**) PAβN was added to the final concentration of 2 µM (Nernstian slope of the electrode is presented in the insert). The concentrated cell suspension was added to obtain OD_{600} of 1. Glucose and Polymyxin B (PMB) were added to the final concentrations of 0.1% and 50 mg/L, correspondingly. In (**b**) LPS additions are shown by the arrows. Numbers indicate the final concentrations of LPS after each of the additions. PMB was added to the final concentration of 50 mg/L, EDTA—to 0.1 mM. Insert shows the relationship between the amount of LPS added and the amount of PAβN bound.

An addition of the intact *S.* Typhimurium wt cells to the inhibitor-containing 100 mM of Tris buffer induced a strong decrease in PAβN concentrations in the medium. In the EDTA-containing Tris buffer, *S.* Typhimurium cells bound considerably higher amounts of

PAβN (Figure 3a). A supplement of glucose to the incubation medium caused a negligible additional binding of PAβN to the cells. Polycationic antibiotic Polymyxin B induced a sharp and very strong increase in PAβN concentrations in the medium, but just after that, the medium concentration of this inhibitor started to decrease. Heat-inactivated cells bound lower amounts of PAβN compared to the intact cells and the addition of glucose did not have any effect. The inactivated cells also bound lower amounts of PAβN in the presence of EDTA and the addition of PMB induced a sharp release of this inhibitor. It should be mentioned that after the PMB addition into the heat-inactivated cell suspensions, PAβN concentrations in the medium increased and remained stable. Another interesting observation is that some amount of PAβN remains bound to the cells in the presence of PMB, which is proportional to the maximal cell-bound amount at different experimental conditions.

The binding of PAβN to heat-inactivated cells and the sharp release of the cell-accumulated inhibitor after the addition of PMB suggest that PAβN binds to the cell surface. *S.* Typhimurium LPS shares an identical lipid A structure with that of *E. coli* [24]. The results of the experiments with the commercial purified LPS from *E. coli* cells supported this anticipation (Figure 3b). The added LPS bound a rather high amount of PAβN and the addition of PMB caused an immediate release of the inhibitor back to the medium. The LPS-bound amount of PAβN was directly proportional to the amount of LPS added (see the insert of Figure 3b) and in an EDTA-containing medium, LPS bound a higher amount of PAβN.

3.4. Saturation in PAβN Binding to S. Typhimurium Cells Can Be Achieved

To better understand the role of EDTA in PAβN interaction with the intact cells, the binding of this inhibitor to *S.* Typhimurium was studied using different PAβN concentrations. The best-pronounced electrode response indicated that relatively, the largest part of PAβN from the medium is accumulated by cells at the lowest PAβN concentration used (2 μM). A total of 1.5×10^{10} cells with an intact OM bound 6.5 nmol of PAβN and the same number of cells in the presence of EDTA—9 nmol of this inhibitor. At 20 μM of PAβN concentration, 1.5×10^{10} cells bound ~25 and ~30 nmol, correspondingly (Figure 4). At 80 μM and higher initial concentrations, the addition of cells caused a negligible change in the medium PAβN concentration and the impact of EDTA was also insignificant. According to our calculations, the maximal amount of PAβN that 1.5×10^{10} cells can bind is 25 or 30 nmol, in the absence or presence of EDTA, which is equal to the amount bound in the 20 μM solution of this inhibitor. For a single cell, this corresponds to 16.7–20 $\times 10^{-19}$ mol or ~1–1.2 $\times 10^{6}$ molecules of PAβN. The results obtained suggest that the high affinity of PAβN to bacterial LPS could be an important factor in strengthening its inhibitory action on efflux pumps in Gram-negative bacteria.

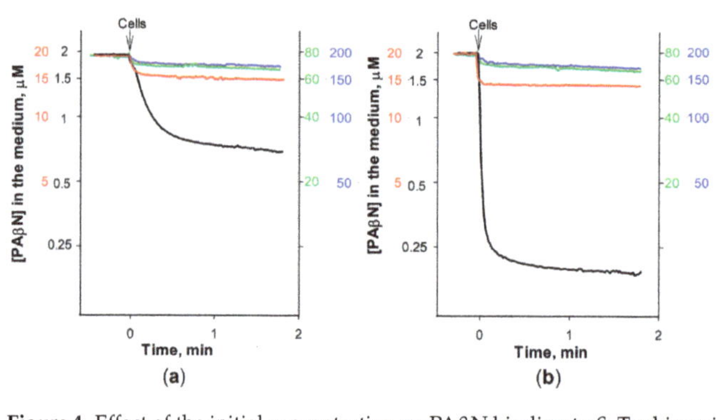

Figure 4. Effect of the initial concentration on PAβN binding to *S.* Typhimurium cells without EDTA (**a**) and in the medium contained 0.1 mM EDTA (**b**). The experiments were performed in 5 mL of 100 mM Tris-HCl, pH 8.0, at 37 °C. The initial PAβN concentration was 2 μM (black), 20 μM (red), 80 μM (green), or 200 μM (blue curves and scales). Arrows indicate the additions of cells.

4. Discussion

Efflux pump inhibitors are considered to be promising tools to combat multidrug-resistant bacteria [6]. Considering the broad range of substrates, it is possible that bacterial efflux inhibitors would also affect the extrusion of lipophilic compounds from the eukaryotic cells. In this context, the knowledge of factors determining the selectivity of efflux inhibitors is of particular importance. PAβN is one of the best-known inhibitors of efflux pumps of the RND superfamily and by using a newly developed PAβN-selective electrode, we demonstrated that this compound with a high affinity binds to *S.* Typhimurium cells and LPSs from *E. coli*. The amount of PAβN bound to the LPS is directly proportional to the amount of LPS added into the solution of this efflux inhibitor, highlighting the specificity and concentration-dependent nature of this interaction.

Schuster et al. [25] showed that PAβN replaces Mg^{2+} in LPSs, and our experiments registered the PAβN binding increasing effect of EDTA, which is consistent with this observation. The stimulatory activity of EDTA on the accumulation of PAβN by the cells could be explained by the removal of divalent cations and the opening of the additional PAβN binding sites on the surface of the OM. The high density of negatively charged residues in LPSs is likely to be of physiological significance, as it concentrates bivalent cations such as Ca^{2+} and Mg^{2+} in the close environment of the cell surface where cations are required for the structural and functional integrity of the outer membrane [24]. The results of our experiments indicate that after the EDTA-induced destruction of the OM permeability barrier, the additional binding of PAβN to the cells is observed, indicating the modulatory role of EDTA in enhancing the binding affinity between PAβN and the LPS. At some experimental conditions, the massive binding of PAβN to the OM could be the reason for the observed permeabilization [26], but this is not an obligatory condition of the action of this efflux inhibitor [13].

The accumulation of PAβN in an anionic periplasm and a negatively charged cytosol could be the additional factors promoting the binding of PAβN to Gram-negative bacteria. Previously, we observed [14] that in an EDTA-containing medium, PAβN induces the efflux of TPP^+ with a higher efficiency than from *S.* Typhimurium cells with an intact OM. On the other hand, a sharp PMB-induced release of PAβN, more prompt than the leakage of cell-accumulated TPP^+, suggests that the accumulation of these cations in bacterial cells differs. First of all, supplementing the incubation medium with glucose induces only a negligible additional accumulation of PAβN, which is not followed by its partial extrusion, as is observed in the case of TPP^+ [13,14]. These observations suggest that PAβN is not just another efflux pump substrate (competitive inhibitor) competing with TPP^+ for the extrusion. According to Kinana et al. [7], PAβN inhibits the efflux of other drugs by binding to the bottom of the distal binding pocket (hydrophobic trap) and interferes with the binding of other drug substrates to the upper part of this pocket. Our data suggest that to get to the hydrophobic pocket, PAβN most probably does not need to cross the PM and enters into the pocket from the periplasm.

Efflux pumps may extrude PAβN from the periplasm, but it is possible that it is not released into the medium and stays bound to the outer layer of the OM. However, PMB, as a compound with a higher affinity to LPSs than PAβN, releases this inhibitor from the cell surface back to the medium. The immediate and very effective release of the bound PAβN upon the addition of PMB implies that PMB disrupts the PAβN–LPS bond, leading to the release of the bound inhibitor. The experimental results, captured in the insert of Figure 3b, offer a clear visual representation of this interplay.

The PAβN binding saturation effect is attained at high concentrations of this compound. The dimensions of *E. coli* and *S.* Typhimurium cells are very similar and one bacterial cell contains approximately 3.5×10^6 of LPS molecules, occupying around three-quarters of the bacterial surface, with the remaining area being filled by proteins [27,28]. Our calculations indicate that one molecule of PAβN binds to ~3 LPS molecules on the cell surface. The results of the experiments with purified LPSs show very similar numbers.

Earlier, PAβN was known as a substrate for the colorimetric determination of peptidase activities. The incubation of intact *E. coli* cells with PAβN results in a time-dependent increase in fluorescence which has an emission spectrum corresponding to that of β-naphthylamine. Kinana et al. have shown [7] that the hydrolysis of PAβN in *E. coli* is nearly entirely dependent on an aminopeptidase, PepN, and the expression of this peptidase in the periplasm. We have found that the concentration of PAβN in the medium starts to decrease after the cell permeabilization with PMB. However, in the case of heat-treated cells, PMB treatment only leads to the release of PAβN back to the medium where the concentration of this compound stays stable.

If the molecular structures of TPP^+ and PAβN are compared, the latter looks more hydrophilic, and a much higher OM penetration rate for TPP^+ could be predicted. The $\Delta\psi$-driven equilibrium distribution of PAβN between the cytosol and the incubation medium should be slower, but the binding of PAβN to cells as well as the release of this inhibitor from the cells to the medium after the cell treatment with PMB are very fast processes, supporting the idea that most of the PAβN does not cross the OM. Also, our results suggest that PAβN is not a proper efflux pump substrate compared to TPP^+. Our earlier published results [13,14], along with the results of the experiments on the interaction of PAβN with the efflux pump mutant ΔacrB (Figure A2), show that at high concentrations, PAβN depolarizes the bacterial PM. However, the mechanism of this depolarization stays unclear if $\Delta\psi$ is not used for the massive efflux of this inhibitor. Although the information is limited, it is known that PAβN also affects the accumulation of antibiotics in Gram-positive bacteria [29,30] and mycobacteria [31]. These results indicate that efflux pumps from other families might also be sensitive to PAβN inhibition. It is possible that the experiments on PAβN binding to Gram-positive bacteria and mycobacteria would allow us to find out how important the accumulation of inhibitors on the cell surface is for the selective and effective blockage of efflux pumps.

5. Conclusions

The developed PAβN-selective electrode led us to the assessment of the *S.* Typhimurium cell affinity of this efflux inhibitor and provided valuable insights into the concentration-dependent dynamics of PAβN action on TPP^+ accumulation, revealing different responses between intact and EDTA-permeabilized cells. This demonstrates the complex interplay of PAβN, and the outer membrane integrity and permeabilizing activity of EDTA, enriching our understanding of this bacterial efflux and its regulation. The electrode's specificity was demonstrated in assessing the interaction of PAβN with *Salmonella enterica* ser. Typhimurium cells. Selective electrodes offer a potent means for real-time monitoring and quantification and enable the monitoring of PAβN at concentrations of 0.2 μM and higher, providing a rapid and stable response.

Author Contributions: Project design and conceptualization, R.D., V.M. and S.S. (Sandra Sakalauskaitė); methodology, V.M. and S.S. (Sandra Sakalauskaitė); investigation, V.M, S.S. (Sandra Sakalauskaitė), S.S. (Simona Sutkuvienė) and R.D.; data curation, R.D.; writing—original draft preparation, V.M., S.S. (Sandra Sakalauskaitė) and S.S. (Simona Sutkuvienė); writing—review and editing, V.M. and R.D.; visualization, V.M. and S.S. (Sandra Sakalauskaitė). All authors have read and agreed to the published version of the manuscript.

Funding: This research was supported by the Research Council of Lithuania, funding grant No MIP-040/2015 to R.D.

Institutional Review Board Statement: Not applicable.

Informed Consent Statement: Not applicable.

Data Availability Statement: The data presented in this study are available on request from the corresponding authors.

Acknowledgments: The authors are grateful to Séamus Fanning for providing *S.* Typhimurium SL1344 and ΔacrB cells.

Conflicts of Interest: The authors declare no conflicts of interest.

Appendix A

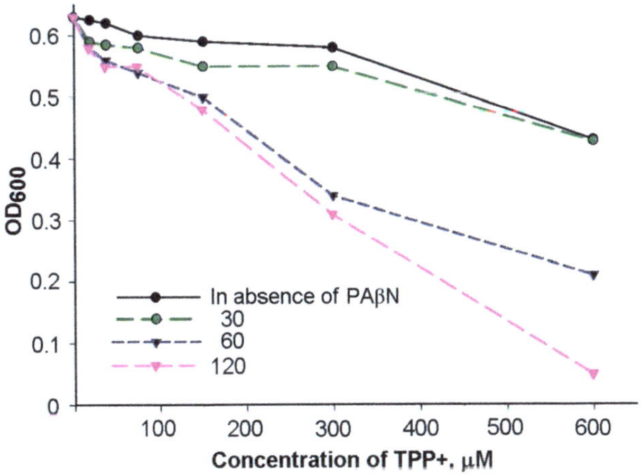

Figure A1. Effect of PAβN on the growth of *S.* Typhimurium cells in TPP^+-containing LB medium. Serial dilutions of TPP^+ were performed in the presence of the figure-indicated supplements of PAβN (μM). The cells were grown overnight in 96-well plates at 37 °C.

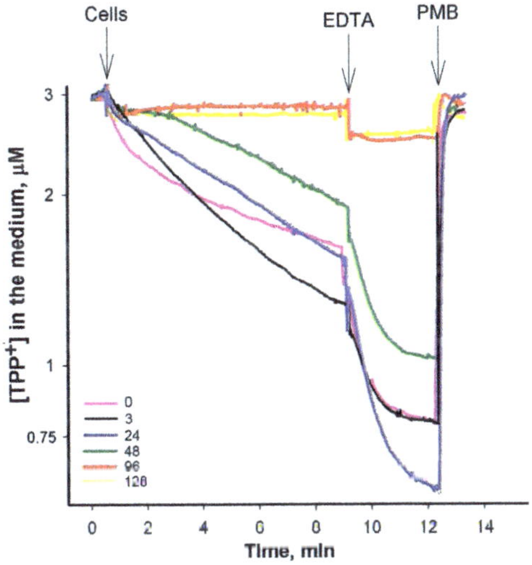

Figure A2. Influence of PAβN concentration on TPP^+ binding to *S.* Typhimurium of ΔacrB mutant cells with intact OM. The experiments were performed at 37 °C in 100 mM Tris/HCl buffer, pH 8.0, containing 0.1% of glucose. The cells were added to obtain OD_{600} of 1, EDTA—to the final concentration of 1 mM, and Polymyxin B (PMB)—to 100 μg/mL. PAβN was added to the medium before the cells to concentrations indicated in the figure (μM).

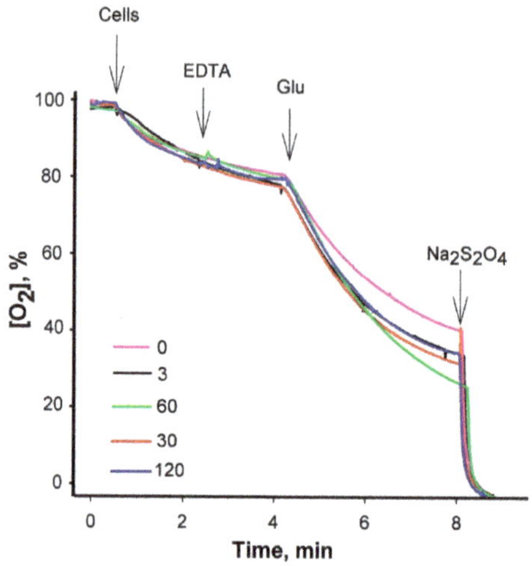

Figure A3. Effect of PAβN on the respiration of *S.* Typhimurium cells. Measurements of the dissolved oxygen concentrations in the medium were performed at 37 °C in 100 mM Tris/HCl buffer, pH 8.0, *S.* Typhimurium wt cells were added to OD_{600} of 1. The final concentrations of the supplements were as follows: EDTA—0.1 mM, glucose (Glu)—0.1%, and $Na_2S_2O_4$—2 mg/mL. The concentrations (μM) of PAβN added to the medium before the cells are indicated in the figure.

References

1. Du, D.; Wang-Kan, X.; Neuberger, A.; van Veen, H.W.; Pos, K.M.; Piddock, L.J.V.; Luisi, B.F. Multidrug efflux pumps: Structure, function and regulation. *Nat. Rev. Microbiol.* **2018**, *16*, 523–539. [CrossRef] [PubMed]
2. Yamasaki, S.; Nagasawa, S.; Hayashi-Nishino, M.; Yamaguchi, A.; Nishino, K. AcrA dependency of the AcrD efflux pump in *Salmonella enterica* serovar Typhimurium. *J. Antibiot.* **2011**, *64*, 433–437. [CrossRef] [PubMed]
3. Pérez, A.; Poza, M.; Fernández, A.; Fernández Mdel, C.; Mallo, S.; Merino, M.; Rumbo-Feal, S.; Cabral, M.P.; Bou, G. Involvement of the AcrAB-TolC efflux pump in the resistance, fitness, and virulence of *Enterobacter cloacae*. *Antimicrob. Agents Chemother.* **2012**, *56*, 2084–2090. [CrossRef] [PubMed]
4. Alav, I.; Sutton, J.M.; Rahman, K.M. Role of bacterial efflux pumps in biofilm formation. *J. Antimicrob. Chemother.* **2018**, *73*, 2003–2020. [CrossRef] [PubMed]
5. Wang, Y.; Venter, H.; Ma, S. Efflux Pump Inhibitors: A Novel Approach to Combat Efflux-Mediated Drug Resistance in Bacteria. *Curr. Drug Targets* **2016**, *17*, 702–719. [CrossRef] [PubMed]
6. Nikaido, H.; Pagès, J.M. Broad-specificity efflux pumps and their role in multidrug resistance of Gram-negative bacteria. *FEMS Microbiol. Rev.* **2012**, *36*, 340–363. [CrossRef] [PubMed]
7. Kinana, A.D.; Vargiu, A.V.; May, T.; Nikaido, H. Aminoacyl β-naphthylamides as substrates and modulators of AcrB multidrug efflux pump. *Proc. Natl. Acad. Sci. USA* **2016**, *113*, 1405–1410. [CrossRef] [PubMed]
8. Lomovskaya, O.; Warren, M.S.; Lee, A.; Galazzo, J.; Fronko, R.; Lee, M.; Blais, J.; Cho, D.; Chamberland, S.; Renau, T.; et al. Identification and characterization of inhibitors of multidrug resistance efflux pumps in *Pseudomonas aeruginosa*: Novel agents for combination therapy. *Antimicrob. Agents Chemother.* **2001**, *45*, 105–116. [CrossRef] [PubMed]
9. Mowla, R.; Wang, Y.; Ma, S.; Venter, H. Kinetic analysis of the inhibition of the drug efflux protein AcrB using surface plasmon resonance. *Biochim. Biophys. Acta Biomembr.* **2018**, *1860*, 878–886. [CrossRef] [PubMed]
10. Reading, E.; Ahdash, Z.; Fais, C.; Ricci, V.; Wang-Kan, X.; Grimsey, E.; Stone, J.; Malloci, G.; Lau, A.M.; Findlay, H.; et al. Perturbed structural dynamics underlie inhibition and altered efflux of the multidrug resistance pump AcrB. *Nat. Commun.* **2020**, *11*, 5565. [CrossRef] [PubMed]
11. Blair, J.M.; Piddock, L.J. How to Measure Export via Bacterial Multidrug Resistance Efflux Pumps. *mBio* **2016**, *7*, e00840-16. [CrossRef] [PubMed]
12. Paixão, L.; Rodrigues, L.; Couto, I.; Martins, M.; Fernandes, P.; de Carvalho, C.C.; Monteiro, G.A.; Sansonetty, F.; Amaral, L.; Viveiros, M. Fluorometric determination of ethidium bromide efflux kinetics in *Escherichia coli*. *J. Biol. Eng.* **2009**, *3*, 18. [CrossRef] [PubMed]
13. Daugelavicius, R.; Buivydas, A.; Sencilo, A.; Bamford, D.H. Assessment of the activity of RND-type multidrug efflux pumps in *Pseudomonas aeruginosa* using tetraphenylphosphonium ions. *Int. J. Antimicrob. Agents* **2010**, *36*, 234–238. [CrossRef] [PubMed]
14. Sutkuvienė, S.; Mikalayeva, V.; Pavan, S.; Berti, F.; Daugelavičius, R. Evaluation of the efficiency of synthesized efflux pump inhibitors on *Salmonella enterica* ser. Typhimurium cells. *Chem. Biol. Drug Des.* **2013**, *82*, 438–445. [CrossRef] [PubMed]

15. Daugelavicius, R.; Bamford, J.K.; Bamford, D.H. Changes in host cell energetics in response to bacteriophage PRD1 DNA entry. *J. Bacteriol.* **1997**, *179*, 5203–5210. [CrossRef] [PubMed]
16. Yasuda, K.; Ohmizo, C.; Katsu, T. Potassium and tetraphenylphosphonium ion-selective electrodes for monitoring changes in the permeability of bacterial outer and cytoplasmic membranes. *J. Microbiol. Methods* **2003**, *54*, 111–115. [CrossRef] [PubMed]
17. Midgley, M. The phosphonium ion efflux system of *Escherichia coli*: Relationship to the ethidium efflux system and energetic studies. *J. Gen. Microbiol.* **1986**, *132*, 3187–3193. [CrossRef] [PubMed]
18. Gros, P.; Talbot, F.; Tang-Wai, D.; Bibi, E.; Kaback, H.R. Lipophilic cations: A group of model substrates for the multidrug-resistance transporter. *Biochemistry* **1992**, *31*, 1992–1998. [CrossRef] [PubMed]
19. Radi, M.S.; Munro, L.J.; Salcedo-Sora, J.E.; Kim, S.H.; Feist, A.M.; Kell, D.B. Understanding Functional Redundancy and Promiscuity of Multidrug Transporters in *E. coli* under Lipophilic Cation Stress. *Membranes* **2022**, *12*, 1264. [CrossRef] [PubMed]
20. Galkina, K.V.; Besedina, E.G.; Zinovkin, R.A.; Severin, F.F.; Knorre, D.A. Penetrating cations induce pleiotropic drug resistance in yeast. *Sci. Rep.* **2018**, *8*, 8131. [CrossRef] [PubMed]
21. Sakalauskaitė, S.; Mikalayeva, V.; Daugelavičius, R. Ethidium Binding to *Salmonella enterica* ser. Typhimurium Cells and Salmon Sperm DNA. *Molecules* **2021**, *26*, 3386. [CrossRef] [PubMed]
22. Wiegand, I.; Hilpert, K.; Hancock, R.E. Agar and broth dilution methods to determine the minimal inhibitory concentration (MIC) of antimicrobial substances. *Nat. Protoc.* **2008**, *3*, 163–175. [CrossRef] [PubMed]
23. Cammann, K. *Working with Ion-Selective Electrodes*; Springer: Berlin/Heidelberg, Germany, 1977; 226p.
24. Erridge, C.; Bennett-Guerrero, E.; Poxton, I.R. Structure and function of lipopolysaccharides. *Microbes Infect.* **2002**, *4*, 837–851. [CrossRef] [PubMed]
25. Schuster, S.; Bohnert, J.A.; Vavra, M.; Rossen, J.W.; Kern, W.V. Proof of an Outer Membrane Target of the Efflux Inhibitor Phe-Arg-β-Naphthylamide from Random Mutagenesis. *Molecules* **2019**, *24*, 470. [CrossRef] [PubMed]
26. Lamers, R.P.; Cavallari, J.F.; Burrows, L.L. The efflux inhibitor phenylalanine-arginine beta-naphthylamide (PAβN) permeabilizes the outer membrane of gram-negative bacteria. *PLoS ONE* **2013**, *8*, e60666. [CrossRef] [PubMed]
27. Nikaido, H.; Vaara, M. Molecular basis of bacterial outer membrane permeability. *Microbiol. Rev.* **1985**, *49*, 1–32. [CrossRef]
28. Neidhardt, F.C.; Curtiss, R. *Escherichia coli and Salmonella: Cellular and Molecular Biology*; ASM Press: Washington, DC, USA, 1996.
29. AlMatar, M.; Albarri, O.; Makky, E.A.; Köksal, F. Efflux pump inhibitors: New updates. *Pharmacol. Rep.* **2021**, *73*, 1–16. [CrossRef] [PubMed]
30. Vecchione, J.J.; Alexander, B., Jr.; Sello, J.K. Two distinct major facilitator superfamily drug efflux pumps mediate chloramphenicol resistance in *Streptomyces coelicolor*. *Antimicrob. Agents Chemother.* **2009**, *53*, 4673–4677. [CrossRef] [PubMed]
31. Balganesh, M.; Dinesh, N.; Sharma, S.; Kuruppath, S.; Nair, A.V.; Sharma, U. Efflux pumps of *Mycobacterium tuberculosis* play a significant role in antituberculosis activity of potential drug candidates. *Antimicrob. Agents Chemother.* **2012**, *56*, 2643–2651. [CrossRef] [PubMed]

biomedicines

MDPI

Article

Resistance Modulation of Individual and Polymicrobial Culture of *S. aureus* and *E. coli* through Nanoparticle-Coupled Antibiotics

Sana Zia [1,2,†], Song Peng [1,†], Arslan Bashir [3], Tasleem Kausar [2], Shanza Rauf Khan [3], Afshan Muneer [4], Attia Nawaz [5], Lina I. Alnajjar [6], Mohd Saeed [7], Nawaf Alshammari [7], Amjad Islam Aqib [8,*] and Kun Li [1,*]

1 MOE Joint International Research Laboratory of Animal Health and Food Safety, Institute of Traditional Chinese Veterinary Medicine, College of Veterinary Medicine, Nanjing Agricultural University, Nanjing 210095, China; sanazia389@gmail.com (S.Z.); payson@stu.njau.edu.cn (S.P.)
2 Department of Zoology, The Government Sadiq College Women University, Bahawalpur 61300, Pakistan; tasleem.kausar@gscwu.edu.pk
3 Department of Chemistry, University of Agriculture, Faisalabad 38000, Pakistan; shanzaraufkhan@gmail.com (S.R.K.)
4 Department of Zoology, Cholistan University of Veterinary and Animal Sciences, Bahawalpur 63100, Pakistan; afshanchudhary9@gmail.com
5 Department of Microbiology, Cholistan University of Veterinary and Animal Sciences, Bahawalpur 63100, Pakistan; attianawaz555@gmail.com
6 Department of Pharmacy Practice, College of Pharmacy, Princess Nourah bint Abdulrahman University, Riyadh 11671, Saudi Arabia; lialnajjar@pnu.edu.sa
7 Department of Biology, College of Science, University of Hail, P.O. Box 2440, Hail 34464, Saudi Arabia; mo.saeed@uoh.edu.sa (M.S.); naib.alshammari@uoh.edu.sa (N.A.)
8 Department of Medicine, Cholistan University of Veterinary and Animal Sciences, Bahawalpur 63100, Pakistan
* Correspondence: amjadislamaqib@cuvas.edu.pk (A.I.A.); lk3005@njau.edu.cn (K.L.)
† These authors contributed equally to this work.

Abstract: Polymicrobial mastitis is now becoming very common in dairy animals, resulting in exaggerated resistance to multiple antibiotics. The current study was executed to find drug responses in individual and mixed Culture of *Staphylococcus aureus* and *Escherichia coli* isolated from milk samples, as well as to evaluate the antibacterial potential of tungsten oxide nanoparticles. These isolates (alone and in mixed culture) were further processed for their responses to antibiotics using the disc diffusion method. On the other hand, tungsten oxide WO_3 (W) nanoparticles coupled with antibiotics (ampicillin, A, and oxytetracycline, O) were prepared through the chemical method and characterized by X-ray diffraction, scanning electron microscopy (SEM), and UV-visible techniques. The preparations consisting of nanoparticles alone (W) and coupled with ampicillin (WA) and oxytetracycline (WO) were tested against individual and mixed Culture through the well diffusion and broth microdilution methods. The findings of the current study showed the highest resistance in *E. coli* was against penicillin (60%) and ampicillin (50%), while amikacin, erythromycin, ciprofloxacin, and oxytetracycline were the most effective antibiotics. *S. aureus* showed the highest resistance against penicillin (50%), oxytetracycline (40%), and ciprofloxacin (40%), while, except for ampicillin, the sensitive strains of *S. aureus* were in the range of 40–60% against the rest of antibiotics. The highest zones of inhibition (ZOI) against mixed Culture were shown by imipenem and ampicillin, whereas the highest percentage decrease in ZOI was noted in cases of ciprofloxacin (−240%) and gentamicin (−119.4%) in comparison to individual Culture of *S. aureus* and *E. coli*. It was noteworthy that the increase in ZOI was not more than 38% against mixed Culture as compared to the individual Culture. On the other hand, there was a significant reduction in the minimum inhibitory concentration (MIC) of nanoparticle-coupled antibiotics compared to nanoparticles alone for individual and mixed-culture bacteria, while MICs in the case of mixed Culture remained consistently high throughout the trial. This study therefore concluded that diverse drug resistance was present in both individual and mixed-culture bacteria, whereas the application of tungsten oxide nanoparticle-coupled antibiotics proved to be an effective candidate in reversing the drug resistance in bacterial strains.

Keywords: mixed culture; *E. coli*; *S. aureus*; tungsten oxide nanoparticles; antibiotics; antibiotics-coupled nanoparticles

1. Introduction

Developing countries like Pakistan experience two-pronged challenges in the form of horizontal expansion of dairy animals and prevailing dairy udder challenges, overall jeopardizing health, and the economy. Such scenarios lead to a daily shortfall of milk for the consumers, e.g., Karachi, a cosmopolitan city of Pakistan, has reached an uncertain supply of up to 4 million liters per day. In addition, it is expected that milk consumption will find a minimum annual growth rate of 5% soon [1]. The dairy sector is currently facing a serious threat in the form of mastitis, which leads to a notable decrease in the yield and quality of milk, along with a sharp increase in treatment expenses and bovine mortality [2]. Furthermore, the augmentation of milk production is hampered by bacterial attacks in the udders of dairy animals, putting both animal and public health at risk [3]. Dairy cows are susceptible to more than 150 types of bacteria that can cause mastitis, including *S. aureus*, *Mycoplasma* species, *Streptococcus uberis*, *Streptococcus dysgalactiae*, *Escherichia coli*, and *Klebsiella pneumoniae* [4], which are responsible for the development of a wide variety of lesions [5]. In veterinary practice, the use of antibiotics often goes unjustified, culminating in development of microbial resistance and hence compromising animal and public health [6]. Microorganisms evolve antimicrobial resistance to survive in continuously changing environments [7]. Recent investigations have documented substantial fluctuations in the efficacy of antibiotics against pathogens such as *E. coli* and *S. aureus*, providing clear indications of their escalating resistance tendencies [8,9].

The wise approach in the current scenario is the five R concept, specifically, responsibility, refinement, reduction, replacement, and review of antimicrobial use (AMU). To reduce the burden of antibiotics or to replace antibiotics, alternatives are needed. As an alternative, nanoparticles are thought to work differently than antibiotics and can help reduce drug resistance by serving as carriers for antibiotics, having synergistic effects with antimicrobials and being antimicrobials themselves. Nanoparticles larger than 10 nm have been found to cause cytotoxicity and cellular disintegration through their interaction with the cellular wall and membrane constituents [9]. The integration of antibacterial agents into biomaterials is a strategic approach aimed at mitigating these challenges. In medicine and pharmaceuticals, nanotechnology plays a crucial role [10]. Apart from their higher reactivity, these particles are unique in their physicochemical properties, which include lower surface-to-volume ratios, greater stability, bioactivity, and bioavailability [11]. Recent studies showed that combining nanoparticles with antibiotics, antimicrobial peptides, and essential oils minimizes the potentially toxic effects of nanoparticles [12]. Tungsten oxide coating has found novel applications in antibacterial coatings, where it helps to improve the antibacterial efficacy of surgical equipment and medical devices [13]. Researchers discovered that WO_3-X nanodots exhibited strong bactericidal activity after they were exposed to membrane stress [14]. WO_3-X nanodots demonstrate a remarkable capability to eliminate both *Escherichia coli* and *Staphylococcus aureus* [15]. Furthermore, previous studies have demonstrated that WO_3 showed promising antimicrobial activity against many bacterial strains, including *E. coli*, *P. multocida*, *B. subtilis* (gram-positive), and *S. aureus* (gram-positive) [16].

The study of nanoparticle coupling with antibiotics targeting *E. coli* and *S. aureus* is of significance because it can helps combat antimicrobial resistance, boost the antimicrobial activity of these drugs, facilitate targeted drug delivery, support combination therapies, advances the diagnosis of various bacterial-related diseases, and expedite the healing process. Hence, the purpose of this research article was to investigate responses of individual and mixed culture *Staphylococcus aureus* and *E. coli* against antibiotics, and to evaluate tungsten oxide nanoparticle-coupled antibiotics as drug resistance modulators.

2. Materials and Methods

2.1. Sample Collection

The sample collection area encompassed districts within the Bahawalpur region, chosen based on accessibility and the consent of dairy farmers. Milk samples were aseptically collected in sterile vials using the convenience statistical method to reach a total sample number of $n = 200$ [17]. The samples were screened at the time of collection for subclinical mastitis following the protocol described by Muhammad et al. [18]. The positive samples were delivered to the laboratory of the Department of Microbiology at Cholistan University of Bahawalpur in a container maintained at a temperature of 4 °C.

2.2. Isolation of E. coli and S. aureus

The milk samples were put into incubation for 24 h at 37 °C, following which centrifugation at a speed of 3634× *g* for 5 min was conducted. The characteristic bacterial colonies were streaked on differential media of both bacteria, i.e., mannitol salt agar for *S. aureus* and MacConkey agar for *E. coli*. The plates were incubated at 37 °C for another 24 h. The mannitol salt agar was transformed into a yellow color with pinpoint round colonies on the media for *S. aureus*, while MacConkey agar was transformed into pink colonies on the media for *E. coli*. These colonies were further subjected to a series of biochemical assays as described by Bergey's manual of determinative bacteriology [19]. The confirmation of *S. aureus* and *E. coli* was determined using pooled information obtained from microbiological and biochemical experiments.

2.3. Antibiotic Susceptibility of E. coli and S. aureus

To determine the susceptibility of *E. coli*, *S. aureus*, and mixed Culture of *E. coli* and *S. aureus*, a total of eight antibiotics (erythromycin, ciprofloxacin, imipenem, amikacin, ampicillin, oxytetracycline, and gentamicin) were tested. The selection of these antibiotics was based on their common usage in clinical laboratories and adherence to the guidelines provided by the Clinical Laboratory and Standard Institute. Some of the antibiotics had previously been used in other studies (unpublished data or under review for publication) of the authors so have been exempted from this study. The Kirby–Bauer disc diffusion method was used to determine zones of inhibition, which were then compared with the standards provided by the Clinical Laboratory and Standard Institute [20]. In brief, fresh *E. coli* and *S. aureus* growth was adjusted at 1–1.5 × 10^8 CFU/mL (colony-forming units/milliliter) on sterile Mueller–Hinton agar and antibiotic discs were put aseptically at equal distances from one another. The agar plates were incubated at 37 °C for 20–24 h, and zones were measured with vernier calipers and compared with the standard zones to identify resistant, intermediate, and sensitive strains of *S. aureus* and *E. coli* [20]. *S. aureus* subspecies *aureus* ATCC 25923™ and *E. coli* ATCC25922 were used as a control strain in this study.

2.4. Response of Mixed Culture against Different Antibiotics

The disc diffusion method was employed to assess the susceptibility of mixed Culture consisting of *S. aureus* and *E. coli* against various antibiotics. For this purpose, a total of three samples each of *S. aureus* (S) and *E. coli* (E) were randomly selected from the previous trial (Section 2.3). These samples were mixed with each other to create three mixed culture samples, namely S1E1, S2E2, and S3E3. Briefly, the 0.5 McFarland solution of each bacterial strain was combined in equal proportion to make a resulting concentration of 0.5 McFarland (equal to 1–1.5 × 10^8 CFU/mL) for the SE combination. This study involved the examination of both individual bacteria and mixed Culture to assess their responses to a total of eight antibiotics with the same protocol as mentioned in Section 2.3 of this manuscript. Zones of inhibition (ZOI) were measured and compared, and comparisons of each individual and mixed culture were made. The percentage increase or decrease in ZOI values of mixed Culture with individual bacteria was calculated to evaluate the extent of variation exhibited by mixed Culture against antibiotics.

2.5. Synthesis and Characterization of Nanoparticle-Coupled Antibiotics

Synthesis of WO_3 Nanoparticles

Sodium tungstate (Na_2WO_4) and cetyltrimethylammonium bromide (CTAB) were individually dissolved in deionized water using a solution of 6 g of Na_2WO_4 and 18 mL of deionized water. Both solutions were mixed, and a small amount of hydrochloric acid (HCl) was added to adjust the pH to within the range of 1–2. This reaction mixture was autoclaved at 80 °C in a 100 mL Teflon vessel for 4 days in a stainless-steel lined autoclave. The hydrothermally precipitated materials were filtered and rinsed with deionized water and ethanol at regular intervals after hydrothermal treatment. After precipitation, the precipitates dried at 120 °C for two hours. A further 3.5 h of calcination at 500 °C was completed. After calcination, the product was ground into powder with a pestle and mortar.

Ampicillin and oxytetracycline were selected for this trial based on their clinical use, resistance profile, and better coupling ability with nanoparticles in a pilot study. These antibiotics were dissolved in 20 mL of deionized water, while 1.5 g of nanoparticles were dispersed in deionized water and stirred for five hours at room temperature in the presence of PVP dissolved in 20 mL of deionized water. The solutions were stored at room temperature for 24 h with regular mixing and stirring. The product was centrifuged for 30 min at 3000 rpm to settle at the bottom. A mortar and pestle were used to grind the product after it was dried for 8 h at 100 °C. The absorbance of coated nanoparticles was measured using a US-1100 double beam UV-visible spectrophotometer from LYNX, LM-56-1001AE in Pakistan. We performed XRD using a Rigaku TTR instrument (Tokyo, Japan) at 40 kV and 300 mA, in the range 2θ between 20 and 80, with Cu kα (λ) radiation around 0.15406 nm, to investigate the crystal structure of nanoparticles. An FTIR analysis was conducted using an FTIR spectrometer (Perkin Elmer Spectrum America) at room temperature within the spectral range of 4000–500 cm^{-1}. Scanning images and elemental analysis were performed on a TESCAN MIRA 3 (Brno, Czech Republic), which is available at the Institute of Space Technology, Islamabad, Pakistan. The Raman spectra were collected using a Peak Seeker Pro-Agiltron Raman spectrometer (USA) with a laser light source of 50 mW at 785 nm. At room temperature, 50 mg of each sample were placed on aluminum slides for examination. Thermal analysis was executed in a controlled nitrogen environment employing a Q600 SDT thermogravimetric analyzer with a heating rate of 10 °C per minute. The analysis of sample mass changes was performed using a POWEREACH JC 2000D2W contact angle tester produced in Pakistan.

2.6. Resistance Modulation by Nanoparitcle Coupled Antibiotics against Individuals and Mixed Culture

An empirical technique was used to estimate the antibacterial activity of WO_3-coated antibiotics. The minimum inhibitory concentrations (MICs) of the preparations were determined using broth microdilutions.

2.6.1. Agar Well Diffusion Method

A fresh culture of individual *E. coli* and individual *S. aureus* was adjusted to 1–1.5 × 10^8 CFU/mL by obtaining turbidity of culture equal to 0.5 McFarland. To make the final adjustment at 1–1.5 × 10^8 CFU/mL, a 1/2:1/2 ratio of both bacteria (each having 0.5 McFarland solutions equaling 1–1.5 × 10^8 CFU/mL) was used to generate the mixed culture for the agar well diffusion method. This was achieved by combining half of an *E. coli* solution (1–1.5 × 10^8 CFU/mL) and half of a *S. aureus* (1–1.5 × 10^8 CFU/mL) solution to make a total of 1 mL (0.5 McFarland). The culture was spread on Mueller Hinton agar homogenously. The well borer was used to make wells (6–8 mm) on sterile Mueller–Hinton agar at equal distances. Tungsten oxide nanoparticles alone and tungsten oxide nanoparticles coupled with antibiotics were poured into the wells (15 μL of 0.01 gm/mL) and incubated at 37 °C for 24 h. The zones of inhibition (ZOI) produced by this preparation against *E. coli*, *S. aureus*, and mixed culture were measured using vernier calipers [21].

2.6.2. Minimum Inhibitory Concentration (MIC)

Sterile broth was poured in all wells of sterile 96-well titration plate. Two-fold serial dilutions, starting from 10,000 µg/mL of each preparation were carried out until the 11th well. A fresh growth of *E. coli* and *S. aureus* adjusted at 1×10^5 CFU/mL was poured in all wells except the negative control. To make the final adjustment at 1×10^5 CFU/mL for mixed Culture, the solutions of *E. coli* (1×10^5 CFU/mL) and *S. aureus* (1×10^5 CFU/mL) were combined in a 1/2:1/2 ratio. Positive control (well containing both broth and culture) and negative control (well containing only broth) were reserved in the 12th column. The optical density at 600 nm wavelength was measured after 0, 4, 20, 24, and 28 h of incubation. In this study, OD values were taken at the 4 h duration both before and after the standard time of incubation, i.e., 24 h. This was intended to find the first minimum dosage if the infection is to be tackled on an immediate basis, as well as the hours around the infection period.

A net OD value was determined by comparing the values taken with those obtained after 0 h of incubation. Various concentrations were tested for inhibition of growth based on the net OD value. The minimum inhibitory concentration (MIC) was determined to be the lowest concentration of preparation that inhibited the growth of bacteria [22].

2.7. Statistical Analysis

Parametric tests, *t*-tests, and analysis of variance (ANOVA) were applied to the data obtained from disc diffusion, well diffusion, and the broth microdilution method. Tukey's test was applied in conjunction with ANOVA to compare the means for statistically significant differences. SPSS version 22 was used to analyze the data at $p < 0.05$.

Formulae Used

$$\begin{aligned}&\%\ \text{Change in ZOI of mixed culture compared with the individual culture bacteria}\\&= \frac{(\text{ZOI of mixed culture bacteria} - \text{ZOI of individual culture bacteria})}{\text{ZOI of mixed culture bacteria}} \times 100\end{aligned}$$

3. Results

3.1. Characterization of Nanoparticles

3.1.1. Tungsten Oxide

The FTIR spectra of tungsten oxide (WO_3) within the range of 1000 to 500 cm^{-1} refer to the characteristic lattice vibration exhibited by tungsten oxide nanoparticles. The stretching vibration of W-O-W and the bending vibration of W-O and W=O are characterized by a prominent peak at around 799 cm^{-1}. The peak at 1600 cm^{-1} is described as a vibration of the symmetrical OH of the hydroxyl group as well as a W-OH phase interaction. The utilization of an acidic medium during the preparation of the sample may yield larger crystallites with higher concentrations of oxygen defects than the sample prepared in less acidic media. Consequently, the symmetrical vibrations of the W-OH, H_2O, and W-OH molecules were affected. A strong stretching W-O-W response in the inorganic compound itself is visible at 799 cm^{-1}. The correlation between vibration and sample preparation was observed to be stronger when using more acidic media, such as a pH value of 2. Figure 1 represents a comprehensive view of the product, where the WO_3 nanoparticles exhibit an array of randomly arranged particles. Upon closer examination, the nanoparticle surface has a smooth texture, characterized by spheroidal or oval morphology. The nanoparticles of WO_3 are oval with rounded ends and their respective ends intersect at different points. Figure 1a displays the scanning electron microscopy (SEM) images depicting the synthesized product WO_3, which was obtained using hydrothermal methods. It was observed that a drug coating had been applied to the WO_3 nanoparticles, resulting in a variation in absorbance with wavelength. However, the absorbance pattern exhibited several peaks.

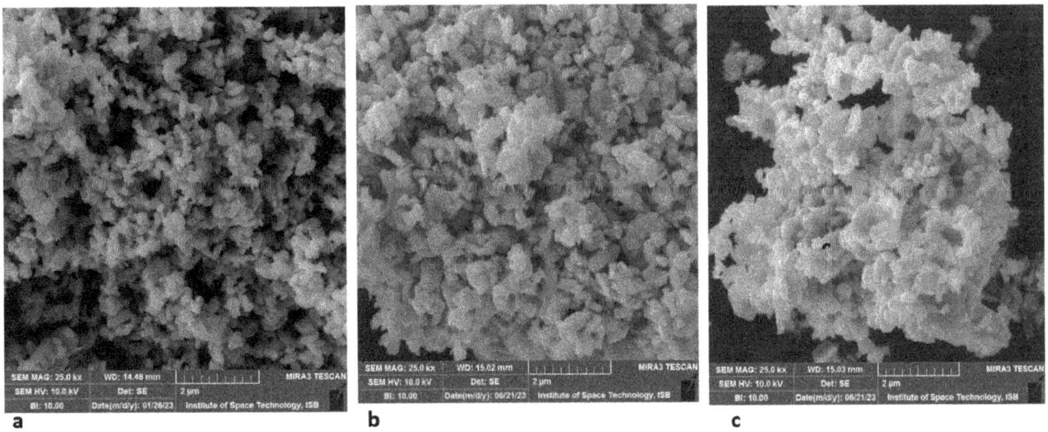

Figure 1. Scanning electron microscopic images of tungsten oxide nanoparticles alone and coated with antibiotics. (**a**) SEM image of tungsten oxide nanoparticle, (**b**) SEM image of tungsten oxide nanoparticles coupled with ampicillin, (**c**) SEM image of tungsten oxide nanoparticles coupled with oxytetracycline.

3.1.2. Tungsten Oxide Coupled Antibiotics

It has been observed in previous studies that the most characteristic bands of oxytetracycline (OTC) fall in the 1100 cm^{-1} to 1700 cm^{-1} region. Peaks at 1653 cm^{-1} to 1522 cm^{-1} were assigned to the -C=O and -NH_2 groups of the amide group in ring A, respectively. The peaks at 1617 cm^{-1} and 1541 cm^{-1} were attributed to the -C=O group in ring C and amide -NH, respectively. The 1457 cm^{-1} peak, on the other hand, was attributed to C=C skeletal vibration. The O-H stretching vibration in the alcohol and phenolic groups is responsible for the intense stretching bond centered between 3300 cm^{-1} to 3500 cm^{-1}. When a tungsten oxide nanoparticle (WO_3) is associated with it, a tungsten oxide appears at 799 cm^{-1}, indicating that the W-O-W stretching vibration is present. These peaks confirmed the presence of pure WO_3 nanoparticles in a sample and their interaction with OTC (Figure 2).

The peak at 799 cm^{-1} indicates that W-O-W stretching vibrations are present in WO_3. The prominent peak at 1550 cm^{-1} indicates that the C-C stretching vibration of the aromatic ring is present in the sample. The presence of an alkene =C-H bend is shown by the peak at 1004 cm^{-1}. The peak from 1680 cm^{-1} to 1630 cm^{-1} reveals the presence of a C=O stretching vibration in the amide group, together with the presence of two alkyl groups. Additionally, the bending vibrations of CH_3 and CH_2 can be observed at 1475 cm^{-1} and 1365 cm^{-1}, respectively. The C=O stretching vibration in the carboxylic group generally exhibits a peak within the range of 1760 cm^{-1} to 1665 cm^{-1} in. The peak at 1585 cm^{-1} corresponds to the vibrational mode associated with the stretching of C-C bonds within the aromatic ring. The peak at 799 cm^{-1} confirms the presence of pure WO_3 nanoparticles in the sample and its interaction with ampicillin (Figure 2).

3.2. Antibiotic Susceptibility of Individual Bacteria

The response of *E. coli* against different antibiotics was more inclined towards the resistant category when compared to *S. aureus* (Table 1). None of the *E. coli* isolates showed less than 20% resistance, however the highest percentage of resistant *E. coli* being 60% against penicillin. In contrast, it was observed that none of the isolates exhibited a sensitivity of less than 30% against any antibiotics tested in this study. The highest percentage of sensitive strains (60%) were noted against imipenem, amikacin, and erythromycin. Subsequently, the percentages of sensitive isolates were noted to be 50% for gentamicin, 40% for erythromycin, penicillin, and ciprofloxacin, and 30% for ampicillin. The largest percentage of intermediate isolates was higher in the case of *S. aureus* as compared to those of *E. coli*. The highest proportion of intermediate susceptible isolates was reported in 40% of *S. aureus* cases against ampicillin, followed by 30% against imipenem, while in *E. coli* cases, the percentage of intermediate susceptible isolates remained between 10 and 20% against different antibiotics.

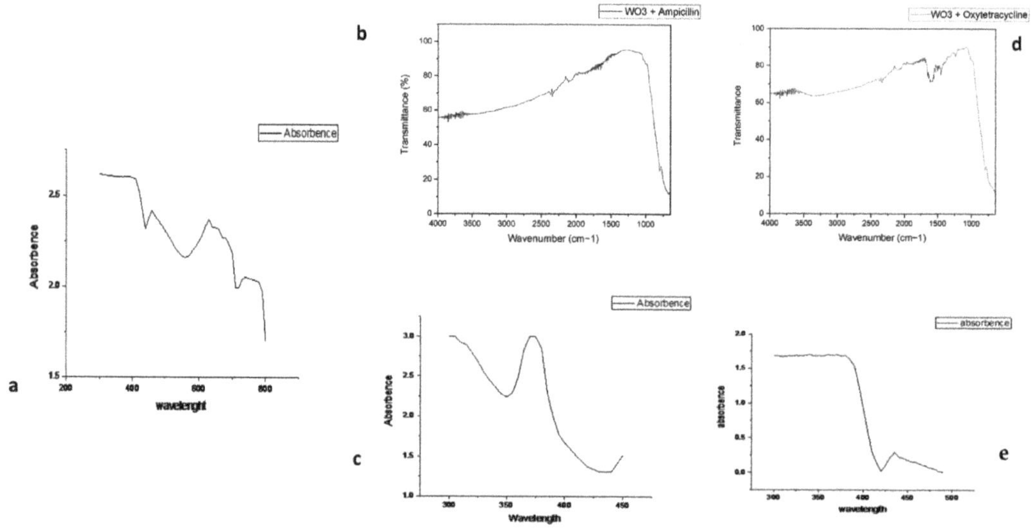

Figure 2. FTIR spectra and UV-visible spectra of tungsten oxide nanoparticles and antibiotics. (**a**) UV-visible spectra of tungsten oxide nanoparticle, (**b**) FTIR spectra of WO_3 + ampicillin, (**c**) UV-visible spectra of oxytetracycline, (**d**) FTIR spectra of WO_3 + ampicillin, (**e**) UV-visible spectra of ampicillin.

Table 1. Antibiotic susceptibility profile of *E. coli* and *S. aureus* against different antibiotics.

Antibiotic	*E. coli*			*S. aureus*		
	R (%)	**I (%)**	**S (%)**	**R (%)**	**I (%)**	**S (%)**
Imipenem	40	20	40	10	30	60
Amikacin	30	10	60	20	20	60
Oxytetracycline	30	20	50	40	20	40
Ampicillin	50	20	30	30	40	30
Gentamicin	40	10	50	30	20	50
Erythromycin	20	20	60	30	10	60
Ciprofloxacin	30	10	60	40	20	40
Penicillin	60	20	20	50	10	40

R = resistant, I = intermediate, S = sensitive.

3.3. *Response of Mixed-Culture Bacteria against Antibiotics*

A significant difference in the ZOI was observed among individual and mixed-culture bacteria when exposed to various antibiotics. The findings of this study challenge the commonly held belief that the reduction in ZOI is a consistent outcome in mixed Culture. Specifically, the results indicate that certain combinations of antibiotics, such as imipenem and ampicillin, exhibited increased susceptibility in terms of higher zones of inhibition compared to individual culture bacteria (Table 1). The variability in the response of mixed-culture bacteria in comparison to individual culture bacteria to different antibiotics was observed, with instances of both elevated and dramatically lowered responses (Figure 3).

This study revealed that the mixed culture exhibited 37.67% and 8.69% higher ZOI compared to the individual *S. aureus* when exposed to ampicillin and imipenem, respectively. However, when tested against other antibiotics, there was a significant decrease in ZOI, except for erythromycin, where no change in ZOI was observed. The most significant reduction in ZOI was observed while testing ciprofloxacin (−240%) followed by gentamicin (−225.94%), penicillin (−70.60%), amikacin (−67.84%), and oxytetracycline (−52.10%) in comparison with individual *S. aureus*. Similarly, the mixed-culture bacteria in comparison to individual *E. coli* showed an increase in percentage ZOI only in the cases of Imipenem (30.4%) and ampicillin (29.51%). Gentamicin caused the highest percentage decrease in ZOI in mixed culture over individual *E. coli* (−119.4%), followed by ciprofloxacin (−100%), penicillin (−55.95%), oxytetracycline (−20.04%), and amikacin (−17.89%). Mixed Culture versus individual *E. coli* showed lesser percentages of decreasing in ZOI when compared to the mixed Culture versus individual *S. aureus*. The findings of this study indicate that the *E. coli* bacteria exhibited a higher level of resistance compared to *S. aureus*. Furthermore, the presence of *E. coli* in polymicrobial (mixed) Culture proved to be the major factor contributing to the observed resistance.

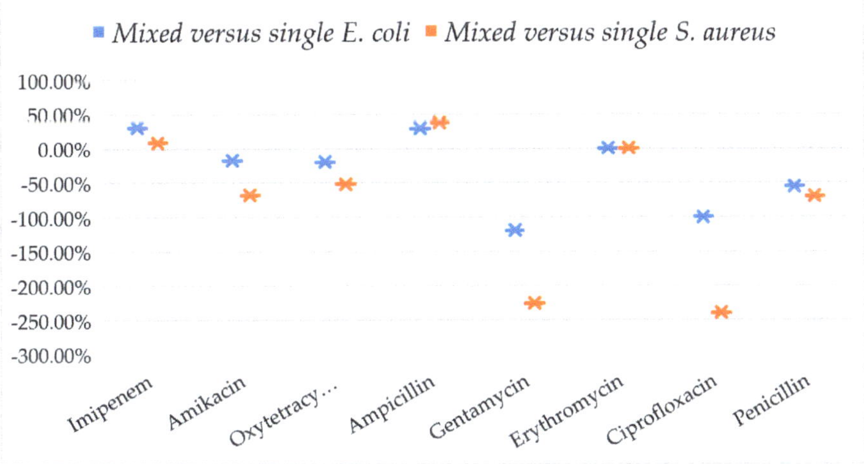

Figure 3. Percentage variation (increase/decrease) in zones of inhibition of mixed culture (*E. coli* plus *S. aureus*) in comparison to individual-culture *E. coli* and individual culture *S. aureus* against different antibiotics. This percentage was measured as "(ZOI of mixed culture − ZOI of individual culture)/mixed culture × 100".

3.4. Antibacterial Potential of Nanoparticle-Coupled Antibiotics against individual and Mixed-Culture Bacteria

The MICs of tungsten oxide (W) nanoparticles alone and tungsten oxide nanoparticle coupled oxytetracycline (WO) and ampicillin (WA) against specific *E. coli* showed significant differences ($p < 0.05$). A similar response was noticed in the cases of individual *S. aureus* and mixed Culture. The highest MIC (2083.33 ± 721.69 µg/mL) was noted in the case of W against mixed culture after 4 h of incubation, while the lowest (19.53 ± 0.00 µg/mL) was noted in the case of WO against individual *S. aureus* after 28 h of incubation (Table 2). However, it was noteworthy that a non-significant difference ($p < 0.05$) in MICs was found between WA and WO at all types of tested incubation periods. The results indicate that the MIC for W in mixed-culture bacteria was found to be higher in comparison to that of individual bacteria (Figures 4 and 5). The MIC of WA against mixed culture, when compared to *E. coli* alone, exhibited elevated values across all types of tested incubation periods except at the 20th and 28th hours of incubation. During this period, no changes in MIC or reduction in MIC, respectively, were observed for the mixed culture in comparison to the individual *E. coli*. Except for the 24th and 28th hours of incubation (where the MIC remained unchanged), there was an observed increase in the MIC of the mixed culture in comparison to that of individual *S. aureus*. With the response of mixed Culture in comparison to those of individual *S. aureus* and individual *E. coli*, it was shown that the application of WO resulted in an increase in MIC at different incubation periods.

Table 2. Comparisons of minimum inhibitory concentrations of nanoparticle-coupled antibiotics against *E. coli*, *S. aureus*, and their mixed culture.

Bacterial Culture Type	Preparation	Minimum Inhibitory Concentration (μg/mL) at Different Incubation Periods			
		4 h	20 h	24 h	28 h
E. coli alone	WA	520.83 ± 180.42^{a}	65.10 ± 22.55^{a}	52.08 ± 22.55^{a}	39.06 ± 0.00^{a}
	WO	312.33 ± 0.29^{a}	65.10 ± 22.55^{a}	39.06 ± 0.00^{a}	32.55 ± 11.28^{a}
	W	1458.33 ± 954.70^{a}	260.42 ± 90.21^{b}	130.21 ± 45.10^{b}	78.12 ± 0.00^{a}
S. aureus alone	WA	416.67 ± 180.42^{a}	65.10 ± 22.55^{a}	52.08 ± 22.55^{ab}	26.04 ± 11.28^{a}
	WO	312.33 ± 0.29^{a}	52.08 ± 22.55^{a}	26.04 ± 11.28^{b}	19.53 ± 0.00^{a}
	W	833.33 ± 360.84^{a}	156.25 ± 0.00^{b}	104.17 ± 45.10^{a}	65.10 ± 22.55^{b}
Mixed (*E. coli* + *S. aureus*)	WA	520.83 ± 180.42^{a}	91.15 ± 59.67^{a}	52.08 ± 22.55^{a}	26.04 ± 11.28^{a}
	WO	520.83 ± 180.42^{a}	104.17 ± 45.10^{a}	65.10 ± 22.55^{a}	39.06 ± 0.00^{a}
	W	2083.33 ± 721.69^{b}	625.00 ± 0.00^{b}	312.50 ± 270.63^{a}	156.25 ± 0.00^{b}

Different alphabetic (a, b) superscripts within the column of incubation for different preparations against each bacterial culture type indicate significant differences ($p < 0.05$); WA = tungsten oxide nanoparticle-coupled ampicillin, WO = tungsten oxide nanoparticle-coupled oxytetracycline, W = tungsten oxide nanoparticles alone.

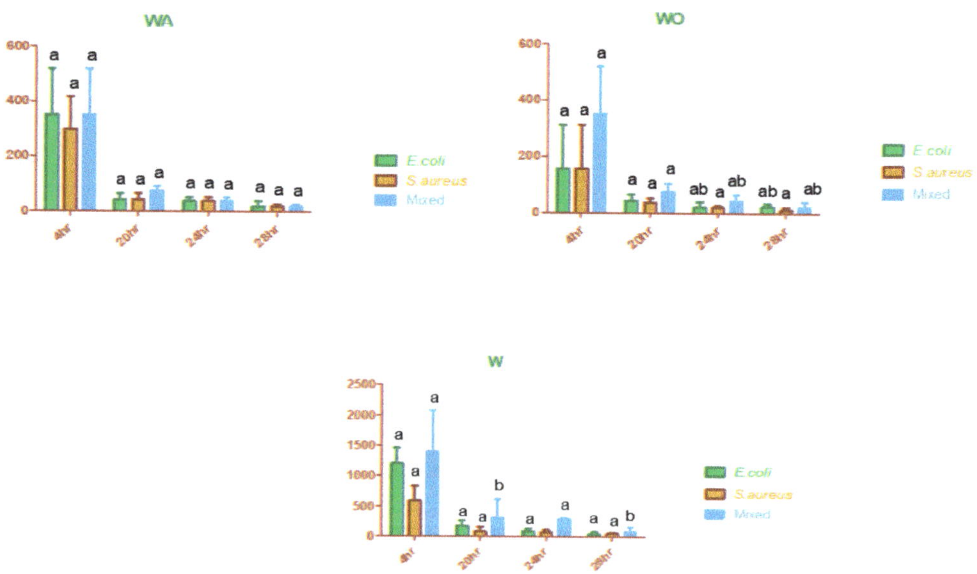

Figure 4. Comparison of responses among individual *E. coli*, individual *S. aureus*, and mixed-culture bacteria against nanoparticles alone and nanoparticle-coupled antibiotics at different incubation periods. Y-axis shows minimum inhibitory concentrations (μg/mL) of *E. coli*, *S. aureus*, and mixed Culture while x-axis shows incubation periods (hours). Different superscripts against each preparation at each incubation interval indicate significant differences ($p < 0.05$), WA = Tungsten oxide nanoparticle coupled ampicillin, WO = Tungsten oxide nanoparticle coupled oxytetracycline, W = Tungsten oxide nanoparticle alone.

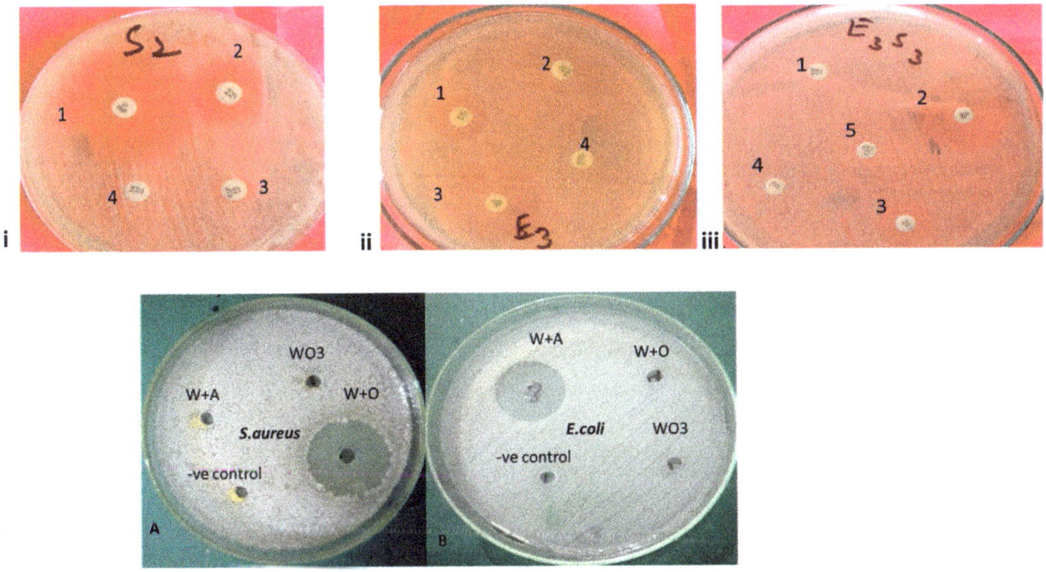

Figure 5. Antibacterial potential of antibiotics through disc diffusion (**i–iii**) and nanoparticle-coupled antibiotics (**A**,**B**) through well diffusion method. 1,2,3,4,5 are indicating different antibiotics, Hand-written S_2 and E_3 are individual bacteria of *S. aureus* and *E. coli*, respectively while E_3S_3 mixed culture, (**i**) = Antibiotic susceptibility of *S. aureus*, (**ii**) = Antibiotic susceptibility of *E. coli*, (**iii**) = Antibiotic susceptibility of mixed culture of *S. aureus* and *E. coli*. Plates A and B show zones of inhibition produced against *S. aureus* (**A**) and *E. coli* (**B**), respectively, by oxide nanoparticles (WO_3), tungsten oxide, tungsten oxide coupled with ampicillin (W + A), tungsten oxide, and tungsten oxide coupled with oxytetracycline (W + O).

There was a statistically significant decrease ($p < 0.05$) in the MIC after 4 h of incubation for each of W, WA, and WO against *S. aureus*, *E. coli*, and mixed-culture bacteria. However, the reduction in MIC remained stable after 24 h without any significant change. The data also indicated that maximum effectiveness in terms of reduced MIC may be attained following a 20 h incubation period. This trend was found to be equally applicable for W, WA, and WO against *S. aureus* alone, *E. coli* alone, and mixed culture (Table S1, Figure 3).

4. Discussion

4.1. Characterization of Nanoparticle-Coupled Antibiotics

There is a clear peak in our current FTIR spectra around 799 cm^{-1}, which is a signal of stretching vibrations in W-O-W and bending vibrations in W-O and O-W. Similarly, studies on the stretching vibration of W-O-W linkages have yielded similar results [23]. The pH of the solution governs the nature of the types of present, and tungsten ions have a strong tendency to agglomerate in an acidic medium to produce a diverse spectrum of polyanions. Other studies have indicated that the broadband appearing in the 600–800 cm^{-1} range has been attributed to the O–W–O stretching modes present in the crystal structure of WO_3 [24].

4.2. Antibiotic Susceptibility

In the past few years, antimicrobial resistance in mastitis pathogens has received attention due to the burden it imposes on lactating cattle in terms of drug toxicity and increased costs of therapeutics. Some studies on antimicrobial resistance were in line with the findings of the current study, but several other studies reported results in contradiction to that of the current study. The difference might be attributed to various factors, such as previous exposure to antibiotics, hygiene, consultation with a veterinarian, implementation of a mastitis control program, or overall animal health. This study contrasts with the findings of Singh et al. [25], who reported the high efficacy of ciprofloxacin against *E. coli* and *S. aureus*. As per the findings of the previous study [26], *E. coli* was found to be sensitive to erythromycin, while in another study it showed resistance to the same antibiotic [27]. In the case of *S. aureus*, other studies reported that gentamicin and enrofloxacin had the highest sensitivity, a finding that contradicts the results of current study [28]. In our study, amikacin, imipenem, and erythromycin exhibited the highest efficacy, whereas ampicillin was found to have the least, which is in accordance with the findings of Verma et al. [29]. The study conducted by León et al. [30] reported penicillin to be highly effective against *S. aureus*, which is in line with the findings of the current study. The utilization of antibiotics for therapeutic and nutritional purposes has resulted in a significant increase in the prevalence of pathogenic bacteria that exhibit resistance to many currently available antimicrobial agents.

4.3. Resistance Modulation by Nanoparticle Coupled Antibiotics

In the current study, nanoparticle-coupled antibiotics showed the highest MIC against *E. coli* and *S. aureus*. Like our findings, other studies on WO_3-X nanodots have also shown significant bactericidal activity attributed to their membrane stress and photocatalytic properties, resulting in a significantly enhanced antibacterial activity [14]. Similarly, Liu et al. [31] showed that increasing the concentration of WS_2 and prolonging the incubation time resulted in the eradication of both the *E. coli* and *S. aureus*. Previous reports have demonstrated that the antibacterial properties of WO_3-X nanodots can be attributed to membrane stress and their photocatalytic properties [32]. A study conducted by Duan et al. [15], showed the remarkable antibacterial efficacy of tungsten oxide nanoparticles by a one-pot synthetic approach against Gram-negative *E. coli* and Gram-positive *S. aureus*. The antibacterial properties of tungsten nanoparticles were also investigated by Syed et al. [33] against *E. coli* and *S. aureus*, resulting in the inhibition of bacterial growth. In a study conducted by Ghasempour et al. [34], it was shown that tungsten oxide nanodots with diameters of 50–90 nm had antibacterial activity against *E. coli* bacteria when exposed to visible light irradiation. Moreover, it has been observed in other experimental findings that nanoscale metal oxides display antimicrobial activity depending on their exposure time to microbial cells, particle size, agglomeration process, and degree of degradation [35].

5. Conclusions

This study concluded diversified trends in drug resistance in individual *S. aureus*, *E. coli*, and mixed-culture bacteria. Tungsten oxide (WO_3), on the other hand, exhibited substantial antibacterial properties against both individual bacterial strains and mixed-culture bacteria. Notably, as compared to non-coupled nanoparticles, WO_3 nanoparticles coupled with antibiotics demonstrated the highest level of antibacterial potential. These nanoparticles exhibited pronounced antibacterial effects during the initial stages of incubation and have shown promise for application in the mitigation of bacterial resistance. However, for the purpose of establishing standardized benchmarks encompassing safety, effectiveness, and stability parameters, it is advisable to undertake further in vivo investigations and field trials.

Supplementary Materials: The following supporting information can be downloaded at: https://www.mdpi.com/article/10.3390/biomedicines11112988/s1, Table S1. Comparative zones of inhibition (mm) produced by bacteria against different antibiotics.

Author Contributions: Conceptualization, T.K., S.R.K., L.I.A., A.I.A. and K.L.; Methodology, S.Z., S.P., A.B., S.R.K., A.M., A.N. and A.I.A.; Software, A.M., A.N., N.A. and K.L.; Validation, S.P., L.I.A. and N.A.; Formal analysis, S.Z., S.P., T.K., S.R.K., L.I.A., M.S., A.I.A. and K.L.; Investigation, M.S. and A.I.A.; Resources, A.N., L.I.A., M.S., A.I.A. and K.L.; Data curation, A.N., M.S., N.A. and A.I.A.; Writing—original draft, S.Z., A.B., S.R.K., A.N., A.I.A. and K.L.; Writing—review & editing, A.B., T.K., S.R.K., A.M., L.I.A., M.S., N.A., A.I.A. and K.L.; Visualization, A.I.A.; Supervision, T.K., N.A.

and A.I.A.; Project administration, T.K.; Funding acquisition, L.I.A., M.S. and N.A. All authors have read and agreed to the published version of the manuscript.

Funding: The research was funded by Princess Nourah bint Abdulrahman University Researchers Supporting Project number (PNURSP2023R341), Princess Nourah bint Abdulrahman University, Riyadh, Saudi Arabia.

Institutional Review Board Statement: Not applicable.

Informed Consent Statement: Not applicable.

Data Availability Statement: Not applicable.

Conflicts of Interest: The authors declare no conflict of interest.

References

1. Sarwar, I.; Ashar, A.; Mahfooz, A.; Aqib, A.I.; Saleem, M.I.; Butt, A.A. Evaluation of antibacterial potential of raw turmeric, nano-turmeric, and NSAIDs against multiple drug resistant *S. aureus* and *E. coli* isolated from animal wounds. *Pak. Vet. J.* **2021**, *41*, 14.
2. Busanello, M.; Rossi, R.S.; Cassoli, L.D.; Pantoja, J.C.F.; Machado, P.F. Estimation of prevalence and incidence of subclinical mastitis in a large population of Brazilian dairy herds. *J. Dairy Sci.* **2017**, *100*, 6545–6553. [CrossRef] [PubMed]
3. Tacconelli, E. *Global Priority List of Antibiotic-Resistant Bacteria to Guide Research, Discovery, and Development*; Infection Control Africa Network: Cape Town, South Africa, 2017.
4. El-Sayed, A.; Kamel, M. Bovine mastitis prevention and control in the post-antibiotic era. *Trop. Anim. Health Prod.* **2021**, *53*, 236. [CrossRef] [PubMed]
5. Akers, R.M.; Nickerson, S.C. Mastitis and its impact on structure and function in the ruminant mammary gland. *J. Mammary Gland. Biol. Neoplasia* **2011**, *16*, 275–289. [CrossRef] [PubMed]
6. Mullard, A. FDA drug approvals. *Nat. Rev. Drug Discov.* **2018**, *17*, 81–86. [CrossRef]
7. Ko, J.H.; Müller, M.A.; Seok, H.; Park, G.E.; Lee, J.Y.; Cho, S.Y.; Peck, K.R. Serologic responses of 42 MERS-coronavirus-infected patients according to the disease severity. *Diagn. Microbiol. Infect. Dis.* **2017**, *89*, 106–111. [CrossRef]
8. Aymen; Naima; Aqib, A.I.; Akram, K.; Majeed, H.; Murtaza, M.; Muneer, A.; Alsayeqh, A.F. Resistance modulation of dairy milk borne *Streptococcus agalactiae* and *Klebsiella pneumoniae* through metallic oxide nanoparticles. *Pak. Vet. J.* **2022**, *42*, 424–428. [CrossRef]
9. Ahmed, A.; Ijaz, M.; Khan, J.A.; Anjum, A.A. Molecular characterization and therapeutic insights into biofilm positive Staphylococcus aureus isolated from bovine subclinical mastitis. *Pak. Vet. J.* **2022**, *42*, 584–590. [CrossRef]
10. Bayda, S.; Adeel, M.; Tuccinardi, T.; Cordani, M.; Rizzolio, F. The history of nanoscience and nanotechnology: From chemical–physical applications to nanomedicine. *Molecules* **2019**, *25*, 112. [CrossRef]
11. Troncarelli, M.Z.; Brandão, H.M.; Gern, J.C.; Guimarães, A.S.; Langoni, H. Nanotechnology and antimicrobials in veterinary medicine. *Formatex* **2013**, *13*, 543–556.
12. Li, X.; Bai, H.; Yang, Y.; Yoon, J.; Wang, S.; Zhang, X. Supramolecular antibacterial materials for combatting antibiotic resistance. *Adv. Mater.* **2019**, *31*, 1805092. [CrossRef]
13. Zhu, W.; Liu, J.; Yu, S.; Zhou, Y.; Yan, X. Ag loaded WO3 nanoplates for efficient photocatalytic degradation of sulfanilamide and their bactericidal effect under visible light irradiation. *J. Hazard. Mater.* **2016**, *318*, 407–416. [CrossRef]
14. Acosta, M.; M'endez, R.A.; Riech, I.; Rodríguez-Pérez, M.; Gattorno, G.R. Structural, optical and photoelectrochemical properties of tungsten oxide thin films grown by non-reactive RF-sputtering, Superlattice. *Superlattices Microstruct.* **2019**, *127*, 123–127. [CrossRef]
15. Duan, G.; Chen, L.; Jing, Z.; De Luna, P.; Wen, L.; Zhang, L.; Zhao, L.; Xu, J.; Li, Z.; Yang, Z.; et al. Robust antibacterial activity of tungsten oxide (WO3-X) nanodots, Chem. *Res. Toxicol.* **2019**, *32*, 1357–1366. [CrossRef]
16. Matharu, R.K.; Ciric, L.; Ren, G.; Edirisinghe, M. Comparative study of the antimicrobial effects of tungsten nanoparticles and tungsten nanocomposite fibres on hospital acquired bacterial and viral pathogens. *Nanomaterials* **2020**, *10*, 1017. [CrossRef]
17. Thrusfield, M. Sampling. In *Veterinary Epidemiology*; Blackwell Science Ltd.: London, UK, 2007; pp. 214–256.
18. Muhammad, G.; Naureen, A.; Asi, M.N.; Saqib, M. Evaluation of a 3% surf solution (surf field mastitis test) for the diagnosis of subclinical bovine and bubaline mastitis. *Trop. Anim. Health Prod.* **2010**, *42*, 457–464. [CrossRef]
19. Sobur, M.A.; Sabuj, A.A.M.; Sarker, R.; Rahman, A.T. Antibiotic-resistant Escherichia coli and Salmonella spp. associated with dairy cattle and farm environment having public health significance. *Vet. World* **2019**, *12*, 984. [CrossRef]
20. *M02-A11*; Performance Standards for Antimicrobial Disk Susceptibility Testing. CLSI: Berwyn, PA, USA, 2018.
21. Anwar, M.A.; Aqib, A.I.; Ashfaq, K.; Deeba, F.; Khan, M.K.; Khan, S.R.; Muzammil, I.; Shoaib, M.; Naseer, M.A.; Riaz, T.; et al. Antimicrobial resistance modulation of MDR *E. coli* by antibiotic-coated ZnO nanoparticles. *Microb. Pathog.* **2020**, *148*, 104450. [CrossRef]

22. Muneer, A.; Kumar, S.; Aqib, A.I.; Khan, S.R.; Shah, S.Q.A.; Zaheer, I.; Rehman, T.U.; Abbas, A.; Hussain, K.; Rehman, A.; et al. Evaluation of Sodium Alginate Stabilized Nanoparticles and Antibiotics against Drug-Resistant Escherichia coli Isolated from Gut of Houbara Bustard Bird. *Oxidative Med. Cell. Longev.* **2022**, *2022*, 7627759. [CrossRef]
23. Sharma, N.; Deepa, M.; Varshney, P.; Agnihotry, S.A. FTIR and absorption edge studies on tungsten oxide based precursor materials synthesized by sol–gel technique. *J. Non-Cryst. Solids* **2002**, *306*, 129–137. [CrossRef]
24. Ingham, B.; Chong, S.V.; Tallon, J.L. Layered tungsten oxide-based organic-inorganic hybrid materials: An infrared and Raman study. *J. Phys. Chem. B* **2005**, *109*, 4936–4940. [CrossRef] [PubMed]
25. Tanzin, T.; Nazir, K.N.H.; Zahan, M.N.; Parvej, M.S.; Zesmin, K.; Rahman, M.T. Antibiotic resistance profile of bacteria isolated from raw milk samples of cattle and buffaloes. *J. Adv. Vet. Anim. Res.* **2016**, *3*, 62–67. [CrossRef]
26. Singh, K.; Chandra, M.; Kaur, G.; Narang, D.; Gupta, D.K. Prevalence and antibiotic resistance pattern among the mastitis-causing microorganisms. *Open J. Vet. Med.* **2018**, *8*, 54. [CrossRef]
27. Mir, A.Q.; Bansal, B.K.; Gupta, D.K. Subclinical mastitis in machine milked dairy farms in Punjab: Prevalence, distribution of bacteria and current antibiogram. *Vet. World* **2014**, *7*, 291–294. [CrossRef]
28. Dhakal, I.P.; Dhakal, P.; Koshihara, T.; Nagahata, H. Epidemiological and bacteriological survey of buffalo mastitis in Nepal. *J. Vet. Med. Sci.* **2007**, *69*, 1241–1245. [CrossRef]
29. Verma, H.; Rawat, S.; Sharma, N.; Jaiswal, V.; Singh, R.; Harshit, V. Prevalence, bacterial etiology and antibiotic susceptibility pattern of bovine mastitis in Meerut. *J. Entomol. Zool. Stud.* **2018**, *6*, 706–709.
30. León-Galván, M.F.; Barboza-Corona, J.E.; Lechuga-Arana, A.A.; Valencia-Posadas, M.; Aguayo, D.D.; Cedillo-Pelaez, C.; Gutierrez-Chavez, A.J. Molecular detection and sensitivity to antibiotics and bacteriocins of pathogens isolated from bovine mastitis in family dairy herds of central Mexico. *BioMed Res. Int.* **2015**, *2015*, 615153. [CrossRef]
31. Liu, X.; Duan, G.; Li, W.; Zhou, Z.; Zhou, R. Membrane destruction-mediated antibacterial activity of tungsten disulfide (WS_2). *RSC Adv.* **2017**, *7*, 37873–37880. [CrossRef]
32. Perelshtein, I.; Lipovsky, A.; Perkas, N.; Gedanken, A.; Moschini, E.; Mantecca, P. The influence of the crystalline nature of nano-metal oxides on their antibacterial and toxicity properties. *Nano Res.* **2015**, *8*, 695–707. [CrossRef]
33. Syed, M.A.; Manzoor, U.; Shah, I.; Bukhari, S.H.A. Antibacterial effects of Tungsten nanoparticles on the Escherichia coli strains isolated from catheterized urinary tract infection (UTI) cases and Staphylococcus aureus. *New Microbiol.* **2010**, *33*, 329–335.
34. Ghasempour, F.; Azimirad, R.; Amini, A.; Akhavan, O. Visible light photoinactivation of bacteria by tungsten oxide nanostructures formed on a tungsten foil. *Appl. Surf. Sci.* **2015**, *338*, 55–60. [CrossRef]
35. Mansoor, A.; Mehmood, M.; Hassan, S.M.U.; Ali, M.I.; Badshah, M.; Jamal, A. Anti-bacterial effect of titanium-oxide nanoparticles and their application as alternative to antibiotics. *Pak. Vet. J.* **2023**, *43*, 269–275. [CrossRef]

biomedicines

Article

Synergy Assessment of Four Antimicrobial Bioactive Compounds for the Combinational Treatment of Bacterial Pathogens

Kevin Masterson [1,*], Ian Major [2], Mark Lynch [1] and Neil Rowan [1]

1 Bioscience Research Institute, Technological University of the Shannon, N37 HD68 Athlone, Ireland; marklynch@tus.ie (M.L.); nrowan@tus.ie (N.R.)
2 PRISM Research Institute, Technological University of the Shannon, N37 HD68 Athlone, Ireland; imajor@tus.ie
* Correspondence: k.masterson@research.ait.ie

Abstract: Antimicrobial resistance (AMR) has become a topic of great concern in recent years, with much effort being committed to developing alternative treatments for resistant bacterial pathogens. Drug combinational therapies have been a major area of research for several years, with modern iterations using combining well-established antibiotics and other antimicrobials with the aim of discovering complementary mechanisms. Previously, we characterised four GRAS antimicrobials that can withstand thermal polymer extrusion processes for novel medical device-based and therapeutic applications. In the present study, four antimicrobial bioactive—silver nitrate, nisin, chitosan and zinc oxide—were assessed for their potential combined use as an alternative synergistic treatment for AMR bacteria via a broth microdilution assay based on a checkerboard format. The bioactives were tested in arrangements of two-, three- and four-drug combinations, and their interactions were determined and expressed in terms of a synergy score. Results have revealed interesting interactions based on treatments against recognised test bacterial strains that cause human and animal infections, namely *E. coli*, *S. aureus* and *S. epidermidis*. Silver nitrate was seen to greatly enhance the efficacy of its paired treatment. Combinations with nisin, which is a lantibiotic, exhibited the most interesting results, as nisin has no effect against Gram-negative bacteria when used alone; however, it demonstrated antimicrobial effects when combined with silver nitrate or chitosan. This study constitutes the first study to both report on practical three- and four-drug combinational assays and utilise these methods for the assessment of established and emerging antimicrobials. The novel methods and results presented in this study show the potential to explore previously unknown drug combination compatibility measures in an ease-of-use- and high-throughput-based format, which can greatly help future research that aims to identify appropriate alternative treatments for AMR, including the screening of potential new bioactives biorefined from various sources.

Keywords: antimicrobial resistance; bioactives; synergy analysis; drug combinations; two-drug combination; three-drug combination; four-drug combination; checkerboard assay; broth microdilution

1. Introduction

Antimicrobial resistance (AMR) has become a topic of academic interest, as it has reached a crisis point that has driven scientists to consider novel appropriate solutions to overcome it [1–3]. An ever-increasing number of antibiotic resistant bacterial species have emerged that pose serious threats to modern medicine, causing a loss in efficacy of critical front-line therapeutics [4]. Antibiotics remain our primary means of eliminating pathological bacterial infections, and while there has been a recent resurgence in the development of novel antibiotic compounds, additional ways of tackling AMR bacteria are urgently needed [5,6]. Research that aims to discover alternative antimicrobials has been a major topic of interest, as it particularly hopes to circumvent the emerging resistance to mainstay antibiotics [7–9]. Additionally, the co-development of methods that can assess the efficacy of appropriate combinations of already established antimicrobial compounds is important to reduce reliance on a single treatment [10–12]. While a vast number of antimicrobial compounds are in use today, many have specific modes of action and, thus, have a narrow effective spectrum in terms of the bacterial species that they can target [13]. This issue can reduce their suitability in medical settings, which ideally require a more broad-spectrum treatment, given the frequent occurrence of co-infections [14]. Furthermore, selecting a narrow-effect spectrum that relies on a singular mechanism of microbial inactivation or inhibition can also make it easier for exposed bacteria to develop unwanted resistance. Thus, the use of two or more treatments in combination to treat bacterial infections represents a highly promising avenue of research. Checkerboard assays are well-documented methods used to assess the effects of different treatments when used in combination, whereby serially diluted concentrations of treatments are combined across a 96-well microtiter plate [15–18]. The resulting effects of combination therapy can be described as synergistic, additive, or antagonistic [16,19,20]. Synergy describes a total effect greater than the sum of the individual effects. An additive effect shows that the combined drugs exhibit a total effect equal to the sum of the individual effects, being no lesser or greater. An antagonistic effect describes combinations in which the total effect is lessened compared to the sum of the individual effects [21]. Combination therapies that result in an overall synergistic effect can allow a much greater impact to result from treatments that would normally hold less or, perhaps no, effect when used alone, such as in the case of AMR bacteria. While co-treatment therapies have been widely used in the treatment of diseases such as cancer, there is a rising interest in the synergistic abilities of previously established antimicrobial compounds [10,12,16].

In a previous study reported by these authors, the individual antibacterial capabilities of four GRAS bioactive compounds—silver nitrate ($AgNO_3$), nisin, chitosan and zinc oxide (ZnO)—were assessed against a number of type-strain bacterial species, as well as AMR wild-type strains [22]. These GRAS bioactives were unusual in the sense that they withstood temperatures used to extrude and process polymers used in the manufacturing of medical devices; thus, these bioactives offer interesting options for new therapeutic research. In the present study, these four bioactives will be assessed regarding their antimicrobial capabilities in combination with one another, using arrangements of two-, three- and four-drug combinations. For this initial combinational study, three standard type strains will be used, namely *Escherichia coli*, *Staphylococcus aureus* and *Staphylococcus epidermidis*. *E. coli* and *S. aureus* were chosen as they represent Gram-negative and Gram-positive bacteria, respectively. *S. epidermidis* was included as it represents opportunistic Gram-positive bacterial pathogens and was observed to hold atypical behaviour against these four compounds in the previous study relative to *S. aureus*. The antimicrobial abilities of the two-drug combinations will be determined via use of a standard broth microdilution protocol in a checkerboard assay format, through which growth will be measured using turbidity absorbance readings. Three- and four-drug combinations will be assessed via use of novel versions of the checkerboard assay developed in the present study. The readings will be used to calculate the % growth of each treatment relative to the 100% growth control. While the checkerboard assay is a method commonly utilised to assess combination effects, there are various methods and programs developed for analysis of results [15–18]. The end results identified in the present study will analysed via use of the recently developed "synergy" python package, which can analyse large amounts of combinations and report their synergy scores [23].

2. Materials and Methods

2.1. Bioactive Solution Preparation

Silver nitrate ($AgNO_3$) (SKU: S8157, CAS: 7761-88-8), nisin, 2.5% (SKU: N5764, CAS: 1414-45-5), chitosan of low molecular weight (SKU: 448869, CAS: 9012-76-4), zinc oxide

(ZnO) and nanopowder of <100 nm in particle size (SKU: 544906, CAS: 1314-13-2) were purchased from Sigma-Aldrich/Merck (Merck Life Science Limited, Arklow, Co. Wicklow, Ireland). Chitosan was dissolved in 1% (*v*/*v*) acetic acid and adjusted to pH 5.5 using 0.4 M sodium hydroxide (NaOH). ZnO was suspended in dH_2O. Nisin was dissolved in a solution of 400 mM sodium chloride (NaCl), which had a pH of 3.25. These solutions were then sterilised through autoclaving. Nisin concentrations were reported in terms of active nisin content, with 1 g of commercial nisin powder containing 25 mg of active nisin. $AgNO_3$ was placed into a solution of 28% (*v*/*v*) Poly (ethylene glycol), which had an average molecular weight of 400 (PEG-400) and 26% (*w*/*v*) d-sorbitol [24]. This solution was then filter sterilised through use of a 0.2-micrometer syringe filter tip.

2.2. Bacterial Cell Culture

The standard strains, namely *E. coli* (ATCC 25922, NCTC 12241) and *S. aureus* (ATCC 29213, NCTC 12973), were purchased from Public Health England (Culture Collections, Public Health England, Salisbury, UK). *S. epidermidis* (ATCC 35984) was purchased from ATCC (LGC Standards, Middlesex, UK). Cultures were prepared via overnight incubation using tryptone soy agar (TSA). Colonies were then suspended in Mueller–Hinton broth (MHB) to 0.5 MacFarland absorbance for use as inoculum [25,26].

2.3. Two-Drug Combinational Broth Microdilution Assay

All steps were conducted under aseptic conditions or in closed systems. The antimicrobial properties of each bioactive solution in combinations of two were assessed in terms of their growth inhibitory capabilities, as determined via use of the broth microdilution method adapted from a previously published protocol [26]. Broth microdilution assays were carried out in flat bottom 96-well plates (untreated) against three chosen bacterial strains, namely *E. coli*, *S. aureus* and *S. epidermidis*. Before use, microplate lids were treated using a hydrophilic coating (20% (*v*/*v*) of isopropyl alcohol (IPA), 0.5% (*v*/*v*) of Triton-X100) [27]. Bacterial inoculums were prepared to give a final in-well concentration of 5×10^5 cfu/mL, as determined via absorbance readings. Two-drug combination assays were prepared in an 8 × 8 checkerboard layout, allowing a total of 64 combinations. Dilutions of drug A and B were prepared in Mueller–Hinton broth (MHB) at a concentration four times higher (4×) than the highest desired final in-well concentration. Serial dilutions (1:2) of drug A and drug B were prepared in separate 96-well plates and combined in the final test plate (1:2 dilution) (See Figures S1 and S2). Each well was then inoculated with the prepared bacterial inoculum (1:2 dilution). Absorbance of the plate was measured using a BioTek® Synergy HT microplate reader and Gen5 Microplate Reader Software (Version 2.01.14) (BioTek® Instruments GmbH, Bad Friedrichshall, Germany). The plate was read using an endpoint absorbance read at 625 nM, and results were recorded as time-point 0 (t = 0) before incubation. This process allowed measurement of any turbidity caused by treatments and was be used as a blank. The plate was placed in a container to help prevent loss of well volume due to evaporation. The container was placed on a rotary incubator at 120 RPM, 37 °C for 18 h. Following incubation, plate absorbance was read (variable shake, and the 1-min endpoint absorbance was read at 625 nm). Results were recorded as timepoint 18 (t = 18). The absorbance values were used to calculate the % inhibition for each treatment well.

2.4. Three-Drug Combinational Broth Microdilution Assay

Three-drug combination assays were carried out as per the two-drug combination assay, albeit using a 6 × 6 checkerboard layout. Six such checkerboards were prepared by combining drug A and drug B, as per a two-drug combination assay, and different concentrations of drug C was added to each individual checkerboard. This setup allowed 6 concentrations of drug A, drug B and drug C to be assessed in combination (6 × 6 × 6), with a total of 216 combinations (See Figure S3 for example layout). The experiment was split across three 96-well plates, allowing two 6 × 6 checkerboards per plate. A separate broth microdilution assay of single treatments was also carried out as a control to ensure that the treatments and bacteria tested performed in a nominal manner. Incubations and absorbance readings were carried out as per two-drug combinational assay.

2.5. Four-Drug Combinational Broth Microdilution Assay

Four-drug combination assays were carried out in a 4 × 4 checkerboard layout, which built on the three-drug layout design. The layout was designed in such a way that four 4 × 4 checkerboards (CBs) were set up within four 96-well plates. Each CB combined drug A with drug B. Each of these four CBs then had a different concentration of drug C added to it. To all CBs within each plate, a difference concentration of drug D will be added. The resulting system will yield a 4 × 4 × 4 × 4 combination (totalling in 256 combinations) (see Figure S4 for example layout).

A separate broth microdilution assay was also carried out using drugs A, B, C and D in tandem with the four-drug combination assay, which was used as a control to ensure that the individual treatments and bacteria being tested performed in a nominal manner. Plate and inoculum preparation, incubations and absorbance readings were all carried out as per the two-drug combinational assay.

2.6. Analysis of Results for the Determination of Synergy/Antagonism

Results from drug combination assays were analysed to determine drug interactions in terms of synergy or antagonism via the "synergy" python package [28]. Input data for synergy were prepared in an excel document using the concentration of each drug (µg/mL) and the % growth. Input data contained an individual column for the concentration of each drug ("drug1.conc", "drug2.conc", "drug3.conc" or "drug4.conc"). The Bliss model was chosen due to its simplicity and ability to analyse four-drug combinations. The reported response was expressed in terms of % growth. The response was input under the column "effect" and expressed as a decimal fraction of 1 (i.e., 100% growth = 1.0, 50% = 0.5, 0% = 0.0). Data were then exported as a .csv file. The synergy package was opened and run using PyCharm (version 2020.2) (JetBrains s.r.o, Prague, Czech Republic), which is a python-integrated development environment (IDE). Following the synergy documentation, input data were imported and analysed using the Bliss model. Results were expressed in terms of a synergy score, with a positive score indicating synergy, a score of 0 representing no effect and negative scores representing antagonism.

3. Results

Due to the number of combinations analysed during this study, only the three highest-scoring interactions of each combination and their average values will be reported and discussed. Synergy scores represent the magnitude of the combination interactions, where a positive score indicates synergy, scores close to 0 indicate an additive effect, a score of 0 represents no effect and negative scores represent antagonism. 2D heat-maps of all 64 combinations of each two-drug combination against each test bacterial species are presented in Figures 1–6 showing each combination's synergy score based on a colour scale, which is shown in the legend. Bar graphs have been prepared and presented in Figures 7–9 for each two-drug, three-drug and four-drug combination respectively, showing the three highest-scoring drug combinations (Combo 1–3 on the x-axis) at the drug concentrations (µg/mL) shown on the left y-axis, with the calculated Bliss synergy score shown on the right y-axis (see Tables S1–S5 for the graphed data).

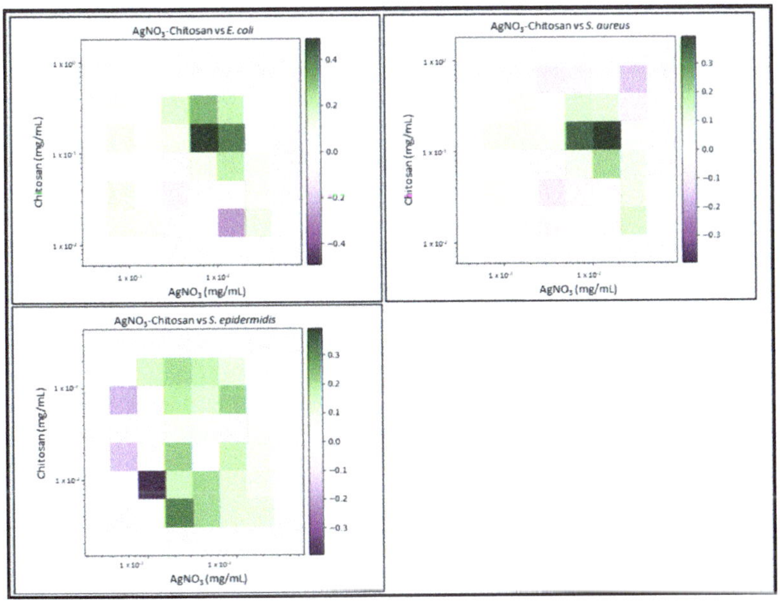

Figure 1. $AgNO_3$–chitosan synergy heat map: Graphs show heat map of synergy between silver nitrate ($AgNO_3$) and Chitosan in inhibiting *E. coli*, *S. aureus* and *S. epidermidis* growth as determined via broth microdilution and absorbance readings. Inhibition results were analysed using the synergy python package using the Bliss synergy model. The synergy python package produced the heatmap graphs of each combination result, giving visual presentations of combinations of synergy (green) or antagonism (purple). $n = 3$.

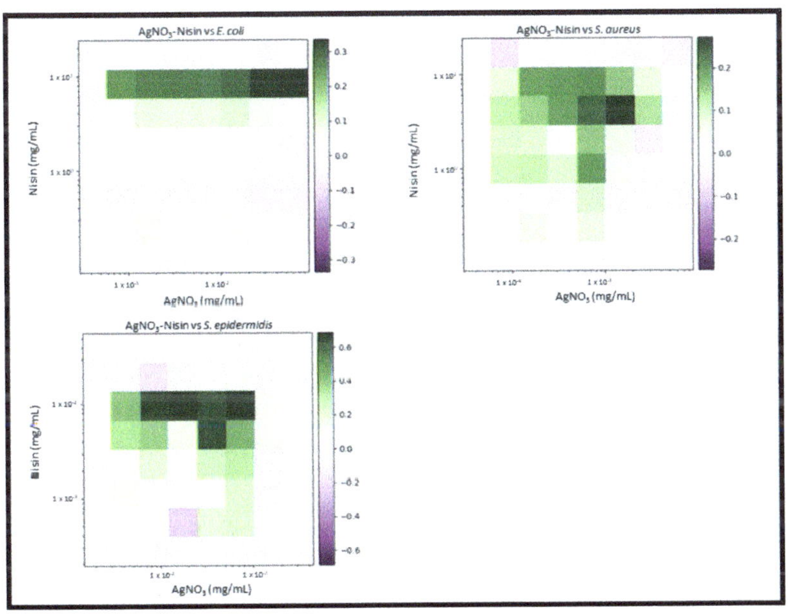

Figure 2. $AgNO_3$–nisin synergy heat map: Graphs show heat map of synergy between silver nitrate ($AgNO_3$) and Nisin in inhibiting *E. coli*, *S. aureus* and *S. epidermidis* growth as determined via broth

microdilution and absorbance readings. Inhibition results were analysed with the synergy python package using the Bliss synergy model. The synergy python package produced the heatmap graphs of each combination result, giving visual presentations of combinations of high (green) and low (purple) synergy. $n = 3$.

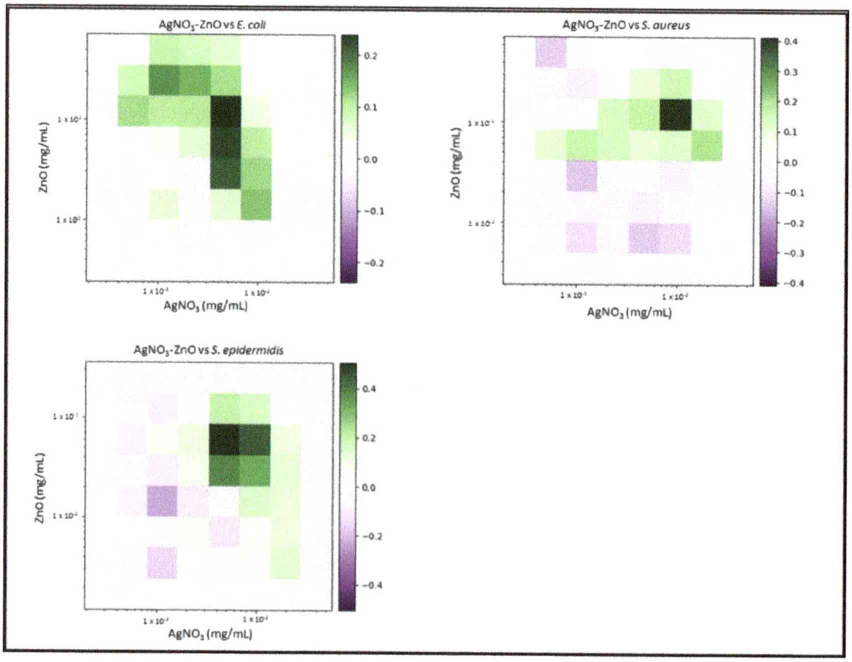

Figure 3. $AgNO_3$–ZnO synergy heat map: Graphs show heat map of synergy between silver nitrate ($AgNO_3$) and zinc oxide (ZnO) in inhibiting *E. coli*, *S. aureus* and *S. epidermidis* growth, as determined via broth microdilution and absorbance readings. Inhibition results were analysed via the synergy python package using the Bliss synergy model. The synergy python package produced the heatmap graphs of each combination result, giving visual presentations of combinations of high (green) and low (purple) synergy. $n = 3$.

3.1. *Two-Drug Combinations*

- $AgNO_3$–Chitosan

$AgNO_3$ and Chitosan reported good synergistic interactions against each bacterial strain. The combination reported the highest average synergy scores against *E. coli* (average 0.4) and *S. aureus* (average 0.32). While the average concentration of chitosan was similar to that of the MIC versus *E. coli*, $AgNO_3$ was reported to be present in lower concentrations. The most effective combination versus *S. aureus* reported concentrations that were 1/2 the MIC, with inhibition being approximately 69%. Results versus *S. epidermidis* reported good overall synergy, as much lower concentrations of each treatment exhibited more effective inhibition, and the second reported combination exhibited 99% inhibition, with 1/3 of the MIC of $AgNO_3$ and less than 1/2 of the MIC of chitosan being used.

- $AgNO_3$–Nisin

$AgNO_3$ and Nisin demonstrated a number of highly synergistic combinations (an average 0.32 versus *E. coli* and an average 0.24 versus *S. aureus*), as well as reporting the highest two-drug score from this study (average 0.68 versus *S. epidermidis*). While the highest-scoring combinations versus *E. coli* did not report inhibition exceeding 70%, there was moderate synergy observed compared to $AgNO_3$ used alone at the same concentrations.

The third highest-scoring combination versus *S. aureus* reported 99% inhibition, using less than 1/4 MIC of $AgNO_3$ and 1/10 MIC of nisin. The three highest-scoring combinations versus *S. epidermidis* indicated that a concentration of 10 μg/mL $AgNO_3$ was most effective in enabling nisin, which was reported to be present in relatively low concentrations, while still having a notable effect upon bacterial growth.

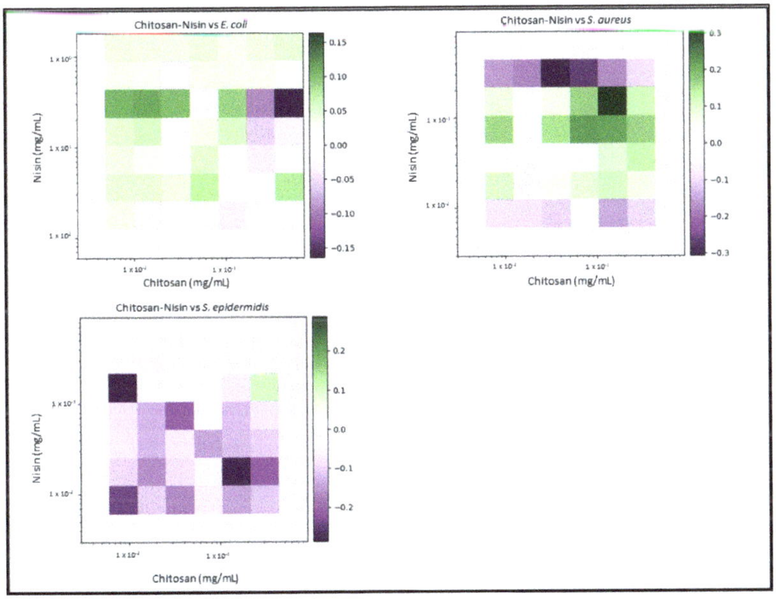

Figure 4. Chitosan–nisin synergy heat map: Graphs show heat map of synergy between Chitosan and Nisin in inhibiting *E. coli*, *S. aureus* and *S. epidermidis* growth, as determined via broth microdilution and absorbance readings. Inhibition results were analysed via the synergy python package using the Bliss synergy model. The synergy python package produced the heatmap graphs of each combination result, giving visual presentations of combinations of high (green) and low (purple) synergy. $n = 3$.

- $AgNO_3$–ZnO

$AgNO_3$ and ZnO reported moderate synergy against *E. coli* (average 0.22) and *S. aureus* (average 0.26), as well as relatively high synergy versus *S. epidermidis* (average 0.44). The highest reported *E. coli* combination exhibited 98.5% growth inhibition at an $AgNO_3$ concentration 1/4MIC and a ZnO concentration 1/2.5MIC, demonstrating a noticeable increase in the efficacy in both treatments. *S. aureus* results reported that lower concentrations of both $AgNO_3$ and ZnO exhibited greater effect when combined. One reported combination exhibited 95.5% growth inhibition using 1/1.8MIC $AgNO_3$ and 1/2.5MIC ZnO. $AgNO_3$ and ZnO demonstrated the second highest-scoring average of all two drug combinations (average 0.44) versus *S. epidermidis*. Reported combinations exhibited effective growth inhibition at much lower concentrations, even reaching 95.4% growth inhibition with 1/1.6MIC $AgNO_3$ and 1/3.33MIC ZnO.

- Nisin–Chitosan

Nisin and chitosan reported mixed results in combination. The highest scoring combinations were identified versus *S. aureus* (average 0.24); however, the highest inhibition of these combinations reached only 50%, with no major reductions being seen in the concentrations of nisin or chitosan. Results versus *E. coli* show that greater concentrations of chitosan were needed to enable nisin; however, these concentrations exceeded the MIC of chitosan, making the combination ineffective. Results versus *S. epidermidis* demonstrated no major interactions, being close a synergy score of 0 in all combinations. Only one combination reported effective synergy, which exhibited 87.3% inhibition with a score of 0.11; however, the concentration of nisin used in this combination exceeded that of its MIC when tested alone.

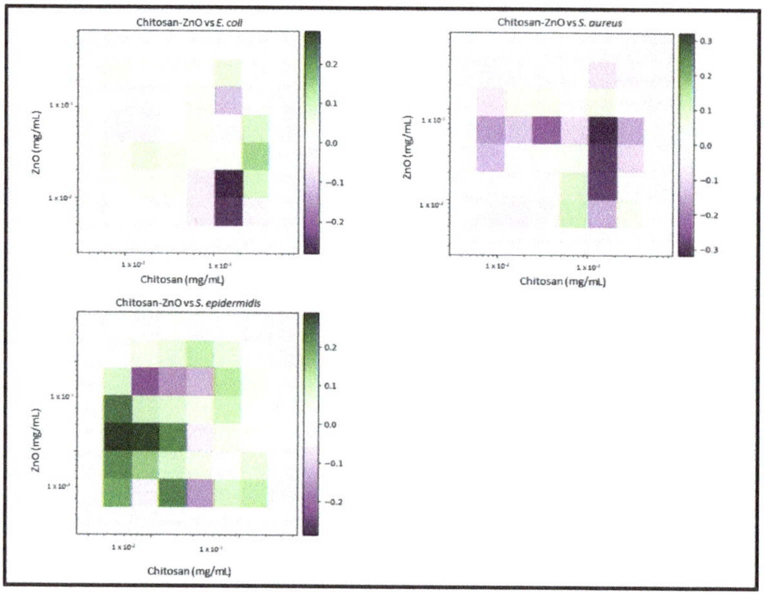

Figure 5. Chitosan–ZnO synergy heat map: Graphs show heat map of synergy between Chitosan and zinc oxide (ZnO) in inhibiting *E. coli*, *S. aureus* and *S. epidermidis* growth, as determined via broth microdilution and absorbance readings. Inhibition results were analysed via the synergy python package using the Bliss synergy model. The synergy python package produced the heatmap graphs of each combination result, giving visual presentations of combinations of high (green) and low (purple) synergy. $n = 3$.

- Chitosan–ZnO

Chitosan and ZnO reported very few synergistic interactions versus *E. coli* (average 0.11), *S. aureus* (average 0.09) and *S. epidermidis* (average 0.27). Analysis of interactions versus *E. coli* show that a high concentration of chitosan was required for synergy to be identified; however, the amount of chitosan was 2 × MIC, and the synergy score was relatively low. While synergy was seen versus *S. aureus* at quite low concentrations of the two highest-scoring combinations, the exhibited growth inhibition was not noteworthy (3.1%, 2.6%, respectively). Combinations versus *S. epidermidis* reported moderate synergy at quite low concentrations of each combination; however, the inhibition did not exceed 31%.

- Nisin–ZnO

Nisin and ZnO reported low synergy versus *E. coli* (average 0.08), *S. aureus* (average 0.06) and *S. epidermidis* (average 0.14). The highest-scoring combination versus *E. coli* (0.09) did not yield noteworthy inhibition, while the next highest-scoring combinations reported concentrations of ZnO that exceed the MIC in order to enable nisin. Highest-scoring combinations versus *S. aureus* reported low concentrations of each treatment; however, they had no noteworthy growth inhibitory effect (3.7–14%). Combinations versus *S. epidermidis* exhibited moderate inhibitory effects; however, the concentrations of ZnO exceeded that of the average MIC, and concentrations of nisin were not much lower that the previously reported MIC average.

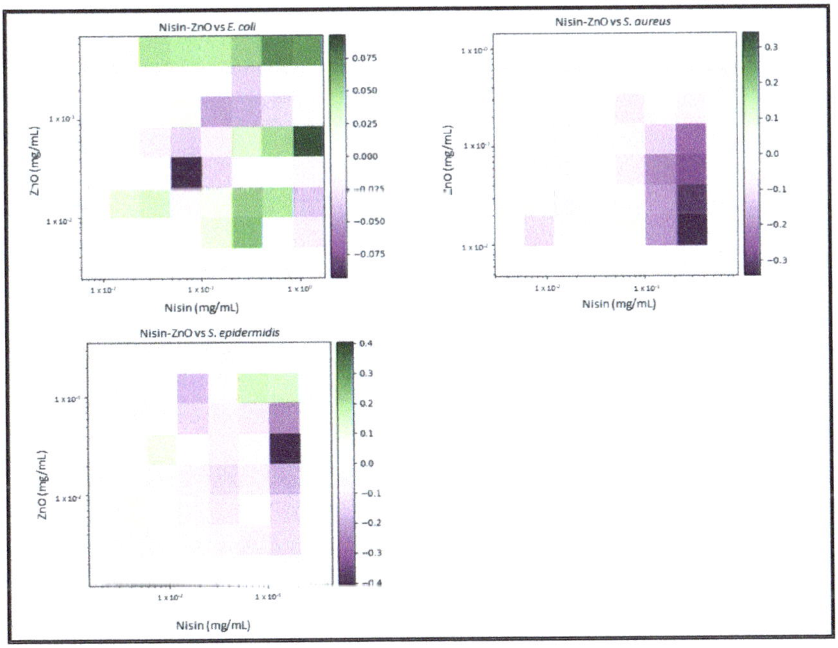

Figure 6. Nisin–ZnO synergy heat map: Graphs that show heat map of synergy between nisin and zinc oxide (ZnO) regarding the inhibition of *E. coli*, *S. aureus* and *S. epidermidis* growth, as determined via broth microdilution and absorbance readings. Inhibition results were analysed via the synergy python package using the Bliss synergy model. The synergy python package produced the heatmap graphs of each combination results, giving visual presentations of combinations of high (green) and low (purple) synergy. $n = 3$.

3.2. Three-Drug Combinations

- Chitosan–$AgNO_3$–Nisin

Chitosan, $AgNO_3$ and nisin reported moderate-to-high synergy in growth inhibition versus *E. coli* (average 0.38), *S. aureus* (average 0.56) and *S. epidermidis* (average 0.43). While the higher scoring combinations versus *E. coli* included high concentrations of chitosan (80–160 µg/mL), reported concentrations of $AgNO_3$ were low (2–4 µg/mL) and had 99% inhibition. Concentrations of nisin were rather high, relative to those of the other test species (3.91–7.81 µg/mL). The highest-scoring combination versus *S. aureus* reported relatively low concentrations of each compound (78.13 µg/mL chitosan, 8 µg/mL $AgNO_3$ and 10.63 µg/mL nisin) and expressed 99% inhibition. The second highest scoring combination showed similar concentrations and levels of inhibition; however, when used twice, the amount of $AgNO_3$ (16 µg/mL) was still less than the previously reported MIC. The third highest-scoring combination reported lower chitosan (39.06 µg/mL), though it did not fully inhibit *S. aureus* growth (71.43%). These combinations versus *S. aureus* reported the second highest average score of all three-drug test combinations. Combinations versus *S. epidermidis* reported near full inhibition (92–96%), as well as good synergy and low concentrations of chitosan (39.06–78.13 µg/mL) and nisin (0.63–1.25 µg/mL); however, concentrations of $AgNO_3$ were near to the MIC (8–16 µg/mL).

- Chitosan–$AgNO_3$–ZnO

Chitosan, $AgNO_3$ and ZnO held moderately low synergy versus *E. coli* (average 0.14), *S. aureus* (average 0.28) and *S. epidermidis* (average 0.35), with combinations versus *E. coli* having the lowest scores of all three-drug combinations. Scores for *E. coli* were quite low, and each of the reported combinations exhibited full inhibition (97.15–100%) at low concentrations of chitosan (40–80 µg/mL), $AgNO_3$ (0.5–4 µg/mL) and ZnO (20 µg/mL). The highest scores versus *S. aureus* reported consistently low concentrations of ZnO (62.5 µg/mL), as well as low concentrations of chitosan (39.06–78.13 µg/mL) and $AgNO_3$ (8–16 µg/mL), while growth inhibition was high (97.55–98.32%). Combinations versus *S. epidermidis* reported good growth inhibition (87.95–98.4%) at relatively low concentrations of chitosan (20–40 µg/mL), $AgNO_3$ (2–4 µg/mL) and ZnO (10 µg/mL).

- Nisin–$AgNO_3$–ZnO

Nisin, $AgNO_3$ and ZnO reported strong synergy versus *E. coli* (average 0.36), *S. aureus* (average 0.38) and *S. epidermidis* (average 0.53). Results versus *E. coli* show that very high concentrations of nisin (31.25 µg/mL) yielded high synergy at low concentrations of $AgNO_3$ (0.5–1 µg/mL) and ZnO (31.25–62.5 µg/mL); however, growth inhibition did not exceed 77%. Combinations versus *S. aureus* reported moderate synergy at low concentrations of nisin (0.63–1.25 µg/mL) and moderate concentrations of $AgNO_3$ (4–16 µg/mL) and ZnO (39.06 µg/mL). Combinations versus *S. epidermidis* reported the third highest average score synergy at low concentrations of nisin (1.25–2.5 µg/mL), $AgNO_3$ (2–4 µg/mL) and ZnO (5 µg/mL); however, these combinations exhibited low-to-moderate growth inhibition (33.6–72.3%).

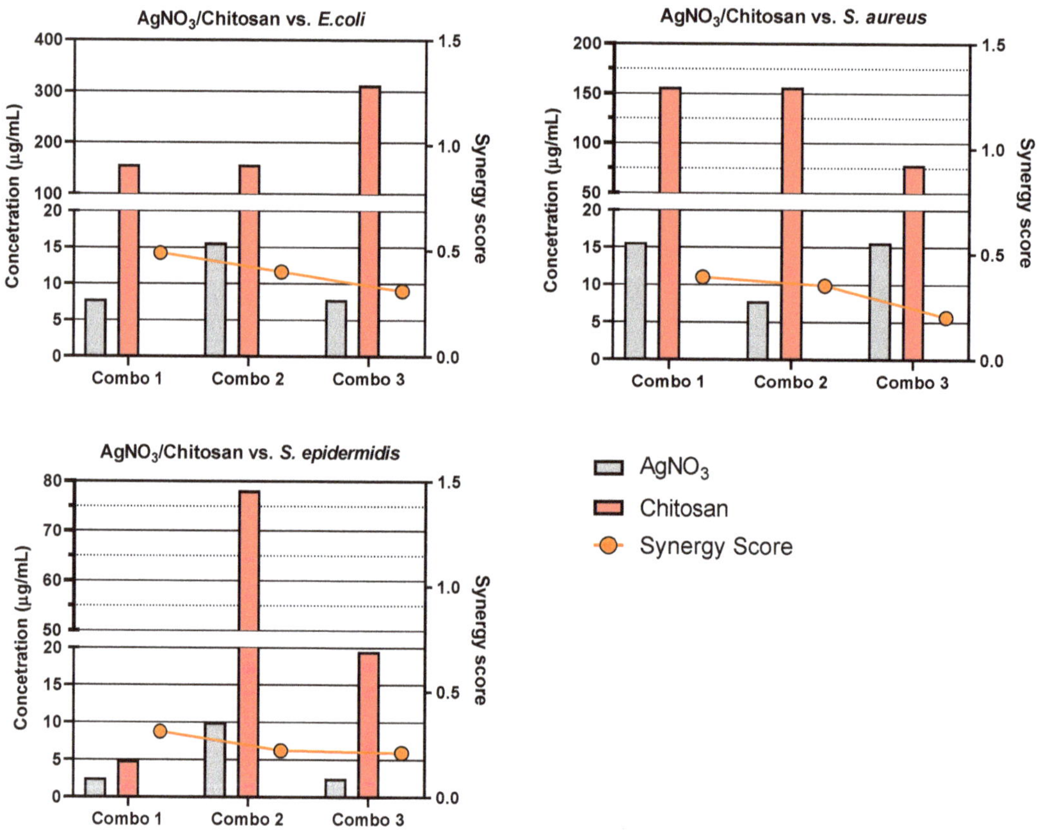

(A)

Figure 7. *Cont.*

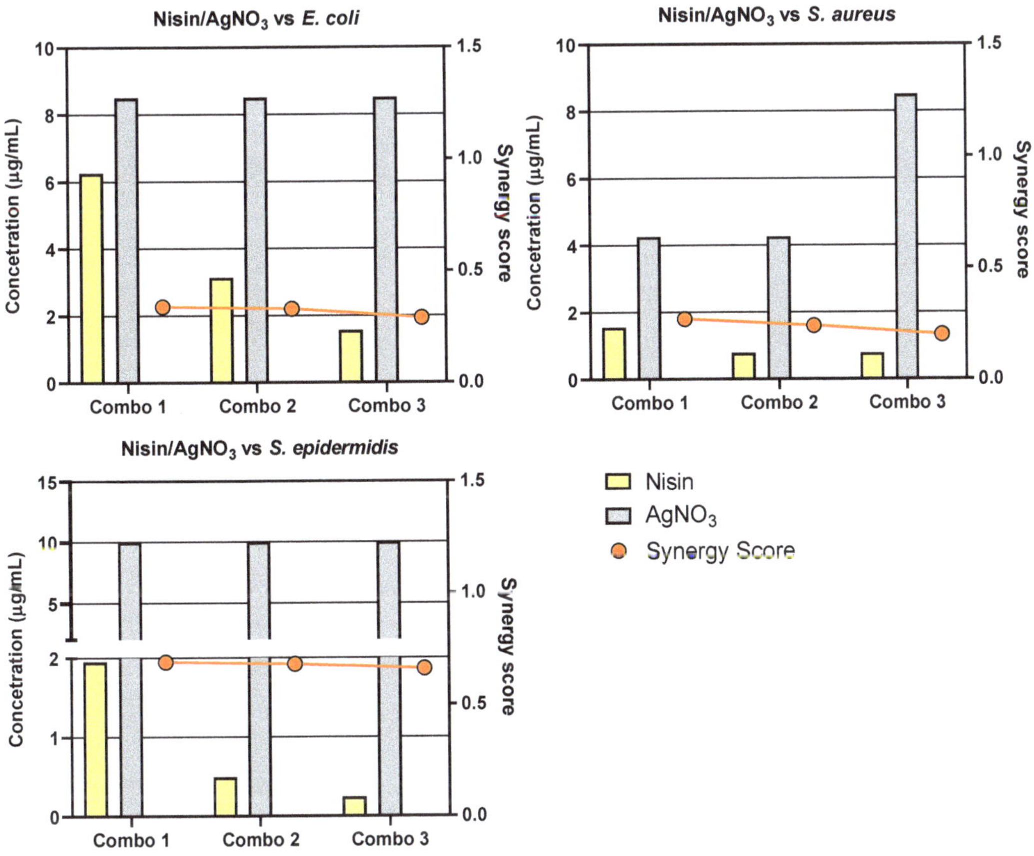

(B)

Figure 7. *Cont.*

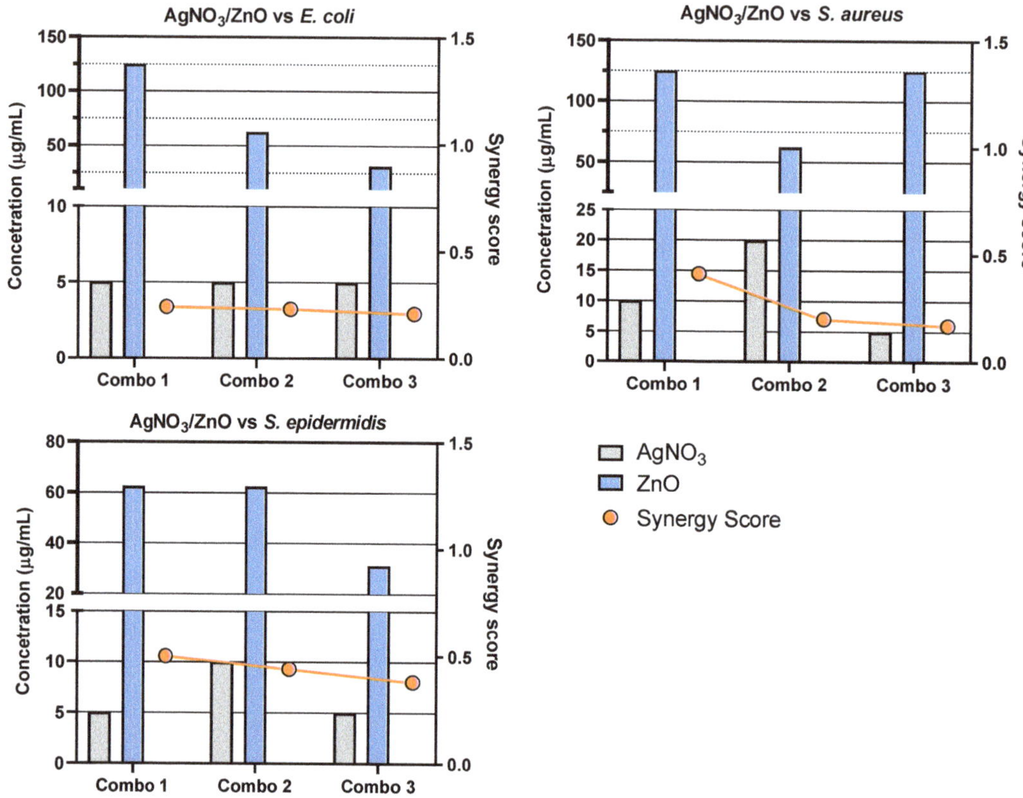

(C)

Figure 7. *Cont.*

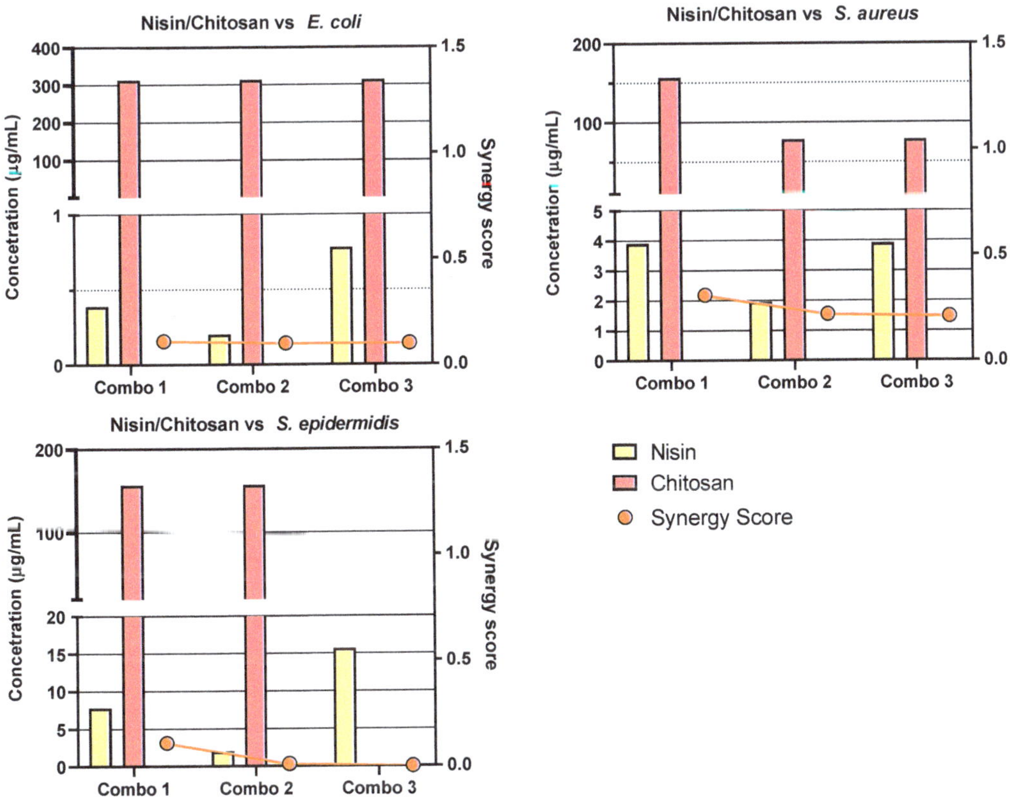

(D)

Figure 7. *Cont.*

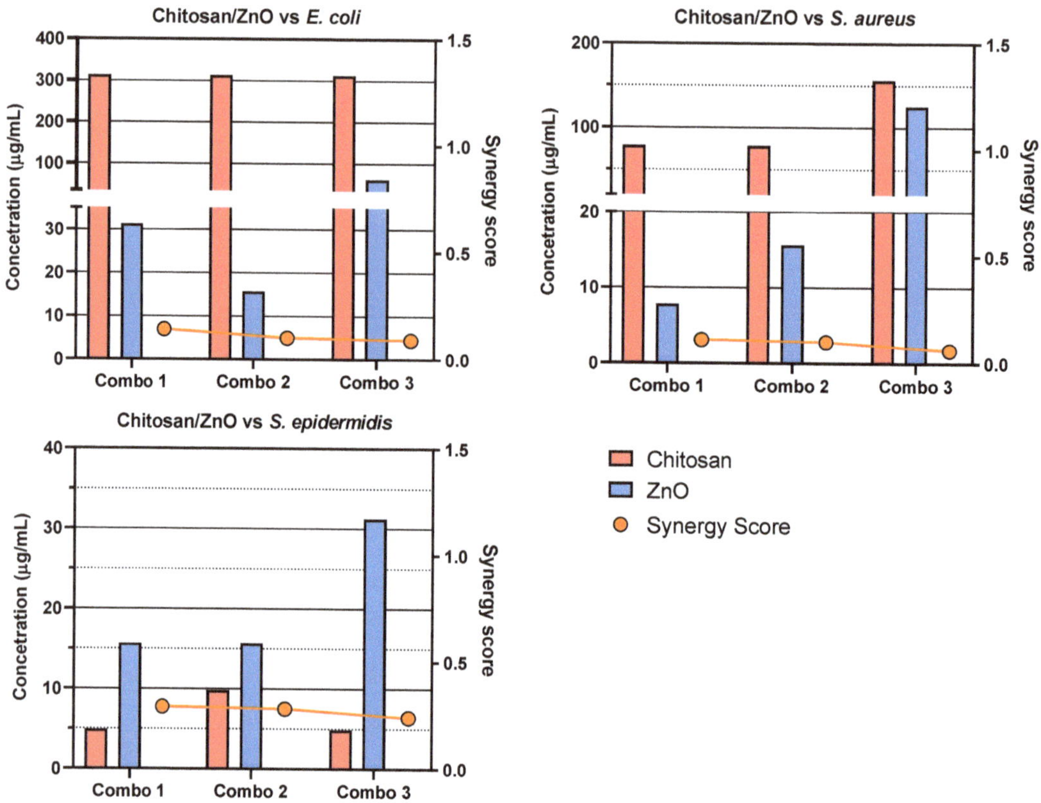

(E)

Figure 7. *Cont.*

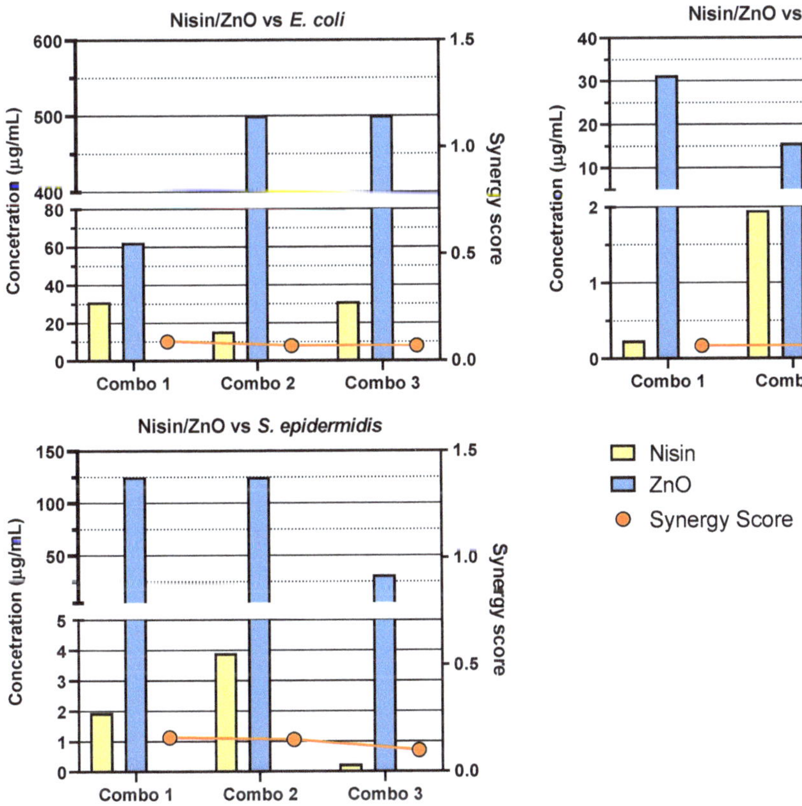

(F)

Figure 7. Synergy score results of the top three two-drug combinations: Bar graphs that present the concentrations of each drug and their respective synergy scores from the top three two-drug combinations (**A–F**) against *E. coli*, *S. aureus* and *S. epidermidis*. (**A**) $AgNO_3$/Chitosan, (**B**) Nisin/$AgNO_3$, (**C**) $AgNO_3$/ZnO, (**D**) Nisin/Chitosan, (**E**) Chitosan/ZnO and (**F**) Nisin/ZnO. Bars show drug concentrations, as indicated on the left *y*-axis, and the line/symbols show each combination's (combo) synergy score, as indicated on the right *y*-axis.

- Nisin–Chitosan–ZnO

Nisin, chitosan and ZnO reported low inhibition synergy versus *E. coli* (average 0.23) and *S. epidermidis* (average 0.21); however, combinations versus *S. aureus* reported the highest synergy score across all three-drug combinations (average 0.83). Concentrations of the reported combinations versus *E. coli* indicated poor synergy between treatments, as high concentrations of chitosan (9.77–312.5 µg/mL) and ZnO (31.25–125 µg/mL) were utilised. Combinations that used low concentrations of each drug exhibited very low inhibition (26.6%).

The highest-scoring combination (0.93) versus *S. aureus* reported low concentrations of nisin (3.91 µg/mL), chitosan (39.06 µg/mL) and ZnO (62.5 µg/mL) with high inhibition (98.7%). The second highest-scoring combination (0.84) also reported low concentrations of nisin (0.977 µg/mL), chitosan (156.25 µg/mL) and ZnO (62.5 µg/mL) with high inhibition (99.4%). While the third highest-scoring combination reported low concentrations of nisin (0.977 µg/mL), chitosan (78.13 µg/mL) and ZnO (62.5 µg/mL), the reported growth inhibition was moderate (71.5%). Moreover, while results versus *S. epidermidis* reported low concentrations of nisin (0.98–1.95 μg/mL), chitosan (39.06 μg/mL) and ZnO (31.25 μg/mL), along with high growth inhibition (99%), there was little synergy observed, as denoted by the two highest scores. The third highest-scoring combination reported a very high concentration of chitosan (625 μg/mL).

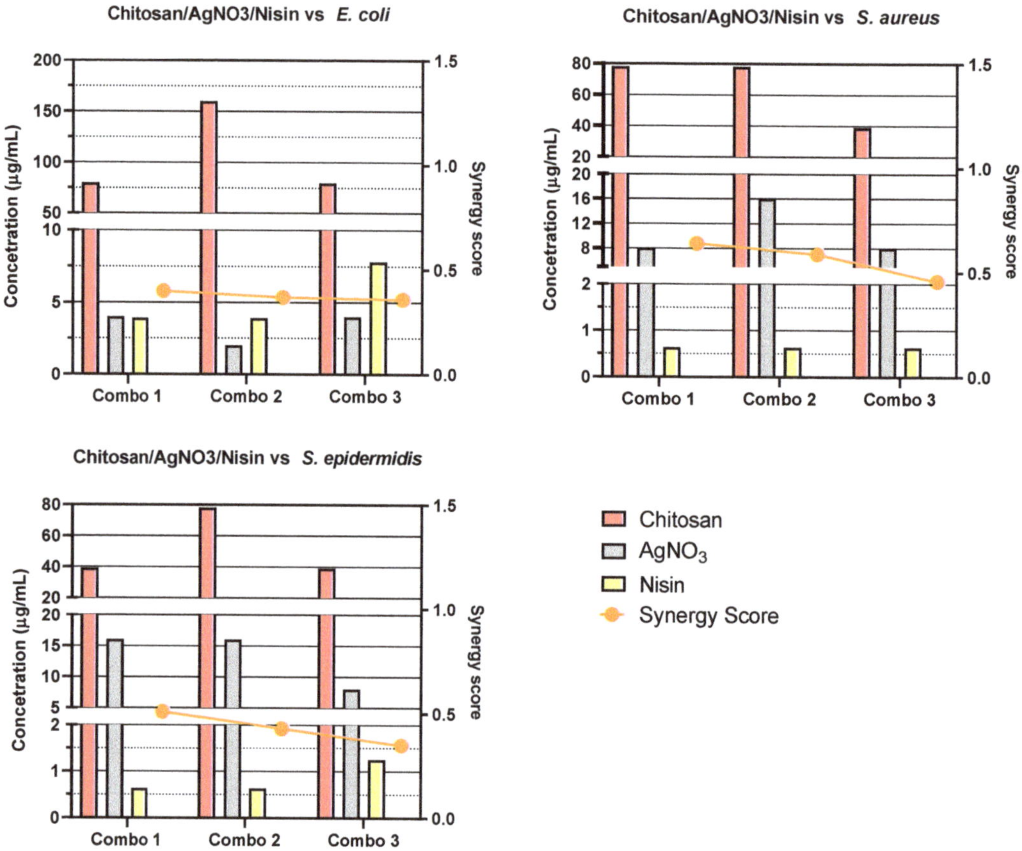

(A)

Figure 8. *Cont.*

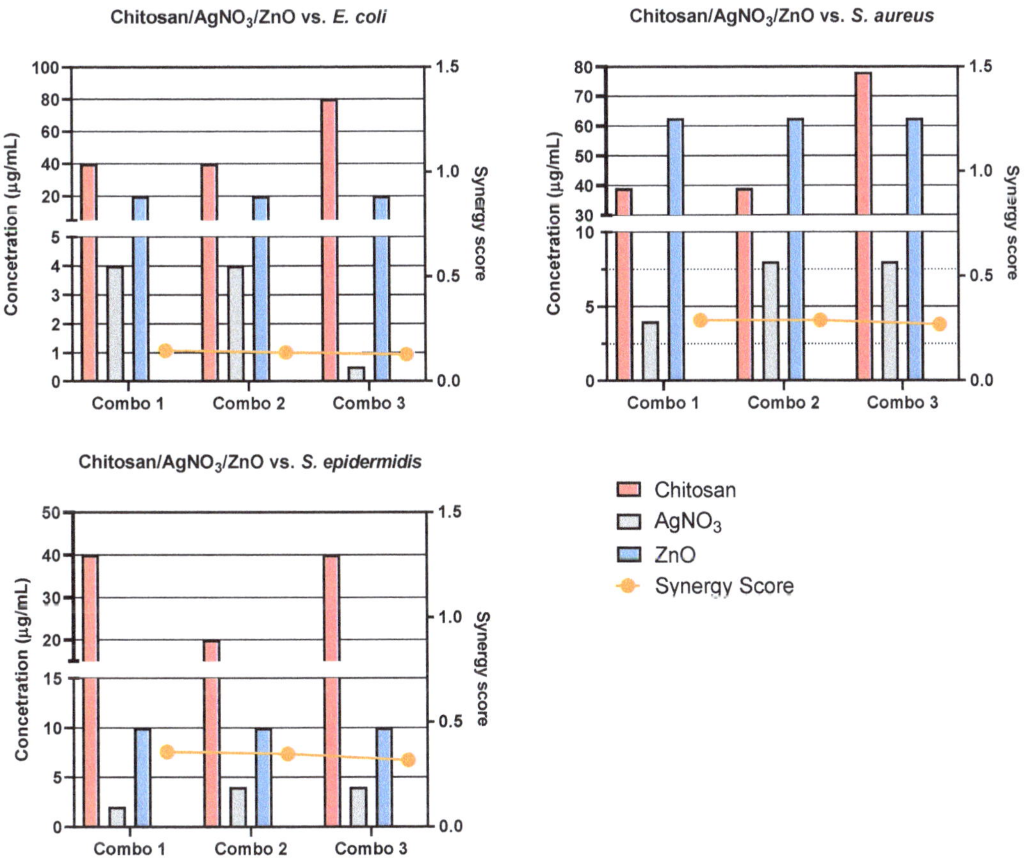

(**B**)

Figure 8. *Cont.*

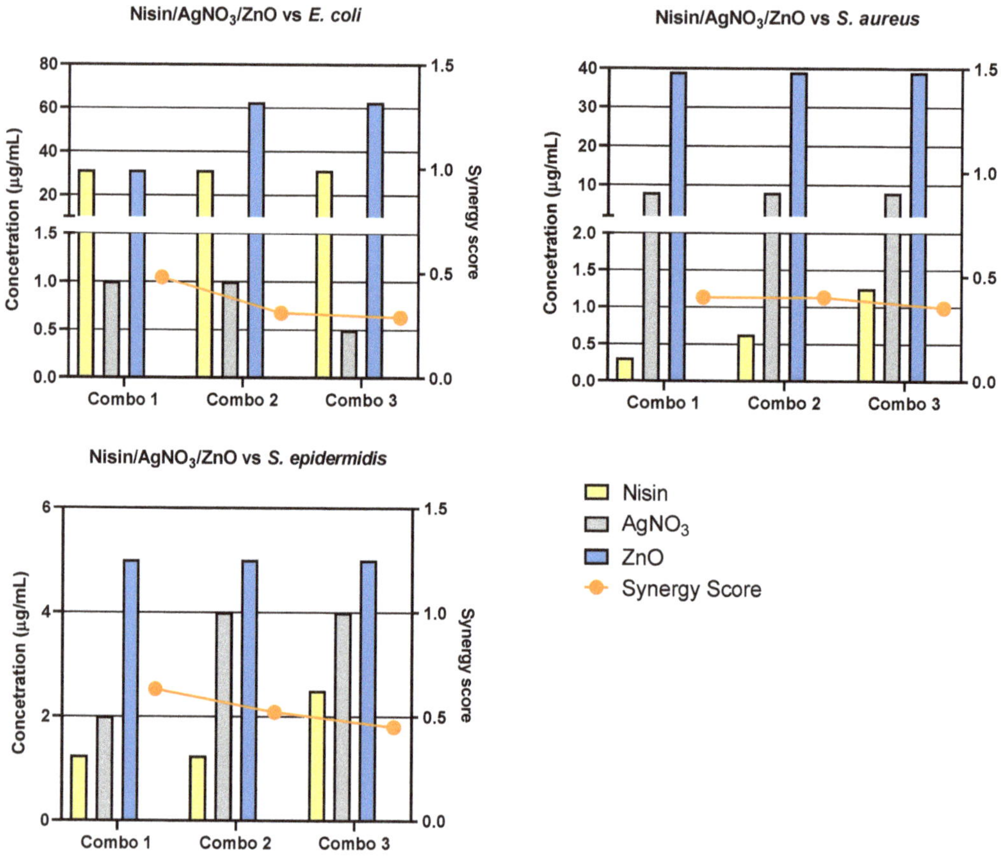

(C)

Figure 8. *Cont.*

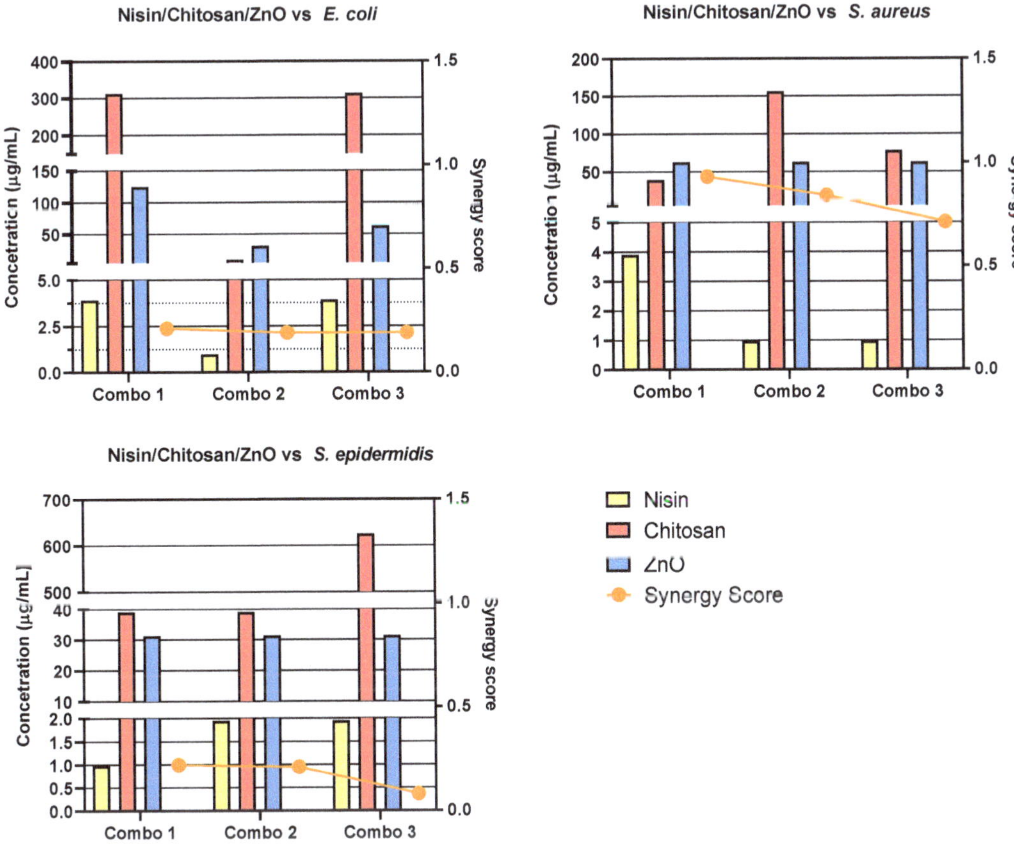

(D)

Figure 8. Synergy score results of top three three-drug combinations: Bar graphs presenting the concentrations of each drug and their respective synergy scores from the top three three-drug combinations (**A**–**D**) against *E. coli*, *S. aureus* and *S. epidermidis*. (**A**) Chitosan/$AgNO_3$/Nisin, (**B**) Chitosan/$AgNO_3$/ZnO, (**C**) Nisin/$AgNO_3$/ZnO and (**D**) Nisin/Chitosan/ZnO. Bars show drug concentrations as indicated on the left *y*-axis and the line/symbols show each combination (combo) synergy score as indicated on the right *y*-axis.

3.3. Four-Drug Combinations

- $AgNO_3$–Nisin–Chitosan–ZnO

Chitosan, nisin, $AgNO_3$ and ZnO exhibited moderately high synergy in combination versus *E. coli* (average 0.36) and very high synergy versus *S. aureus* (average 0.91) and *S. epidermidis* (average 1.11). While the compound's average synergy score versus *E. coli* is lower than that versus the other two test species, results indicate the positive contributions of each treatment at low concentrations of chitosan (80 µg/mL), nisin (1.95–31.25 µg/mL), $AgNO_3$ (8 µg/mL) and ZnO (10–40 µg/mL), which showed effective inhibition (69–98.9%). The highest-scoring combination, which exhibited 98.9% inhibition, reported a very high concentration of nisin (31.25 µg/mL), indicating that nisin had a strong influence within the combination.

Reported synergy scores versus *S. aureus* are quite high, with the top scoring combinations exhibiting effective inhibition (97.5–99.5%) at low concentrations of chitosan (19.53–78.13 µg/mL), nisin (0.39–1.56 µg/mL) and $AgNO_3$ (4 µg/mL) and moderate concentrations of ZnO (62.50 µg/mL).

Combinations versus *S. epidermidis* reported the highest synergy scores identified within the present study at low concentrations of chitosan (20–80 µg/mL), nisin (1.25 µg/mL), $AgNO_3$ (8 µg/mL) and ZnO (10–40 µg/mL). While the top combination reported a very high synergy score (1.3), the reported inhibition greatly deviated (stdev 69.18), having an average value of 24.96%. The second highest (1.13) and third highest (0.9) scoring combinations exhibited stable inhibition (98.7–99.1%) at similarly low concentrations of each treatment.

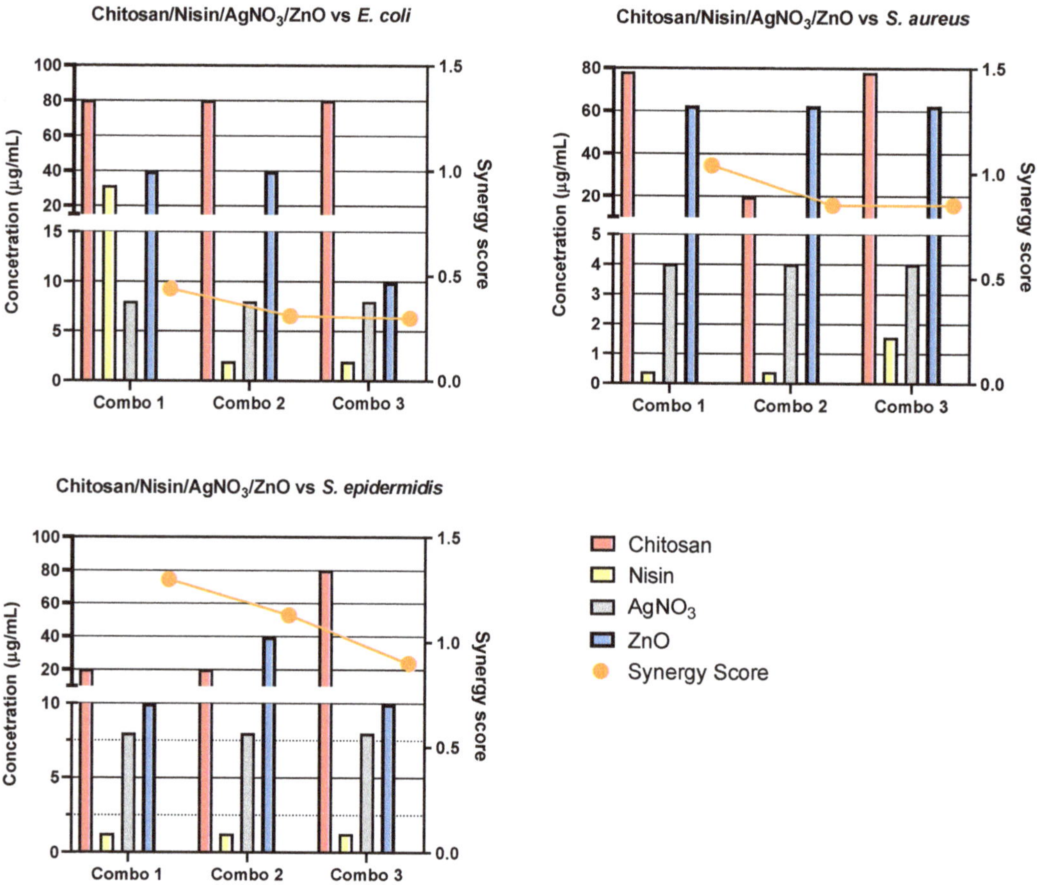

Figure 9. Synergy score results of the top three four-drug combinations: Bar graphs that present the concentrations of each drug and their respective synergy scores from the top three four-drug combinations (Chitosan/Nisin/$AgNO_3$/ZnO) against *E. coli*, *S. aureus* and *S. epidermidis*. Bars show drug concentrations, as indicated on the left *y*-axis, and the line/symbols show each combination (combo) synergy score, as indicated on the right *y*-axis.

4. Discussion

Antimicrobial synergy holds great promise as a solution for use in meeting the AMR crisis for several reasons. While a bacterial species may hold or even develop resistance to a single therapeutic agent, co-treatment with an alternative compound that exhibits alternative modes of antimicrobial action could help to alleviate this issue. Additionally, certain groups of bacteria hold intrinsic metabolic or physical characteristics that can prevent certain classes of antimicrobials from exhibiting their effect. Co-treatment using a compound that can disrupt these characteristics would allow the primary treatment to carry out its effect unimpeded. Gram-negative bacteria are an example of one such group, as they have an additional outer membrane that can act to prevent compounds from reaching their target ligands. Following this example, nisin is a poly-cyclic lantibiotic that targets the inner-membrane-bound lipid II molecule. Due to the presence of an outer membrane, nisin is prevented from reaching its target, thus rendering it ineffective [29,30]. However, in theory, it would be possible to enable nisin by combining it via treatment with an additional compound that targets the outer membrane. By removing the outer membrane or compromising its integrity, nisin could freely interact with its lipid-II target. While this interaction can be clearly deemed to be synergistic, it is not enough on its own to observe a positive end result from the combination. While it would stand to reason that combining two or more already well-known and effective treatments would produce a greater gross effect than that of each individual treatment, previous studies of drug combinations have shown this predicted outcome to be incorrect, as have the results presented in this study [16,20,31]. To determine the synergistic abilities of two or more compounds, it is necessary to assess an array of various concentrations in different combinations. It is not important to determine the highest effect of combined treatments; the concept is to instead determine combinations that express a higher effect than that of the individual drugs at an identical concentration. The aim is to more easily discern the ratio of each drug required to enable another's mechanism of action, thus giving the most efficient synergy.

4.1. Inhibition and Synergy

Previously, $AgNO_3$ was shown to be the most effective bacterial growth inhibitor of the tested bacterial species [22]. Nisin was shown to have very efficient inhibitory effects on test Gram-positive bacterial species, while having no effect on Gram-positive bacteria. Both compounds differ majorly in their modes of action, with $AgNO_3$ permeating bacterial membranes through reactive silver ions (Ag^{2+}), while nisin has specific binding affinity to the lipid-II molecules bound in the inner bacterial membrane. Nisin's inability to affect Gram-negative bacteria is based on its inability to breach its outer membrane and interact with the lipid-II ligand. By combining both $AgNO_3$ and nisin, it was hypothesised that the reactive Ag^{2+} ions of $AgNO_3$ breach the Gram-negative outer bacterial membrane, allowing nisin to reach its target ligand [32–35]. Similar hypotheses were devised regarding ZnO and nisin, as ZnO had efficacy against Gram-negative and Gram-positive bacteria, as well as a similar mode of action wherein it destabilises membranes through release of Zn^{2+} ions and reactive oxygen species (ROS) [36–39]. Chitosan also had a noteworthy effect on all test strains, though it also had an alternate mechanism via which it targeted the bacterial cell wall [40,41]. The varying mechanisms had great significance for combinational studies and allowed us to observe whether effects unlock one other treatments' drawbacks ($AgNO_3$–nisin, ZnO–nisin), stack upon against them ($AgNO_3$ ZnO) or complement them (chitosan–nisin, chitosan–$AgNO_3$, chitosan–ZnO).

In this study, the combinatory compatibility of four chosen bioactives was successfully established, as were the magnitude of their interactions with one another. The checkerboard assay was utilised to screen the inhibitory effect of bioactive combinations against each test bacterial strain. The checkerboard assay was a well-established method used to screen drug combinations in various areas of clinical research [12,15–18]. Through the use of the synergy python package and the Bliss model, the synergy score of each test combination was successfully determined. The results of this study presented interesting interactions between the bioactives, many of which were predictable, though others were unanticipated.

4.2. Two-Drug Combination Synergy

The results of two-drug combination studies carried out against the Gram-negative bacteria *E. coli* have yielded varying results. Nisin, which is a lantibiotic that targets the inner-membrane-bound lipid II molecule, is hindered by the outer membrane found in Gram-negative bacteria, which prevents nisin from carrying out its mechanism of action. It was hypothesised that combining nisin with a compound capable of penetrating the outer membrane, such as $AgNO_3$ or ZnO, would enable nisin, with the resulting interaction being marked as synergetic. Combinations of nisin–$AgNO_3$ exhibited moderate-to-high synergy (average 0.32), showing a consistent concentration of $AgNO_3$ (8.49 μg/mL) to be the most accommodating compound for varying concentrations of nisin. While the inhibition ranged between 64 and 68% for this combination, it shows that nisin was able to have an effect upon a previously unaffected target. In contrast, ZnO was not found to enable nisin; rather, it appeared that nisin was antagonizing ZnO, as the concentrations of ZnO in the most synergistic combinations were higher than that of its previously determined MIC. Combinations of nisin–chitosan also exhibited undesired results, with higher concentrations of chitosan being utilised to observe an inhibitory effect. While such results are unfavourable, they still present a promising observation, showing that nisin had an effect on Gram-negative bacteria. Combinations of $AgNO_3$–chitosan exhibited a strong synergistic interaction, with effects being evident at lower concentrations of $AgNO_3$, which would indicate chitosan's ability to enable it. Chitosan has also shown to enable ZnO, which also exhibited lower concentrations; however, these combinations scored quite low, which reflects the fact that the concentration of chitosan was quite high.

Two-drug combinations used to inhibit *S. aureus* and *S. epidermidis* growth presented some moderate-to-strong synergistic combinations; however, there was a pattern of ZnO not effectively combining with nisin or chitosan. $AgNO_3$ demonstrates itself to be the most effect bioactive, enabling all other bioactives with which it is combined, reporting lower concentrations with higher inhibition responses. The highest overall scoring two-drug combination involved nisin–$AgNO_3$ versus *S. epidermidis*. Chitosan also demonstrated notable synergy with most bioactives, though it only effectively combined with ZnO against *S. epidermidis*, as much lower concentrations of both gave a greater response; however, the inhibition response was weak.

4.3. Three-Drug Combinations against Gram-Negative Bacteria

Increasing the combination number can further alter the effect exhibited by treatments, as is evident from three-drug combinations. Combinations that included nisin were shown to demonstrate high synergy versus *E. coli* with near full inhibition. Following the two-drug analysis, it was predictable that chitosan–$AgNO_3$–nisin would effectively synergise, presenting the highest-scoring combination versus *E. coli*. Furthermore, the relatively high concentrations of nisin in this combination showed that it had an active effect on *E. coli*, as it can be presumed to be heavily involved in enabling resistance (i.e., a concentration close to 0 would indicate little-to-no input). A more unpredictable result was seen regarding combinations that involved ZnO, as two-drug combinations demonstrated ZnO to be a poor component in combination, though three-drug combinations showed opposing results. Nisin–ZnO was the lowest scoring combination versus *E. coli*; however, with the inclusion of $AgNO_3$ or chitosan, these combinations were the second and third highest-scoring three-drug combinations against *E. coli*, respectively. Most interestingly, the nisin–$AgNO_3$–ZnO combination highlights how little $AgNO_3$ was reported in the higher scoring combinations, while relatively high concentrations of nisin were reported. This result again indicates nisin's active role in the combination, whereas $AgNO_3$ is at too low a concentration to have an inhibitory effect. This result could also demonstrate the ability of $AgNO_3$ to enable the mechanism of nisin. While chitosan–$AgNO_3$–ZnO reported low scoring combinations, the results seemed to be promising, with low concentrations of all three bioactives and nearly full inhibition reported. This result, once again, does not follow the patterns observed in two-drug combinations of the same bioactives.

4.4. Three-Drug Combinations against Gram-Positive Bacteria

Three-drug combinations versus *S. aureus* and *S. epidermidis* offered interesting points of comparison. The chitosan–$AgNO_3$–nisin combination scored highly versus both bacterial strains; however, concentrations of $AgNO_3$ were quite high at low concentrations of nisin. Scores from combinations that involved ZnO also proved to be quite unpredictable versus Gram-positive bacteria. Combinations of nisin–ZnO and chitosan–ZnO against *S. aureus* scored quite poorly; however, nisin–chitosan–ZnO reported the highest score of all three-drug combinations. In contrast, this combination had the second lowest score against *S. epidermidis*. The individual concentrations were quite low, and the reported synergy scores were also quite low. An interesting observation of this combination is that it was also predictable based on the two-drug combinations of nisin–chitosan, nisin–ZnO and chitosan–ZnO, which produced synergy scores that very closely averaged that of the nisin–chitosan–ZnO synergy score. Concentrations versus *S. aureus* indicate that ZnO enabled the effects of chitosan and $AgNO_3$; however, concentrations versus *S. epidermidis* did not indicate that any single bioactive enabled another bioactive, highlighting the even distribution of activity between the three bioactives.

Nisin–$AgNO_3$–ZnO demonstrated strong synergy versus both Gram-positive bacteria. While the reported synergy was particularly high against *S. epidermidis*, the reported inhibitory effects were quite low. Two-drug reports show that nisin–ZnO combinations interacted very poorly, which implies that the influence of $AgNO_3$ caused the three-drug combinations to more favourably interact, which was also predictable given the synergy scores of nisin–$AgNO_3$ and $AgNO_3$–ZnO.

4.5. Four-Drug Combinations

While a four-drug combination gave valuable insights into interactions between all four treatments at once, the 4×4 sized checkerboard had some disadvantages relative to larger 6×6 or 8×8 checkerboards, primarily the fact that it could not accommodate enough combinations to generate a full model of the possible combinational interactions. However, if determined via a two- or three-drug assay, key concentrations can be selected and utilised within the four-drug assay for further investigation. As such, the present four-drug assay layout should be used as a follow-on study, rather than as an initial combinational study, due to such limitations. Likewise, with the chosen three-drug assay layout, the four-drug assay layout allowed reduced experimental size and a faster setup.

Four-drug combinations reported a marked increase in the efficacy of all four bioactives relative to their individual capabilities against each bacterial strain. The combination of chitosan–nisin–$AgNO_3$–ZnO against *E. coli* exhibited some predictable results, with synergy scores comparable to scores derived from two- and three-drug combinations. While concentrations of each bioactive in the highest-scoring combinations were lower than their individual MICs, concentrations of chitosan and $AgNO_3$ were still quite moderate. Furthermore, only the highest-scoring combination reported complete inhibition while using a high concentration of nisin, which was expected due to nisin's inability to target Gram-negative species.

The reported four-drug synergy scores against *S. aureus* and *S. epidermidis* were very high relative to other scores determined during this study. Concentrations versus *S. aureus* were notably lower than their individual MICs, thus having strong inhibitory effects. While *S. epidermidis* reported the highest synergy score of this study, its highest-scoring combination reported low inhibition. Compared to the other two reported combinations, a slight increase in either chitosan or ZnO was sufficient to push the effects toward complete inhibition, albeit remaining well below their individual MICs. From analysis of the most effective combinations against each individual bacterial strain, the most effective concentrations of each drug in combination were determined, which gave a four-drug combination that could cause complete inhibition against all tested bacterial strains, while the amount of each drug used was limited (See Table 1).

Table 1. Most effective concentrations of the four bioactive compounds in combination against *E. coli*, *S. aureus* and *S. epidermidis*. These concentrations were established by evaluating the highest-scoring combinations against each bacterial species and determining the lowest concentrations of each compound that would cause complete inhibition of all three bacterial species.

Bioactive	Most Effective Concentration (μg/mL)
Chitosan	80
Nisin	2
$AgNO_3$	8
ZnO	60

5. Conclusions

Combinational antimicrobial bioactive studies hold great potential in many areas of clinical, pharmaceutical and medical research. The discovery of cross-treatment synergism could potentially unlock many new avenues of therapy for various pathogens and conditions. While drug synergy is a key target of combinational studies, drug antagonism is also a well-documented occurrence in pharmaceuticals, and while there are several models under development for its prediction, in many cases, it is difficult to determine which treatments may interact negatively without performing pre-clinical or clinical studies. Though it is important to find compatible combinations of drugs, another key goal is finding combinations in which the individual drugs are more effective within a combination than they are when acting alone. Determining synergy scores is an efficient method of screening many combinations of treatments and deducing the most effective option. It is evident from results presented here that treatment interactions cannot be accurately predicted and can differ greatly between bacterial strains. Furthermore, increasing the combination number has also been shown to have an unpredictable effect, as two-drug combinations cannot predict the effects of three-drug combinations of the same components, and likewise, two- and three-drug combinations cannot predict the effects of four-drug combinations. Current findings show that models previously used to predict drug combinations cannot be wholly trusted, as there are aberrant results presented in this study that contradict predictive models. While such results may not be considered advantageous, they provide knowledge critical to the development of combinational treatments. Within the scope of investigating new or previously unsuitable compounds as alternative antimicrobial treatments, the possibility of using as little of each compound possible while still holding an antimicrobial effect holds great potential. Identification of antimicrobial compounds based on complementary modes of action also provides an additional avenue for the discovery of novel treatments that combat AMR bacteria, wherein the activity of one treatment may enable the activity of another treatment that is naturally ineffective against the bacteria, as was the case of nisin against Gram-negative species in the present study.

There is also a pressing need to use effective screening methods to evaluate alternative bioactives that are biorefined via various environmental and food waste streams to help address the shortage in appropriate antimicrobials and the development of AMR [42,43]. Interestingly, there is increased interest in exploring new sources of antimicrobials, such as marine, peatland and food waste streams, that may present stressful environments that favor the production of unique antimicrobial bioactives [44]. This simple mass-throughput screening-based approach to evaluating combinational bioactives will also help address the surge in resistance to anti-fungal drugs among problematical fungi that cause significant human and animal infections [45,46]. There is also a proportionate interest in progressing interdisciplinary research through Quadruple Helix Hub frameworks (combining academia, industry, society and regulators), which use shared access to specialist equipment and subject-matter experts across disciplines to overcome these challenges [47].

The four chosen bioactives, namely $AgNO_3$, ZnO, nisin and chitosan, which have previously been characterised in terms of their individual antimicrobial abilities, have now been characterised for their combinational interactions in two-, three-, and four-drug arrangements. Using this data, we accurately determined the most effective concentrations of each compound required for most effective microbial growth inhibition, limiting the amount of each compound needed to inhibit bacterial growth while having a broader spectrum of effect. Using this data, it is clear that the use of these compounds in combination produces a much more effective inhibitor of microbial growth, as the concentration of each individual bioactive is much lower than that of their MIC alone. As the current study relied on combinations of serial dilutions, more extensive analyses must be carried out regarding the most synergistic combinations, with further testing used to determine the most ideal concentration combinations and verify their efficacy. Additionally, the methods and analysis procedures presented have been shown to produce detailed and high-throughput assessments of drug combinations, and as such, they should be carried forward to evaluate additional treatments and combinations. The methodology could also be adjusted accordingly to allow studies of pharmaceuticals in other fields of research.

Supplementary Materials: The following supporting information can be downloaded via the following link: https://www.mdpi.com/article/10.3390/biomedicines11082216/s1, Figure S1. Drug combination setup, Figure S2. Two-drug combination assay final layout, Figure S3. Three-drug combination assay final layout, Figure S4. Drug combination assay final layout; Table S1. Three highest Bliss scoring two-drug combinations against *E. coli*, Table S2. Three highest Bliss scoring two-drug combinations against *S. aureus*. Table S3. Three highest Bliss scores of each two-drug combination versus *S. epidermidis*. Table S4. Three highest Bliss synergy scores of each three-drug combination versus *E. coli*, *S. aureus* and *S. epidermidis*. Table S5. Three highest Bliss synergy scores and the average of each four-drug combination versus *E. coli*, *S. aureus* and *S. epidermidis*.

Author Contributions: Conceptualisation, K.M., I.M. and N.R.; methodology, K.M.; investigation, K.M.; writing—original draft preparation, K.M.; writing—review and editing, all authors; supervision, N.R., I.M. and M.L.; funding acquisition, N.R. All authors have read and agreed to the published version of the manuscript.

Funding: The authors acknowledge funding support provided by DAFM-2022-PSS125 and Athlone Institute of Technology (grant No. AIT-2017-008) for this work.

Institutional Review Board Statement: Not applicable.

Informed Consent Statement: Not applicable.

Data Availability Statement: Not applicable.

Conflicts of Interest: The authors declare no conflict of interest.

References

1. Hsia, Y.; Lee, B.R.; Versporten, A.; Yang, Y.; Bielicki, J.; Jackson, C.; Newland, J.; Goossens, H.; Magrini, N.; Sharland, M. Use of the WHO Access, Watch, and Reserve classification to define patterns of hospital antibiotic use (AWaRe): An analysis of paediatric survey data from 56 countries. *Lancet Glob. Health* **2019**, *7*, e861–e871. [CrossRef] [PubMed]
2. Dadgostar, P. Antimicrobial resistance: Implications and costs. *Infect. Drug Resist.* **2019**, *12*, 3903–3910. [CrossRef] [PubMed]
3. Interagency Coordination Group on Antimicrobial Resistance. No Time to Wait: Securing the Future from Drug-Resistant Infections Report to the Secretary-General of the United Nations. 2019. Available online: https://www.who.int/antimicrobial-resistance/interagency-coordination-group/IACG_final_report_EN.pdf?ua=1 (accessed on 29 May 2020).
4. Talebi Bezmin Abadi, A.; Rizvanov, A.A.; Haertlé, T.; Blatt, N.L. World Health Organization Report: Current Crisis of Antibiotic Resistance. *Bionanoscience* **2019**, *9*, 778–788. [CrossRef]
5. Koulenti, D.; Xu, E.; Yin Sum Mok, I.; Song, A.; Karageorgopoulos, D.E.; Armaganidis, A.; Lipman, J.; Tsiodras, S. Novel Antibiotics for Multidrug-Resistant Gram-Positive Microorganisms. *Microorganisms* **2019**, *7*, 270. [CrossRef]
6. Tacconelli, E.; Carrara, E.; Savoldi, A.; Harbarth, S.; Mendelson, M.; Monnet, D.L.; Pulcini, C.; Kahlmeter, G.; Kluytmans, J.; Carmeli, Y.; et al. Discovery, research, and development of new antibiotics: The WHO priority list of antibiotic-resistant bacteria and tuberculosis. *Lancet Infect. Dis.* **2018**, *18*, 318–327. [CrossRef] [PubMed]
7. Rima, M.; Rima, M.; Fajloun, Z.; Sabatier, J.M.; Bechinger, B.; Naas, T. Antimicrobial peptides: A potent alternative to antibiotics. *Antibiotics* **2021**, *10*, 1095. [CrossRef]
8. Girard, M.; Bee, G. Invited review: Tannins as a potential alternative to antibiotics to prevent coliform diarrhea in weaned pigs. *Animal* **2020**, *14*, 95–107. [CrossRef]

9. Alekish, M.; Ismail, Z.B.; Albiss, B.; Nawasrah, S. In vitro antibacterial effects of zinc oxide nanoparticles on multiple drug-resistant strains of *Staphylococcus aureus* and *Escherichia coli*: An alternative approach for antibacterial therapy of mastitis in sheep. *Vet. World* **2018**, *11*, 1428–1432. [CrossRef]
10. Duss, F.R.; Garcia De La Mària, C.; Croxatto, A.; Giulieri, S.; Lamoth, F.; Manuel, O.; Miró, J.M. Successful treatment with daptomycin and ceftaroline of MDR *Staphylococcus aureus* native valve endocarditis: A case report. *J. Antimicrob. Chemother.* **2019**, *74*, 2626–2630. [CrossRef]
11. Gonzales, P.R.; Pesesky, M.W.; Bouley, R.; Ballard, A.; Biddy, B.A.; Suckow, M.A.; Wolter, W.R.; Schroeder, V.A.; Burnham, C.A.D.; Mobashery, S.; et al. Synergistic, collaterally sensitive β-lactam combinations suppress resistance in MRSA. *Nat. Chem. Biol.* **2015**, *11*, 855–861. [CrossRef]
12. Fatsis-Kavalopoulos, N.; Roemhild, R.; Tang, P.C.; Kreuger, J.; Andersson, D.I. CombiANT: Antibiotic interaction testing made easy. *PLoS Biol.* **2020**, *18*, e3000856. [CrossRef]
13. Acuña, L.; Morero, R.D.; Bellomio, A. Development of Wide-Spectrum Hybrid Bacteriocins for Food Biopreservation. *Food Bioprocess Technol.* **2011**, *4*, 1029–1049. [CrossRef]
14. Singh, S.B.; Young, K.; Silver, L.L. What is an "ideal" antibiotic? Discovery challenges and path forward. *Biochem. Pharmacol.* **2017**, *133*, 63–73. [CrossRef]
15. Mataraci, E.; Dosler, S. In vitro activities of antibiotics and antimicrobial cationic peptides alone and in combination against methicillin-resistant *Staphylococcus aureus* biofilms. *Antimicrob. Agents Chemother.* **2012**, *56*, 6366–6371. [CrossRef]
16. Torres, N.S.; Montelongo-Jauregui, D.; Abercrombie, J.J.; Srinivasan, A.; Lopez-Ribot, J.L.; Ramasubramanian, A.K.; Leung, K.P. Antimicrobial and antibiofilm activity of synergistic combinations of a commercially available small compound library with colistin against *Pseudomonas aeruginosa*. *Front. Microbiol.* **2018**, *9*, 2541. [CrossRef]
17. Cokol-Cakmak, M.; Cetiner, S.; Erdem, N.; Bakan, F.; Cokolid, M. Guided screen for synergistic three-drug Combinations. *PLoS ONE* **2020**, *15*, e0235929. [CrossRef]
18. Meletiadis, J.; Pournaras, S.; Roilides, E.; Walsh, T.J. Defining fractional inhibitory concentration index cutoffs for additive interactions based on self-drug additive combinations, Monte Carlo simulation analysis, and in vitro-in vivo correlation data for antifungal drug combinations against *Aspergillus fumi*. *Antimicrob. Agents Chemother.* **2010**, *54*, 602–609. [CrossRef]
19. Bollenbach, T. Antimicrobial interactions: Mechanisms and implications for drug discovery and resistance evolution. *Curr. Opin. Microbiol.* **2015**, *27*, 1–9. [CrossRef]
20. Yadav, B.; Wennerberg, K.; Aittokallio, T.; Tang, J. Searching for Drug Synergy in Complex Dose-Response Landscapes Using an Interaction Potency Model. *Comput. Struct. Biotechnol. J.* **2015**, *13*, 504–513. [CrossRef]
21. Tang, J.; Wennerberg, K.; Aittokallio, T. What is synergy? The Saariselkä agreement revisited. *Front. Pharmacol.* **2015**, *6*, 181. [CrossRef]
22. Masterson, K.; Meade, E.; Garvey, M.; Lynch, M.; Major, I.; Rowan, N.J. Development of a low-temperature extrusion process for production of GRAS bioactive-polymer loaded compounds for targeting antimicrobial-resistant (AMR) bacteria. *Sci. Total Environ.* **2021**, *800*, 149545. [CrossRef] [PubMed]
23. Wooten, D.J.; Meyer, C.T.; Lubbock, A.L.R.; Quaranta, V.; Lopez, C.F. MuSyC is a consensus framework that unifies multi-drug synergy metrics for combinatorial drug discovery. *Nat. Commun.* **2021**, *12*, 4607. [CrossRef] [PubMed]
24. Boekema, B. How to Prevent Silver Nitrate ($AgNO_3$) Precipitating in Growth Broth?. ResearchGate. 2018. Available online: https://www.researchgate.net/post/How_to_prevent_Silver_Nitrate_AgNO3_precipitating_in_growth_broth (accessed on 25 January 2018).
25. Harrison, J.J.; Stremick, C.A.; Turner, R.J.; Allan, N.D.; Olson, M.E.; Ceri, H. Microtiter susceptibility testing of microbes growing on peg lids: A miniaturized biofilm model for high-throughput screening. *Nat. Protoc.* **2010**, *5*, 1236–1254. [CrossRef] [PubMed]
26. Wiegand, I.; Hilpert, K.; Hancock, R.E.W. Agar and broth dilution methods to determine the minimal inhibitory concentration (MIC) of antimicrobial substances. *Nat. Protoc.* **2008**, *3*, 163–175. [CrossRef]
27. Brewster, J.D. A simple micro-growth assay for enumerating bacteria. *J. Microbiol. Methods* **2003**, *53*, 77–86. [CrossRef]
28. Wooten, D.J.; Albert, R. Synergy: A Python library for calculating, analyzing and visualizing drug combination synergy. *Bioinformatics* **2021**, *37*, 1473–1474. [CrossRef]
29. He, L.; Zou, L.; Yang, Q.; Xia, J.; Zhou, K.; Zhu, Y.; Han, X.; Pu, B.; Hu, B.; Deng, W. Antimicrobial Activities of Nisin, Tea Polyphenols, and Chitosan and their Combinations in Chilled Mutton. *J. Food Sci.* **2016**, *81*, 1466–1471. [CrossRef]
30. Ruhr, E.; Sahl, H.G. Mode of action of the peptide antibiotic nisin and influence on the membrane potential of whole cells and on cytoplasmic and artificial membrane vesicles. *Antimicrob. Agents Chemother.* **1985**, *27*, 841–845. [CrossRef]
31. Chen, D.; Liu, X.; Yang, Y.; Yang, H.; Lu, P. Systematic synergy modeling: Understanding drug synergy from a systems biology perspective. *BMC Syst. Biol.* **2015**, *9*, 56. [CrossRef]
32. Gut, I.M.; Blanke, S.R.; Van Der Donk, W.A. Mechanism of inhibition of bacillus anthracis spore outgrowth by the Lantibiotic Nisin. *ACS Chem. Biol.* **2011**, *6*, 744–752. [CrossRef]
33. Pandian, S.R.K.; Deepak, V.; Kalishwaralal, K.; Viswanathan, P.; Sangiliyandi, G. Mechanism of bactericidal activity of silver nitrate—A concentration dependent bi-functional molecule. *Braz. J. Microbiol.* **2010**, *41*, 805–809. [CrossRef]
34. Prabhu, S.; Poulose, E.K. Silver nanoparticles: Mechanism of antimicrobial action, synthesis, medical applications, and toxicity effects. *Int. Nano Lett.* **2012**, *2012*, 32. [CrossRef]

35. Van Heusden, H.E.; De Kruijff, B.; Breukink, E. Lipid II induces a transmembrane orientation of the pore-forming peptide lantibiotic nisin. *Biochemistry* **2002**, *41*, 12171–12178. [CrossRef]
36. Espitia, P.J.P.; Soares, N.d.F.F.; Coimbra, J.S.d.R.; de Andrade, N.J.; Cruz, R.S.; Medeiros, E.A.A. Zinc Oxide Nanoparticles: Synthesis, Antimicrobial Activity and Food Packaging Applications. *Food Bioprocess Technol.* **2012**, *5*, 1447–1464. [CrossRef]
37. Fiedot-Toboła, M.; Ciesielska, M.; Maliszewska, I.; Rac-Rumijowska, O.; Suchorska-Woźniak, P.; Teterycz, H.; Bryjak, M. Deposition of zinc oxide on different polymer textiles and their antibacterial properties. *Materials* **2018**, *11*, 707. [CrossRef]
38. Pasquet, J.; Chevalier, Y.; Couval, E.; Bouvier, D.; Noizet, G.; Morlière, C.; Bolzinger, M.A. Antimicrobial activity of zinc oxide particles on five micro-organisms of the Challenge Tests related to their physicochemical properties. *Int. J. Pharm.* **2014**, *460*, 92–100. [CrossRef]
39. Pasquet, J.; Chevalier, Y.; Pelletier, J.; Couval, E.; Bouvier, D.; Bolzinger, M.A. The contribution of zinc ions to the antimicrobial activity of zinc oxide. *Colloids Surfaces A Physicochem. Eng. Asp.* **2014**, *457*, 263–274. [CrossRef]
40. Kunjachan, S.; Jose, S. Understanding the mechanism of ionic gelation for synthesis of chitosan nanoparticles using qualitative techniques. *Asian J. Pharm.* **2010**, *4*, 148. [CrossRef]
41. Qin, C.; Li, H.; Xiao, Q.; Liu, Y.; Zhu, J.; Du, Y. Water-solubility of chitosan and its antimicrobial activity. *Carbohydr. Polym.* **2006**, *63*, 367–374. [CrossRef]
42. Rowan, N.J.; Galanakis, C.M. Unlocking challenges and opportunities presented by COVID-19 pandemic for cross-cutting disruption in agri-food and green deal innovations: Quo Vadis? *Sci. Total Environ.* **2020**, *748*, 141362. [CrossRef]
43. Rowan, N.J.; Pogue, R. Editorial overview: Green new deal era—Current challenges and emerging opportunities for developing sustaining and disruptive innovation. *Curr. Opin. Environ. Sci. Health* **2021**, *22*, 100294. [CrossRef]
44. O'Neill, E.A.; McKeon Bennett, M.; Rowan, N.J. Peatland-based innovation can potentially support and enable the sustainable development goals of the United Nations: Case study from the Republic of Ireland. *Case Stud. Chem. Environ. Eng.* **2022**, *6*, 100251. [CrossRef]
45. Garvey, M.; Meade, E.; Rowan, N.J. Effectiveness of front line and emerging fungal disease prevention and control interventions and opportunities to address appropriate eco-sustainable solutions. *Sci. Total Environ.* **2022**, *851*, 158284. [CrossRef] [PubMed]
46. Garvey, M.; Rowan, N.J. Pathogenic Drug Resistant Fungi: A Review of Mitigation Strategies. *Int. J. Mol. Sci.* **2023**, *24*, 1584. [CrossRef]
47. Rowan, N.J.; Casey, O. Empower Eco multiactor HUB: A triple helix 'academia-industry-authority' approach to creating and sharing potentially disruptive tools for addressing novel and emerging new Green Deal opportunities under a United Nations Sustainable Development Goals framework. *Curr. Opin. Environ. Sci. Health* **2021**, *21*, 100254. [CrossRef]

biomedicines

MDPI

Article

Computational Guided Drug Targets Identification against Extended-Spectrum Beta-Lactamase-Producing Multi-Drug Resistant Uropathogenic *Escherichia coli*

Harpreet Kaur, Vinay Modgil, Naveen Chaudhary, Balvinder Mohan and Neelam Taneja *

Department of Medical Microbiology, Post Graduate Institute of Medical Education and Research, Chandigarh 160012, India
* Correspondence: drneelampgi@yahoo.com; Tel.: +91-01722755160; Fax: +91-01722744401

Abstract: Urinary tract infections (UTIs) are one of the most frequent bacterial infections in the world, both in the hospital and community settings. Uropathogenic *Escherichia coli* (UPEC) are the predominant etiological agents causing UTIs. Extended-spectrum beta-lactamase (ESBL) production is a prominent mechanism of resistance that hinders the antimicrobial treatment of UTIs caused by UPEC and poses a substantial danger to the arsenal of antibiotics now in use. As bacteria have several methods to counteract the effects of antibiotics, identifying new potential drug targets may help in the design of new antimicrobial agents, and in the control of the rising trend of antimicrobial resistance (AMR). The public availability of the entire genome sequences of humans and many disease-causing organisms has accelerated the hunt for viable therapeutic targets. Using a unique, hierarchical, in silico technique using computational tools, we discovered and described potential therapeutic drug targets against the ESBL-producing UPEC strain NA114. Three different sets of proteins (chokepoint, virulence, and resistance genes) were explored in phase 1. In phase 2, proteins shortlisted from phase 1 were analyzed for their essentiality, non-homology to the human genome, and gut flora. In phase 3, the further shortlisted putative drug targets were qualitatively characterized, including their subcellular location, broad-spectrum potential, and druggability evaluations. We found seven distinct targets for the pathogen that showed no similarity to the human proteome. Thus, possibilities for cross-reactivity between a target-specific antibacterial and human proteins were minimized. The subcellular locations of two targets, ECNA114_0085 and ECNA114_1060, were predicted as cytoplasmic and periplasmic, respectively. These proteins play an important role in bacterial peptidoglycan biosynthesis and inositol phosphate metabolism, and can be used in the design of drugs against these bacteria. Inhibition of these proteins will be helpful to combat infections caused by MDR UPEC.

Keywords: extended-spectrum beta-lactamase; urinary tract infections; *Escherichia coli*; multi-drug resistant; subtractive genomics; drug targets

1. Introduction

The World Health Organization (WHO) has listed AMR as one of the top ten worldwide public health concerns to humanity in the twenty-first century [1]. Recently, AMR has been referred to as the "silent tsunami facing modern medicine" [2]. The heightened use/misuse of antibiotics in human medicine and animal agriculture primarily contributes to this phenomenon. An alarming increase of AMR in bacteria causes community or hospital-acquired infections. Of particular interest is the multi-drug resistant (MDR) pathogen *Escherichia coli (E. coli)*, which is the most common causative pathogen in many types of infections, especially in countries with poor healthcare systems [3]. *E. coli* are part of normal human and animal gut flora, but are also the predominant cause of community and hospital-acquired UTIs. AMR among UPEC has significantly grown recently, restricting the choice of treatment options. Special notes are ESBL-producing *E. coli* (ESBL-EC),

which are responsible for severe human morbidity and mortality, with significant economic losses and disease burdens [4]. ESBLs are a fast-growing class of beta-lactamases that can hydrolyze oxy-imino cephalosporins (ceftriaxone, cefuroxime, ceftazidime, cefepime, and cefotaxime) and monobactams which results in drug resistance [5]. Thus, these bacteria have become resistant to many available antibiotics, and may also be resistant to drugs of last resort such as carbapenems [6]. The use of carbapenems has dramatically increased over the previous few years, leading to carbapenem-resistant *Enterobacteriaceae* (CRE), against which hardly any antibiotics are available, except for colistin [6]. ESBLs are rampant in low and middle-income countries (LMICs) [4]. WHO has put ESBL-EC on its priority list against which new therapeutics are being developed [7].

The accessibility of both the human and microbial genomes has made it easier to apply comparative and subtractive genomics approaches, in which the genomes of the human host and pathogens are compared and homologous host proteins are identified as non-targets. Pathogen-specific proteins with various therapeutic properties have been discovered [8–10]. Such approaches have been used to identify non-homologous proteins as potential drug targets in many pathogens, including *Staphylococcus saprophyticus, Mycobacterium avium, Acinetobacter baumannii, Pseudomonas aeruginosa, Burkholderia pseudomallei,* and *Streptococcus pneumoniae* [9,11–13]. These approaches, combined with the advancements in computational biology and the availability of diverse bioinformatics tools, are revolutionizing fields of drug discovery and design by minimising time and expense of wet-lab screening [13,14].

In this study, we applied a subtractive genomics approach using an ESBL- producing MDR UPEC strain NA114. This UPEC strain was identified in Pune, India, in the urine of a 70-year-old man who had prostatitis. It is an MDR pathogen and ESBL-producing strain, resistant to amoxicillin, co-trimoxazole, tetracycline, gentamicin, nalidixic acid, ciprofloxacin, ceftazidime, and nitrofurantoin [15]. To identify and select the proteins that are crucial for the pathogen's survival, we made use of a variety of computational techniques, software, and web databases. Only proteins specific to pathogens were shortlisted, thereby avoiding host homology. In order to identify possible broad-spectrum therapeutic targets, we considered all functional metabolic pathways, virulence genes, and resistance genes and discovered seven distinct targets that are unique to UPEC. Furthermore, as they showed no similarity to the human proteome, the possibilities of cross-reactivity between a new drug that uses one of these targets and human proteins were minimized. The subcellular locations of two targets, ECNA114_0085 and ECNA114_1060, were predicted as cytoplasmic and periplasmic, respectively. These proteins play an important role in peptidoglycan biosynthesis and inositol phosphate metabolism. Therefore, the development of new drugs against these targets could be promising steps towards eliminating UTIs caused by *E. coli*.

2. Material and Methods

Figure 1 displays the full workflow that was employed in this study.

2.1. Phase I: Comparative Analysis of Pathogen and Human Proteome

Step1: Non-Homology Analysis

The Kyoto Encyclopedia of Genes and Genomes (KEGG) database was used to compare the pathways of the host *Homo sapiens* (*H. sapiens*) and the UPEC strain NA114 [16]. Pathways that are not present in the human host were considered to be distinct pathways. All proteins in the pathways were searched using the protein basic local alignment search tool (BLASTp) against the non-redundant database. Non-homologous proteins were chosen based on rigorous measurements showing no similarity [17].

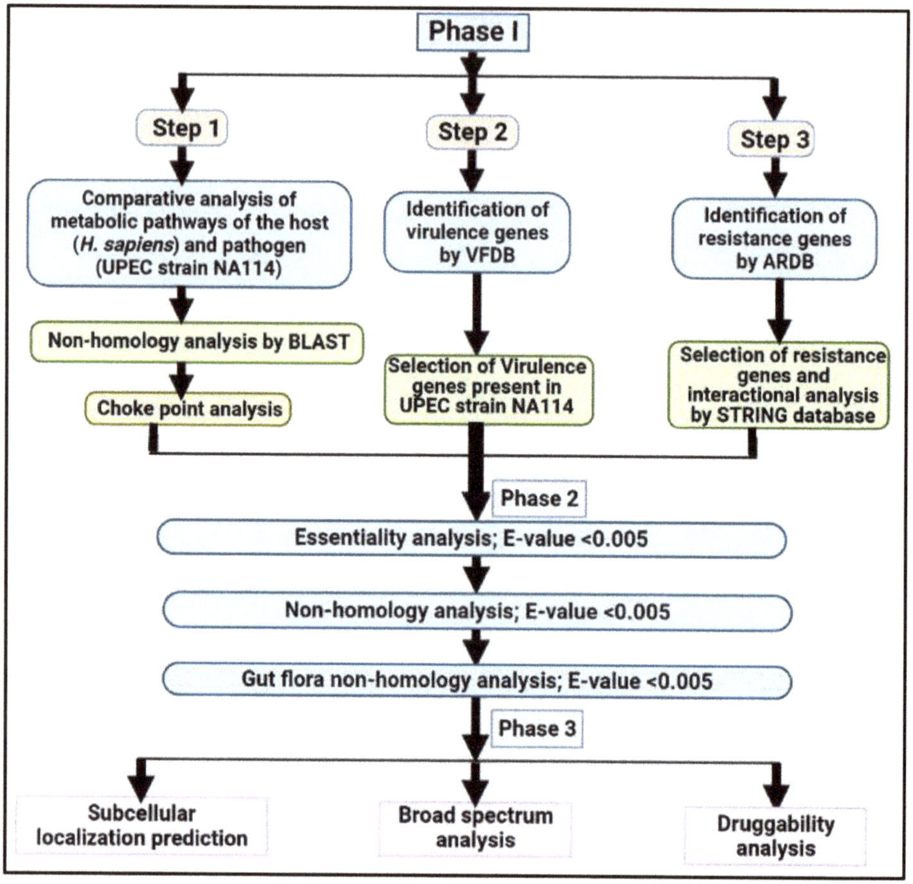

Figure 1. Complete workflow of the study.

2.2. Chokepoint Analysis

In a metabolic network, a "chokepoint reaction" is a reaction that consumes or creates a single substrate or product. The inhibition of enzymes involved in chokepoint reaction will hamper the important cell function [18]. Chokepoint analysis is carried out using simple statistics. The confidence level (CL) of all non-homologous proteins (enzymes) has been calculated by using the simple formula [19],

$$Confidence\ level\ of\ an\ enzyme = \\ No.\ of\ pathways\ in\ which\ an\ enzyme\ is\ found\ as\ a\ choke\ point / \\ Total\ no.\ of\ pathways\ of\ the\ enzyme \times 100.$$

Proteins with confidence scores below 50% were removed, while those with confidence scores above 50% were chosen for further study.

2.2.1. Step 2: Analysis of Virulence Genes

Virulence factors (VF) have been identified as prospective therapeutic targets in drug development. The pathogen would become avirulent if these virulence proteins were inhibited, as these proteins are essential for the establishment and severity of infection [20]. The virulence factor database (VFDB) contains virulence genes from different bacterial species, including *E. coli* [21]. The VFDB was used to compile a comprehensive list of virulence genes, and corresponding sequences of these genes were taken from the national center for biotechnology information (NCBI)/KEGG database.

2.2.2. Step 3: Analysis of Resistance Genes

In a survey in India, resistance rates of UPEC to various antibiotics were reported for gentamicin (58.2%), beta-lactams (57.4%), quinolones (74.5%), amikacin (33.4%), nalidixic acid (77.7%), co-trimoxazole (48.5%) and cefuroxime (56%) [22]. The genes that produce resistance to popular antibiotics were found using the antibiotic resistance database (ARDB). High-confidence interactors were predicted using the Search Technique for the Retrieval of Interacting Genes/Proteins (STRING) version 10 for each resistant protein [23]. Further, the proteins with high interactions were manually screened for their presence in the human host.

2.3. Phase II: Subtractive Analysis

2.3.1. Analysis of Essential Genes

We identified potential therapeutic drug targets for powerful essential proteins using a BLASTp search against the database of essential genes (DEG) [24]. DEG, a database of essential genes (genes required for an organism's survival), contains essential genes from both eukaryotic and prokaryotic organisms. Protein alignments with an e-value of 0.005 or less were considered more meaningful. The resultant proteins were reconstituted and used as input for the non-homology analysis.

2.3.2. Non-homology Analysis

The goal of the non-homology study is to identify proteins specific to the pathogen that are non-homologous to the human host. This process is important because it prevents the drug from binding to any homologous host proteins and reduces the drug's undesired cross-reactivity [25]. All short-listed proteins were analysed through a BLASTp against a non-redundant *H. sapiens* database [17,25,26]. Proteins not present in the human host were chosen for the following phase.

2.3.3. Human Gut Flora Non-homology Analysis

Shortlisted proteins from the prior step were compared to the proteome of human gut microflora. A normal healthy human's gastrointestinal tract contains approximately 1014 bacterial species [27]. The gut microbiota has a symbiotic relationship with the host, and it plays an important role in metabolism by digesting food particles and protecting the gut against harmful bacteria invasion [28]. Unintentional inhibition of gut flora proteins can degrade the microbiota, resulting in negative consequences for the host. To avoid pharmacological cross-reactivity with gut microflora proteins, the target proteins were homologized against the gut flora proteome using BLASTp with an expected value (e-value) of 0.005 [26]. The analysis included a list of gut microorganisms reported in the literature [29]. The proteins with no more than ten matches on their own were chosen. The potential target proteins were further analyzed for their quantitative characterization.

2.4. Phase III: Quantitative Characterization of Putative Drug Targets

2.4.1. Subcellular Localization Prediction

E.coli is a gram-negative bacterium that possesses an outer cell membrane. Therefore, proteins after synthesis were localized in five possible locations, namely, cytoplasm, periplasm, and the plasma, extracellular, and outer membranes. The localization analysis aims to identify a protein as a potential therapeutic or vaccination target. Surface membrane proteins can be employed as vaccination targets, while cytoplasmic/periplasmic proteins can be exploited as therapeutic drug targets [30]. Once the potential targets were selected, identification of the location of these proteins was attempted to know their functional assignment. For the identification of sub-cellular locations, the Cell Ontology-based Classification (Cello) tool [31], PSORTb (https://www.psort.org/psortb/) [32], and a program for identification of sub-cellular localization of bacterial proteins (ProtCompB) were used.

2.4.2. Broad-Spectrum Analysis

For the identification of broad-spectrum targets, potential therapeutic targets were examined using a BLASTp search against a large number of pathogenic microorganisms [29]. A list of pathogenic bacteria involved in UTIs reported in the virulence factor database (VFDB), and pathogenic bacteria from the literature, as reported by Shanmugham et al. 2013 were considered in this analysis [21]. According to the homology analysis for each pathogen, close homologs are more likely to represent a "promising broad-spectrum target" if they are found in a larger number of pathogenic species.

2.4.3. Druggability Analysis

Further, we performed a BLASTp search against the DrugBank database with an e-value of 0.005 to assess each protein's drug-ability potential. The Drug Bank database is a one-of-a-kind resource that combines pharmacological data with information on drug targets at the sequencing, structure, and pathway levels [33].

3. Results

3.1. Phase I: Comparative Analysis

The metabolic pathways of the host *H. sapiens* and the UPEC have been compared *in silico*. Pathways that are not found in humans but are found in pathogens are referred to as distinct pathways. The 119 metabolic pathways considered in this study are linked to nucleotide, carbohydrate, amino acid, or vitamin metabolism, or cofactor biosynthesis. Thirty four of the 119 pathways have been identified as distinct. Non-homologous proteins were identified in 30 of the 34 pathways.

3.1.1. Chokepoint Enzymes

Chokepoint analysis was investigated using non-homologous proteins from several metabolic pathways. Thirty-one of the 62 proteins were identified as choke point enzymes, with a confidence level of more than 50%, and were chosen for further analysis (Table 1).

Table 1. List of chokepoint enzymes from different pathways.

Enzyme ID	Total No. of Pathways of the Enzyme
ECNA114_0085	ena00473, ena00550, ena01100, ena01502
ECNA114_3261	ena00550
ECNA114_1004, ECNA114_3778	ena00540, ena01100
ECNA114_2045	ena01053
ECNA114_0627	ena00010, ena00030, ena00052, ena00230, ena00500, ena00520, ena00521, ena01100, ena01110, ena01120, ena01130
ECNA114_2137, ECNA114_2136	ena00521, ena00523, ena01100, ena01130
ECNA114_2904, ECNA114_2317	ena00071, ena00280, ena00310, ena00362,00380, ena00620, ena00630, ena00640,ena00650, ena00900, ena01100, ena01110, ena01120, ena01130, ena01200, ena01220, ena01212
ECNA114_1060	ena00562, ena00627, ena01120
ECNA114_3742	ena00910, ena01120, ena02020
ECNA114_1411	ena00010, ena00071, ena00350, ena00625, ena00626, ena00650, ena01100, ena01110, ena01120, ena01130, ena01220

Table 1. *Cont.*

Enzyme ID	Total No. of Pathways of the Enzyme
ECNA114_1698, ECNA114_1052, ECNA114_3080	ena00633, ena01120
ECNA114_3742, ECNA114_3652	ena00010, ena0071, ena00350,ena00625, ena00626, ena01100, ena01110, ena01120, ena01130, ena01150
ECNA114_4463	ena02060, ena00500
ECNA114_2504	ena00520, ena02060
ECNA114_1862	ena00051, ena00520, ena01100, ena02060
ECNA114_4175, ECNA114_4173, ECNA114_4172, ECNA114_2735, ECNA114_3748, ECNA114_2977	ena00051, ena02060
ECNA114_3218, ECNA114_3216, ECNA114_3217, ECNA114_3224,	ena00052, ena02060

3.1.2. Virulence Factors Analysis

Several types of virulence factors, including adherence, iron uptake-aerobactin, and hemolysin are enlisted in VFDB. In UPEC, 69 virulence genes were found. Out of 69, nineteen were found in UPEC strain NA114 (Table 2).

Table 2. List of virulence genes.

Virulence Factors UPEC	Genes Name	Found in UPEC Strain NA114 and Not in Human
Iron uptake	iutA, iucA, iucB, iucC, iucD	iucA, iucB, iucD, iucC
Chu (*E. coli* hemin uptake)	chuA, chuS, chuT, chuU, chuV, chuW, chuX, and chuY	chuU, chuW
Enterobactin	entA, entB, entC, entD, entE, entF, fepA, fepB, fepC, fepD, fepE, and fepG	fepA, fepB, fepC, fepD, fepG, entE, entA, entB, entF, entC
IroN	iroN	iroN
Hemolysin	hlyA, hlyB, hlyC, and hlyD	hlyB, hlyD

3.1.3. Resistance Gene Analysis

Resistance genes were taken from the ARDB. In the UPEC strain used in this study, we found 1805 resistance genes. These proteins were further screened from the STRING database, which resulted in 20 genes with high interactions. Out of these 20, six proteins were present in UPEC and not found in humans (Table 3). Protein sequences of these resistance-causing proteins were taken from KEGG/NCBI database. Generally, resistance-causing proteins and their related proteins are thought to be 'promising therapeutic targets', because inhibiting them could stop the drug resistance mechanism from working.

Table 3. List of selected resistance genes.

Resistance Gene Found in String after ARDB Tool	Found UPEC Strain NA114	Found in Human	Enzyme No.
acrB	Yes	No	c0580
acrA	Yes	No	c0581
macB	Yes	No	c1016
arnA	Yes	No	c2797
tolC	Yes	No	c3781
bacA	Yes	No	c3807

3.2. *Phase II: Subtractive Channel of Analysis*

3.2.1. Analysis of Essential Genes

Using the DEG server, and a threshold e-value of 0.005, the combined proteins from phase 1 were further screened for essential genes. The most important condition for a prospective therapeutic target is that it must be a protein that the organism needs to survive. Fifteen of the 52 input proteins identified in phase I were found to be required for the pathogen's survival and growth (Table 4). Proteins that did not match against any in the DEG database were regarded as non-essential and excluded from the study. The selectivity/specificity of the proteins were analyzed by finding proteins that are non-homologous to the human and gut microbiota proteome.

Table 4. List of selected essential genes.

Sr. No.	Query Protein	No. of a Homolog in DEG	DEG Accession Number
1	ECNA114_0085	1	DEG10180021
2	ECNA114_1004	2	DEG10190079, DEG10180150
3	ECNA114_3778	3	DEG10480267,DEG10180536, DEG10190246
4	ECNA114_2045	1	DEG10180357
5	ECNA114_2137	1	DEG10180210
6	ECNA114_3652	1	DEG10180338
7	ECNA114_1052	1	DEG10180159
8	ECNA114_1862	1	DEG10180464
9	ECNA114_3218	1	DEG10180032
10	entD	1	DEG10190060
11	entE	1	DEG10180357
12	fepB	1	DEG10180106
13	fepC	2	DEG10190239; DEG10480100
14	arnAc2797	3	DEG10180489, DEG10480227, DEG10190203
15	ECNA114_0580	1	DEG10480311

3.2.2. Non-homology Analysis

The use of proteins that are homologous to the host as therapeutic targets could hurt the host's metabolism. As a result, several in silico drug-target-discovery algorithms involving filtering out proteins homologous to the human proteome were utilized as the initial step. Non-homology analysis was performed on the short-listed protein datasets. BLASTp was used to look for similarities between the fifteen UPEC proteins and the entire proteome of *H. sapiens* (host). Five proteins were found to be homologous to a human host and excluded from the study. The remaining ten non-homologous proteins were used in further studies.

3.2.3. Gut Flora Non-homology Analysis

The resulting list of proteins that are non-homologous to *H. sapiens* was screened against the whole proteome microbiota and other gut flora organisms found in the literature using the BLASTp search [29]. From gut flora non-homology analysis, three proteins were found as homologs in microbiota and other gut flora organisms and excluded from the study. The remaining seven proteins were further assessed as putative drug targets.

3.3. Phase III

In this phase, the putative drug targets were further analyzed for their properties, namely, cellular location, broad-spectrum, and druggability analysis.

3.3.1. Subcellular Localization Prediction of Putative Targets

The sub-cellular location was predicted by using ProtCompB, Cello, and PSORTb tools. ProtCompB predicts the localization of proteins at the subcellular level in gram-negative bacteria; by using these, five of the seven proteins were found in the inner membrane, one in the periplasm, and one in the cytoplasm. Cello predicts the localization of protein domains using a multi-class support vector machine (SVM) system based on the physicochemical properties of proteins and predicted the location of one protein in the cytoplasm, one in the periplasm, and five in the inner membrane. PSORTb is another tool for the prediction of protein location for gram-positive strains, gram-negative strains, and archaeal sequences. In our dataset, it returned five proteins as cytoplasmic-membrane, and two as cytoplasm. By comparing the results from different tools, the location of five proteins was found in the inner membrane, one in the cytoplasm and one in the periplasm (Table 5).

Table 5. Predicted subcellular locations of putative drug targets.

Protein Enzyme Code	Cello	PSORTb	ProtCompb	Subcellular Location
ECNA114_4463	Innermembrane	Cytoplasmic Membrane	Innermembrane	Innermembrane
ECNA114_4172	Innermembrane	Cytoplasmic membrane	Innermembrane	Innermembrane
ECNA114_2735	Innermembrane	Cytoplasmic membrane	Innermembrane	Innermembrane
ECNA114_3216	Innermembrane	Cytoplasmic membrane	Innermembrane	Innermembrane
ECNA114_3224	Innermembrane	Cytoplasmic membrane	Innermembrane	Innermembrane
ECNA114_0085	Cytoplasmic	Cytoplasmic	Cytoplasmic	Cytoplasmic
ECNA114_1060	Periplasmic	Cytoplasmic	Periplasmic	Periplasmic

3.3.2. Broad-Spectrum Analysis

Comparative genomic analysis of the screened targets, using a clinically significant *E. coli* strain that causes UTIs described in the VFDB as reference, allowed for an efficient assessment of potential broad-spectrum therapy targets. The broad-spectrum investigation used a list of 240 pathogenic pathogens. A promising broad-spectrum target was discovered using a BLASTp homology search against the entire proteome of each of these bacterial pathogens. The homology search revealed that the screened targets had close homologs in multiple pathogenic species, indicating that *E. coli* proteins are multispecies. Therapeutic compounds that inhibit such broad-spectrum targets could help to eradicate UPEC infections and could be used as possible UPEC-specific drug targets. Such pathogen-specific targets could help to lower the possibility of treatment resistance in a variety of infections. All proteins had more than 70 matches, indicating that the therapeutic targets have a broad spectrum.

3.3.3. Druggability Analysis

The druggability of the short-listed candidate drug targets was assessed by utilizing a sequence-similarity search against the Drug Bank target database. All seven potential drug targets were searched against the Drug Bank target database manually, which includes US Food and Drug Administration (FDA)-approved drugs, revealing that ECNA114_0085 is homologous to a known target, D-Alanine-D-Alanine-ligase. None of the other targets exhibited any matches to the drug target database (Table 6).

Table 6. List of identified putative drug targets.

Putative Targets ID's	Function	Pathways Involved	Druggability Analysis	Subcellular Location
ECNA114_4463	treB, PTS system,	Phosphotransferase system, Starch and sucrose metabolism	No	Innermembrane
ECNA114_4172	PTS system	Mannose and fructose metabolism, Metabolic pathways, Phosphotransferase system	No	Innermembrane
ECNA114_2735	srlE, PTS system	Mannose and fructose metabolism, Metabolic pathways, Phosphotransferase system	No	Innermembrane
ECNA114_3216	agaW, component of PTS system	Galactose metabolism, Metabolic pathways, Phosphotransferase system	No	Innermembrane
ECNA114_3224	agaD, a component of the PTS system	Galactose metabolism, Metabolic pathways, Phosphotransferase system	No	Innermembrane
ECNA114_0085	D-alanine-D-alanine ligase	D-alanine metabolism, Metabolic pathways, Vancomycin resistance, Peptidoglycan biosynthesis	Yes	Cytoplasmic
ECNA114_1060	appA, Phosphoanhydride phosphohydrolase	Inositol phosphate and riboflavin metabolism, Metabolic pathways	No	Periplasmic

4. Discussion

In this study, we used a broad approach and included all functional biosynthetic pathways, such as the metabolism of carbohydrates, energy, amino acids, vitamins, terpenoids, and ketides, as well as those for the biodegradation of xenobiotics. We identified distinct non-homologous pathways that are only found in pathogens and not in human hosts, and the proteins found in these pathways were classified as distinct proteins [34]. Out of 34 considered pathways, non-homologous proteins were found in 30 pathways. However, the fact that they are distinct from, or unrelated to human hosts does not qualify them as effective therapeutic drug targets, as isoenzymes or paralogs exist [35]. To remove this bias, a chokepoint analysis was done. A chokepoint reaction consumes or creates a particular product [18]. Therefore, inhibiting an enzyme that uses a harmful substrate may result in the accumulation of toxic products. Alternatively, if the inhibited enzyme is involved in the creation of a special product, inhibiting it would starve the cell, which may ultimately impair vital cell processes [36]. In light of this, enzymes implicated in chokepoint reactions may be crucial for infectivity and thus, may represent a potential therapeutic target for drug discovery. Thirty-one enzymes from the 62 non-homologous enzymes identified in various metabolic pathways were recognized as chokepoint enzymes. In this work, we have also taken UPEC virulence factors into consideration. Targeting virulence factors has two benefits: first, it will reduce pathogenicity; and second, selection pressure will be minimized, preventing the development of drug resistance. Cinnamaldehyde, baicalein, naringenin, and catechin are a few of the recently identified anti-virulence substances that act by preventing the initiation of quorum sensing (QS) and biofilm production [37,38]. In our study, UPEC strain NA114 was found to contain 19 virulence factors, including those involved in lysine degradation, ABC transporters, biosynthesis of siderophores, two-component pathways, and bacterial secretion systems. By stymieing the actions of proteins linked to antibiotic resistance proteins, drug resistance can also be avoided. To identify potential therapeutic targets for UPEC NA114, we thus included antibiotic resistance genes as targets in the current investigation. Similar combinatorial approaches have been used for the identification of putative drug targets for *Mycobacterium abscessus* where 40 targets were identified [29]. Out of 15 proteins found in the DEG database deemed to be essential for the survival of UPEC, further non-homology and gut flora non-homology analysis yielded seven putative drug targets (Table 3). Since the eligibility and effectiveness of a protein as a therapeutic target depends on where it is located in the cell, in silico tools were employed to predict the subcellular location of the protein [13,39]. Proteins found in the membrane are not ideal as therapeutic targets due to their difficulty in purification and testing. We found two potent druggable proteins (ECNA114_0085 and ECNA114_1060) as promising targets in UPEC strain NA114.

The cytoplasmic ECNA114_0085 protein, identified as a potential target here, is a D-alanine-D-alanine ligase (Ddl) involved in the synthesis of peptidoglycan [9,40]. The peptidoglycan precursor UDPMurNAc-pentapetide contains the terminal dipeptide D-Ala-D-Ala, which is an essential building component for peptidoglycan cross-linking, which provides stability to the cell wall. Therefore, blocking the enzyme Ddl can dramatically reduce the strength of the cell wall, which causes the bacteria to be osmotically lysed. According to a prior study, Ddl is an attractive target for the discovery of drugs in *Mycobacterium tuberculosis* [41]. Vancomycin, which acts only on gram-positive bacteria, binds to Ddl, which inhibits glucosyltransferase, thereby preventing the polymerization and synthesis of N-acetylglucosamine (NAG) and N-acetylmuramic acid (NAM) throughout the peptidoglycan layer [42]. The essential penicillin-binding proteins (PBPs), which are enzymes involved in the final stages of peptidoglycan cross-linking in both gram-negative as well as gram-positive bacteria, covalently bind to beta-lactam antibiotics, making them bactericidal substances that prevent the formation of bacterial cell walls. Early intracellular phases of cell wall synthesis have received little attention as prospective therapeutic targets so far, in contrast to the extracellular stages of peptidoglycan biosynthesis, which are inhibited by beta-lactam and glycopeptide antibiotics [43]. Currently, only two drugs are in use to block intracellular peptidoglycan synthesis: D-cycloserine, which blocks both Ddl and alanine racemase, and fosfomycin, a MurA ligase inhibitor. The internal peptidoglycan precursor UDPMurNAc-pentapeptide is assembled by the mur ligases MurC, MurD, MurE, and MurF by the sequential addition of L-Ala, D-Glu, m-Dpm or L-Lys, and D-Ala-D-Ala to UDP-MurNAc. Ddl is responsible for supplying the MurF substrate, D-alanyl-D-alanine [43]. Because cross-linking of peptidoglycan chains takes place between the penultimate D-Ala in a second pentapeptide strand and the C6-NH2-group of meso diaminopimelic acid in gram-negative bacteria and the NH2-group of pentaglycine in gram-positive *Staphylococcus*, this terminal dipeptide is crucial in the construction of the bacterial cell wall. In both situations, the intrastrand D-Ala-D-Ala bond is broken and a new interstrand peptide bond is formed [43]. This enzyme is common in prokaryotes, but absent in eukaryotes, making it a likely candidate target for antibiotic development. It is possible to design inhibitors against this enzyme, crucial to the peptidoglycan synthesis process, that could cause a loss in the structural integrity of bacterial cell walls and osmotic lysis of pathogenic bacteria.

The periplasmic protein appA (ECNA114_1060) involved in the metabolism of inositol phosphate was found to be another attractive target. Gram-negative bacteria are extremely resistant to the penetration of antimicrobials due to their double-layered cell envelope and a variety of efflux pumps. Proteins in the periplasm are easier to target, since only penetration of the outer membrane is required, whereas cytoplasmic targets are difficult to reach [44]. In addition to assisting in the production of virulence factors, adhesion molecules, and signaling molecules, periplasmic proteins are crucial for sustaining cell viability, cell division, and membrane integrity. All eukaryote genomes contain the inositol polyphosphate production enzymes, yet standard search techniques frequently fail to uncover the amino acid sequence homology of these enzymes [45]. Similarity searching between microbial and human inositol phosphate kinases is restricted to fewer catalytically important residues, which include proline, aspartic acid, lysine, glycine, serine, leucine, tyrosine, isoleucine, histidine, threonine, glutamic acid, and phenylalanine [45]. Recent studies of the Inositol phosphate metabolic pathways in pathogenic protozoa (*Trypanosome brucei*) and fungi (*Cryptococcus neoformans*) are being exploited as promising antimicrobial targets [45]. Inositol phosphate metabolism has been identified as a putative drug target for many bacterial pathogens including *Staphylococcus saprophyticus, Klebsiella pneumoniae,* and *Streptococcus pneumoniae* [11,46,47]. Despite the benefits and optimistic future of comparative genomics in the identification of possible therapeutic targets, there are still certain limitations. Despite the fact that comparative genomics is frequently utilized in the creation of medications to combat drug-resistant bacteria, the failure rate of existing antibiotics is substantially higher than the rate at which new antibiotics are developed. Additionally, antibiotics are transient treatments for infections. The use of comparative genomics in the creation of antibiotics is a long-debated topic due to the fact that their value is significantly lower than medications used in the treatment of chronic diseases [48]. A further problem is that, despite the fact that comparative genomics can decrease the number of experimental targets and identify some appealing proteins as prospective therapeutic targets, the range of potential targets that can be screened by this method is still quite broad and the cost and time used in the development process is prohibitive. The majority of these putative targets identified by comparative genomics are unlikely to be validated through experiments. Therefore, it would be advantageous to integrate network-based approaches with comparative genomics to further limit the range of experimental targets. This will use less time and material, and cut costs during the early stages of drug research and development.

5. Conclusions

Using a unique hierarchical in silico technique, we discovered and described two potent druggable proteins as prospective therapeutic drug targets against ESBL-producing MDR UPEC strain NA114. The identified drug targets are distinct and have the potential to be exploited for designing new antimicrobial agents against UPEC. The computer-aided drug design method can also be utilized to identify homologous compounds for these targets. The search for possible therapeutic targets in ESBL-producing MDR UPEC strain NA114 was directed by comparative and subtractive genomics. Essentiality, non-homology to the human host, availability in the drug bank, and sub- cellular location were used to narrow down the list of targets in this study. Our research uncovered prospective therapeutic targets (ECNA114_0085 and ECNA114_1060) that are critical in the treatment of *E. coli* urinary tract infections.

Author Contributions: H.K. and N.T., Conceptualization. H.K., Data curation. H.K., Formal analysis. H.K., Investigation. H.K., V.M. and N.C., Methodology. H.K., Software. B.M. and N.T., Supervision. H.K. and N.T., Validation. H.K. and N.T. Writing—original draft. H.K. and N.T., Writing—review & editing. All authors have read and agreed to the published version of the manuscript.

Funding: This research received no external funding.

Institutional Review Board Statement: Not applicable.

Informed Consent Statement: Not applicable.

Data Availability Statement: Not applicable.

Acknowledgments: We gratefully acknowledge to the Director, PGIMER Chandigarh for providing facilities to carry out this work. The Indian Council of Medical Research (ICMR), New Delhi support in the form of a fellowship (ICMR/RA/Fellowship/2021/11125) to the first author is highly acknowledged.

Conflicts of Interest: The authors declare no conflict of interest.

References

1. Murray, C.J.; Ikuta, K.S.; Sharara, F.; Swetschinski, L.; Robles Aguilar, G.; Gray, A.; Han, C.; Bisignano, C.; Rao, P.; Wool, E.; et al. Global Burden of Bacterial Antimicrobial Resistance in 2019: A Systematic Analysis. *Lancet* **2022**, *399*, 629–655. [CrossRef]
2. Exner, M.; Bhattacharya, S.; Christiansen, B.; Gebel, J.; Goroncy-Bermes, P.; Hartemann, P.; Heeg, P.; Ilschner, C.; Kramer, A.; Larson, E.; et al. Antibiotic Resistance: What Is so Special about Multidrug-Resistant Gram-Negative Bacteria ? Antibiotikare-sistenz: Was Ist so Besonders an Den Gram-Negativen. *GMS Hyg. Infect. Control* **2017**, *12*, Doc05.
3. Kayastha, K.; Dhungel, B.; Karki, S.; Adhikari, B.; Banjara, M.R.; Rijal, K.R.; Ghimire, P. Extended-Spectrum β-Lactamase-Producing Escherichia Coli and Klebsiella Species in Pediatric Patients Visiting International Friendship Children's Hospital, Kathmandu, Nepal. *Infect. Dis. Res. Treat.* **2020**, *13*, 117863372090979. [CrossRef]
4. Taneja, N.; Rao, P.; Arora, J.; Dogra, A. Occurrence of ESBL & Amp-C b -Lactamases & Susceptibility to Newer Antimicrobial Agents in Complicated UTI. *Indian J Med Res.* **2008**, *127*, 85–88.
5. Shaikh, S.; Fatima, J.; Shakil, S.; Rizvi, S.M.D.; Kamal, M.A. Antibiotic Resistance and Extended Spectrum Beta-Lactamases: Types, Epidemiology and Treatment. *Saudi J. Biol. Sci.* **2015**, *22*, 90–101. [CrossRef] [PubMed]
6. Kaza, P.; Mahindroo, J.; Veeraraghavan, B.; Mavuduru, R.S.; Mohan, B.; Taneja, N. Evaluation of Risk Factors for Colistin Resistance among Uropathogenic Isolates of Escherichia Coli and Klebsiella Pneumoniae: A Case–Control Study. *J. Med. Microbiol.* **2019**, *68*, 837–847. [CrossRef] [PubMed]
7. Asokan, G.V.; Ramadhan, T.; Ahmed, E.; Sanad, H. WHO Global Priority Pathogens List: A Bibliometric Analysis of Medline-Pubmed for Knowledge Mobilization to Infection Prevention and Control Practices in Bahrain. *Oman. Med. J.* **2019**, *34*, 184–193. [CrossRef] [PubMed]
8. Chakrabarty, R.P.; Alam, A.S.M.R.U.; Shill, D.K.; Rahman, A. Identification and Qualitative Characterization of New Therapeutic Targets in Stenotrophomonas Maltophilia through in Silico Proteome Exploration. *Microb. Pathog.* **2020**, *149*, 104293. [CrossRef] [PubMed]
9. Kaur, H.; Kalia, M.; Taneja, N. Identification of Novel Non-Homologous Drug Targets against Acinetobacter Baumannii Using Subtractive Genomics and Comparative Metabolic Pathway Analysis. *Microb. Pathog.* **2021**, *152*, 104608. [CrossRef]
10. Kaur, H.; Kalia, M.; Singh, V.; Modgil, V.; Mohan, B.; Taneja, N. In Silico Identification and Characterization of Promising Drug Targets in Highly Virulent Uropathogenic Escherichia Coli Strain CFT073 by Protein-Protein Interaction Network Analysis. *Inf. Med. Unlocked* **2021**, *25*, 100704. [CrossRef]
11. Shahid, F.; Ashfaq, U.A.; Saeed, S.; Munir, S.; Almatroudi, A.; Khurshid, M. In Silico Subtractive Proteomics Approach for Identification of Potential Drug Targets in Staphylococcus Saprophyticus. *Int. J. Environ. Res. Public Health* **2020**, *17*, 3644. [CrossRef] [PubMed]
12. Solanki, V.; Tiwari, V. Subtractive Proteomics to Identify Novel Drug Targets and Reverse Vaccinology for the Development of Chimeric Vaccine against Acinetobacter Baumannii. *Sci. Rep.* **2018**, *8*, 9044. [CrossRef] [PubMed]
13. Dutta, A.; Singh, S.K.; Ghosh, P.; Mukherjee, R.; Mitter, S.; Bandyopadhyay, D. In Silico Identification of Potential Therapeutic Targets in the Human Pathogen Helicobacter Pylori. *Silico Biol.* **2006**, *6*, 43–47.
14. Lin, X.; Li, X.; Lin, X. A Review on Applications of Computational Methods in Drug Screening and Design. *Molecules* **2020**, *25*, 1375. [CrossRef]
15. Avasthi, T.S.; Kumar, N.; Baddam, R.; Hussain, A.; Nandanwar, N.; Jadhav, S.; Ahmed, N. Genome of Multidrug-Resistant Uropathogenic Escherichia Coli Strain NA114 from India. *J. Bacteriol.* **2011**, *193*, 4272–4273. [CrossRef] [PubMed]
16. Kanehisa, M.; Goto, S.; Kawashima, S.; Nakaya, A. The KEGG Databases at GenomeNet. *Nucleic Acids Res.* **2002**, *30*, 42–46. [CrossRef]
17. Anishetty, S.; Pulimi, M.; Pennathur, G. Potential Drug Targets in Mycobacterium Tuberculosis through Metabolic Pathway Analysis. *Comput. Biol. Chem.* **2005**, *29*, 368–378. [CrossRef]
18. Singh, S.; Malik, B.K.; Sharma, D.K. Choke Point Analysis of Metabolic Pathways in E.Histolytica: A Computational Approach for Drug Target Identification. *Bioinformation* **2007**, *2*, 68–72. [CrossRef]
19. Sc, S.K.M.; Phil, M. Protocol of Rice Genome Annotation through Comparative Functional Genomics Approach. *Genome* **2009**, *4*, 1–8.
20. Baron, C.; Coombes, B. Targeting Bacterial Secretion Systems: Benefits of Disarmament in the Microcosm. *Infect. Disord. Drug Targets* **2007**, *7*, 19–27. [CrossRef]

21. Chen, L.; Zheng, D.; Liu, B.; Yang, J.; Jin, Q. VFDB 2016: Hierarchical and Refined Dataset for Big Data Analysis--10 Years On. *Nucleic Acids Res.* **2016**, *44*, D694–D697. [CrossRef] [PubMed]
22. Niranjan, V.; Malini, A. Antimicrobial Resistance Pattern in Escherichia Coli Causing Urinary Tract Infection among Inpatients. *Indian J. Med. Res.* **2014**, *139*, 945–948. [PubMed]
23. Szklarczyk, D.; Franceschini, A.; Wyder, S.; Forslund, K.; Heller, D.; Huerta-Cepas, J.; Simonovic, M.; Roth, A.; Santos, A.; Tsafou, K.P.; et al. STRING V10: Protein–Protein Interaction Networks, Integrated over the Tree of Life. *Nucleic Acids Res.* **2015**, *43*, D447–D452. [CrossRef] [PubMed]
24. Luo, H.; Lin, Y.; Gao, F.; Zhang, C.T.; Zhang, R. DEG 10, an Update of the Database of Essential Genes That Includes Both Protein-Coding Genes and Noncoding Genomic Elements. *Nucleic Acids Res* **2014**, *42*, D574–D580. [CrossRef] [PubMed]
25. Sarkar, M.; Maganti, L.; Ghoshal, N.; Dutta, C. In Silico Quest for Putative Drug Targets in Helicobacter Pylori HPAG1: Molecular Modeling of Candidate Enzymes from Lipopolysaccharide Biosynthesis Pathway. *J. Mol. Model* **2012**, *18*, 1855–1865. [CrossRef]
26. Altschul, S.F.; Gish, W.; Miller, W.; Myers, E.W.; Lipman, D.J. Basic Local Alignment Search Tool. *J. Mol. Biol.* **1990**, *215*, 403–410. [CrossRef]
27. Fujimura, K.E.; Slusher, N.A.; Cabana, M.D.; Lynch, S.V. Role of the Gut Microbiota in Defining Human Health. *Expert Rev. Anti. Infect. Ther.* **2010**, *8*, 435–454. [CrossRef]
28. Rabizadeh, S.; Sears, C. New Horizons for the Infectious Diseases Specialist: How Gut Microflora Promote Health and Disease. *Curr. Infect. Dis. Rep.* **2008**, *10*, 92–98. [CrossRef]
29. Shanmugham, B.; Pan, A. Identification and Characterization of Potential Therapeutic Candidates in Emerging Human Pathogen Mycobacterium Abscessus: A Novel Hierarchical In Silico Approach. *PLoS ONE* **2013**, *8*, e59126. [CrossRef]
30. Barh, D.; Tiwari, S.; Jain, N.; Ali, A.; Santos, A.R.; Misra, A.N.; Azevedo, V.; Kumar, A. In Silico Subtractive Genomics for Target Identification in Human Bacterial Pathogens. *Drug Dev. Res.* **2011**, *72*, 162–177. [CrossRef]
31. Yu, C.; Lin, C.; Hwang, J.-K. Predicting Subcellular Localization of Proteins for Gram-Negative Bacteria by Support Vector Machines Based on n -Peptide Compositions. *Proteins Sci.* **2004**, *13*, 1402–1406. [CrossRef] [PubMed]
32. Yu, N.Y.; Wagner, J.R.; Laird, M.R.; Melli, G.; Rey, S.; Lo, R.; Dao, P.; Cenk Sahinalp, S.; Ester, M.; Foster, L.J.; et al. PSORTb 3.0: Improved Protein Subcellular Localization Prediction with Refined Localization Subcategories and Predictive Capabilities for All Prokaryotes. *Bioinformatics* **2010**, *26*, 1608–1615. [CrossRef] [PubMed]
33. Wishart, D.S.; Knox, C.; Guo, A.C.; Cheng, D.; Shrivastava, S.; Tzur, D.; Gautam, B.; Hassanali, M. DrugBank: A Knowledgebase for Drugs, Drug Actions and Drug Targets. *Nucleic Acids Res.* **2008**, *36*, D901–D906. [CrossRef]
34. Butt, A.M.; Nasrullah, I.; Tahir, S.; Tong, Y. Comparative Genomics Analysis of Mycobacterium Ulcerans for the Identification of Putative Essential Genes and Therapeutic Candidates. *PloS ONE* **2012**, *7*, e43080. [CrossRef] [PubMed]
35. Ferdous, S.; Akter, A. Identification of Potential Drug Targets by Subtractive Genome Analysis of Escherichia Coli O157: H7: An in Silico Approach. *Adv. Appl. Bioinform. Chem.* **2015**, *8*, 49–63.
36. Yeh, I. Computational Analysis of Plasmodium Falciparum Metabolism: Organizing Genomic Information to Facilitate Drug Discovery. *Genome Res.* **2004**, *14*, 917–924. [CrossRef]
37. Martínez, O.F.; Cardoso, M.H.; Ribeiro, S.M.; Franco, O.L. Recent Advances in Anti-Virulence Therapeutic Strategies with a Focus on Dismantling Bacterial Membrane Microdomains, Toxin Neutralization, Quorum-Sensing Interference and Biofilm Inhibition. *Front. Cell Infect. Microbiol.* **2019**, *9*, 74. [CrossRef]
38. Topa, S.H.; Palombo, E.A.; Kingshott, P.; Blackall, L.L. Activity of Cinnamaldehyde on Quorum Sensing and Biofilm Susceptibility to Antibiotics in Pseudomonas Aeruginosa. *Microorganisms* **2020**, *8*, 455. [CrossRef]
39. Butt, A.M.; Tahir, S.; Nasrullah, I.; Idrees, M.; Lu, J.; Tong, Y. Mycoplasma Genitalium: A Comparative Genomics Study of Metabolic Pathways for the Identification of Drug and Vaccine Targets. *Infect. Genet. Evol.* **2012**, *12*, 53–62. [CrossRef]
40. Prosser, G.A.; Carvalho, L.P.S. De Kinetic Mechanism and Inhibition of Mycobacterium Tuberculosis D—Alanine: D—Alanine Ligase by the Antibiotic D—Cycloserine. *FEBS J.* **2013**, *280*, 1150–1166. [CrossRef]
41. Bruning, J.B.; Murillo, A.C.; Chacon, O.; Barletta, R.G.; Sacchettini, J.C. Structure of the Mycobacterium Tuberculosis D-Alanine:D-Alanine Ligase, a Target of the Antituberculosis Drug D-Cycloserine. *Antimicrob. Agents Chemother.* **2011**, *55*, 291–301. [CrossRef] [PubMed]
42. Howden, B.P.; Davies, J.K.; Johnson, P.D.R.; Stinear, T.P.; Grayson, M.L. Reduced Vancomycin Susceptibility in Staphylococcus Aureus, Including Vancomycin-Intermediate and Heterogeneous Vancomycin-Intermediate Strains: Resistance Mechanisms, Laboratory Detection, and Clinical Implications. *Clin. Microbiol. Rev.* **2010**, *23*, 99–139. [CrossRef]
43. Kovač, A.; Konc, J.; Vehar, B.; Bostock, J.M.; Chopra, I.; Janežič, D.; Gobec, S. Discovery of New Inhibitors of D-Alanine:D-Alanine Ligase by Structure-Based Virtual Screening. *J. Med. Chem.* **2008**, *51*, 7442–7448. [CrossRef] [PubMed]
44. Pandeya, A.; Ojo, I.; Alegun, O.; Wei, Y. Periplasmic Targets for the Development of Effective Antimicrobials against Gram-Negative Bacteria. *ACS Infect. Dis.* **2020**, *6*, 2337–2354. [CrossRef] [PubMed]
45. Saiardi, A.; Azevedo, C.; Desfougères, Y.; Portela-Torres, P.; Wilson, M.S.C. Microbial Inositol Polyphosphate Metabolic Pathway as Drug Development Target. *Adv. Biol. Regul.* **2018**, *67*, 74–83. [CrossRef] [PubMed]
46. Ramesh, P.; Nagendrappa, J.H.; Shivashankara, S.K.H. Comparative Analysis of Rosetta Stone Events in Klebsiella Pneumoniae and Streptococcus Pneumoniae for Drug Target Identification. *Beni. Suef. Univ. J. Basic Appl. Sci.* **2021**, *10*, 37. [CrossRef]

47. Pirruccello, M.; Nandez, R.; Idevall-Hagren, O.; Alcazar-Roman, A.; Abriola, L.; Berwick, S.A.; Lucast, L.; Morel, D.; De Camilli, P. Identification of Inhibitors of Inositol 5-Phosphatases through Multiple Screening Strategies. *ACS Chem. Biol.* **2014**, *9*, 1359–1368. [CrossRef]
48. Liao, J.; Wang, Q.; Wu, F.; Huang, Z. In Silico Methods for Identification of Potential Active Sites of Therapeutic Targets. *Molecules* **2022**, *27*, 7103. [CrossRef]

biomedicines

MDPI

Article

Coupling CRISPR/Cas9 and Lambda Red Recombineering System for Genome Editing of *Salmonella* Gallinarum and the Effect of *ssaU* Knock-Out Mutant on the Virulence of Bacteria

Hamza Tahir [1,†], Abdul Basit [1,2,†], Hafsa Tariq [1], Zulquernain Haider [1], Asim Ullah [1,3], Zafar Hayat [4,5] and Shafiq Ur Rehman [1,*]

1 Institute of Microbiology and Molecular Genetics, University of the Punjab, Lahore 54590, Pakistan
2 School of Biology, University of St Andrews, St Andrews KY16 9AJ, UK
3 Division of Infection and Immunity, The Roslin Institute, University of Edinbrugh, Edinburgh EH8 9YL, UK
4 Department of Animal Nutrition, University of Veterinary and Animal Sciences, Lahore 54000, Pakistan
5 Department of Animal Sciences, University of Sargodha, Sargodha 40100, Pakistan
* Correspondence: shafiq.mmg@pu.edu.pk; Tel.: +92-3214905423
† These authors contributed equally to this work.

Abstract: The poultry industry in developing countries still faces a significant threat from fowl typhoid, a disease caused by *Salmonella* Gallinarum that has been well contained in more economically developed countries. In addition to the virulence exhibited by large virulence plasmid (85 kb), Salmonella Pathogenicity Island 2 in *S.* Gallinarum plays a key role in mediating disease through its type III secretion systems (TTSS). The TTSS secrete effector protein across the *Salmonella* containing vacuoles and mediate the internalization of bacteria by modulating vesicular passage. In this study, candidate virulent *ssaU* gene (~1 kb) encoding type III secretion system was successfully deleted from indigenously isolated *S.* Gallinarum genome through homology-directed repair using CRISPR/Cas9 and lambda recombination systems. CRISPR/Cas9-based genome editing of poultry-derived *Salmonella* Gallinarum has not been previously reported, which might be linked to a lack of efficiency in its genetic tools. This is the first study which demonstrates a complete CRISPR/Cas9-based gene deletion from this bacterial genome. More importantly, a poultry experimental model was employed to assess the virulence potential of this mutant strain (ΔssaU_SG18) which was unable to produce any mortality in the experimentally challenged birds as compared to the wild type strain. No effect on weight gain was observed whereas bacteria were unable to colonize the intestine and liver in our challenge model. This in vivo loss of virulence in mutant strain provides an excellent functionality of this system to be useful in live vaccine development against this resistant and patho genic bacteria.

Keywords: CRISPR/Cas9; lambda recombineering; genome editing; type three secretion system; virulent genes; *Salmonella* Gallinarum; poultry experimental model

1. Introduction

Salmonella enterica sub sp. *enterica* serovar Gallinarum biovar Gallinarum (*S.* Gallinarum) is a crucial pathogen of poultry causing fowl typhoid (FT), an acute septicemic disease leading to catarrhal enteritis, hepatitis, pericarditis and splenomegaly [1]. With a mortality rate of 40–90%, the disease causes havoc in terms of decreased weight gain in broilers and low egg production in layer birds [2]. This disease spreads through both horizontal and vertical transmission [3]. Currently, live attenuated strain (9R) with a semi-rough lipo-polysaccharide structure has been used as vaccine to control FT in poultry [4,5]. In the last two decades, various outbreaks of FT have been reported in India, Nigeria, Korea and Japan [4–7]. In addition, the emerging resistance of *S.* Gallinarum against ciprofloxacin, amoxicillin, erythromycin, tetracycline and doxycycline reported in recent studies demonstrates an alarming situation in the future for poultry industry [8]. This urges the undermining of the virulence factors and molecular pathways that contribute to SG's pathogenicity. The enormous 85 kb plasmid of *Salmonella* Gallinarum is crucial for virulence, and various plasmid-borne virulence genes have been characterized for their virulence potential [9]. Salmonella pathogenicity island 2 (SPI-2) has been shown to play a crucial role in disease transmission by *Salmonella* enterica via type III secretion system apparatus (TTSS) [10].

The SPI-2 encoded TTSS gets activated and interfere with the maturation of phagosome once the *Salmonella* is engulfed. This results in the formation of specified *Salmonella* Containing vacuole (SCV), leading to intracellular survival and replication of *Salmonella* [11].

A number of steps precede the SCV formation, including endocytic pathway hence inhibiting the delivery of hydrolytic enzymes; relocation of SCV to perinuclear area and development of Salmonella-induced-filaments (Garcia-del Portillo et al., 1993). Due to bacterial intervention with many cellular processes these events proceeds to the survival, colonization, and spread of *Salmonella*. Hence SPI-2-encoded TTSS is crucial to orchestrate SCV formation and intracellular survival of *Salmonella* [12].

The genes belonging to SPI-2 including *ssaG, ssaV, SsaU* and *ssrA* have been classified as potential virulent factors for *Salmonella* infections [13]. These TTSS-associated genes regulate the trafficking of several virulence-related effector proteins from bacteria to host cells [14]. Translocator proteins are involved in the trafficking of another set of proteins directly into the host cytoplasm through their needle like injectisome. After translocation, these "effector proteins" act as the virulence mediator affecting the changes in the host cells [15].

In this study, complete CRISPR/Cas9-based knockout ΔssaU_SG18 strain was employed for a poultry experimental infection model to investigate the effect of the SPI-2 TTSS on the virulence of S. Gallinarum infections. Here, we demonstrate that a functional SPI-2 TTSS is necessary for the development of fowl typhoid and that ΔssaU_SG18 is incapable of colonization and death in chickens.

The CRISPR system which was discovered as a bacterial immune system usually consists of the (CRISPR-associated) Cas protein (typically Cas9) (type IIA), an RNA-dependent endonuclease which cleaves the target DNA determined by the CRISPR RNA (crRNA) dictating the region of double stranded break (DSB) through its spacer sequence (20 nucleotide) and a trans-activating (tracrRNA) which forms a complex structure with crRNA to guide the Cas nuclease [16]. Nowadays this crRNA and tracrRNA can be fused together to transcribe a functional single chimeric guide RNA (sgRNA) [17], which form a complex with Cas9 protein directing it to specified DNA locus by base pairing with protospacer adjacent motif (PAM). PAM motif consists of a few base pairs with their specific sequence and position which vary with defined CRISPR/Cas9 system [18]. In the type II system of the S.pyogenes, PAM refers to a –NGG- consensus sequence comprising of two (G:C) base pairs which occur one base pair downstream of the spacer binding site, within the target DNA [16]. Hence after the recognition, this complex mediates a double stranded break (DSB), three base pairs upstream of the PAM sequence. Since, DSB is lethal, cells during their cycle of repair can be exploited to acquire specific modification via homologous recombination or integrated homology-directed repair (HDR) via a PCR product, plasmid or oligonucleotide [19].

Phage-derived lambda recombination system has emerged as a useful and alternate method for the chromosomal genome engineering [20,21]. The method relies upon the phage recombination proteins which utilize short homology region at 5′ end of linearized DNA fragment to commence the homology directed recombination. The use of a phage-derived recombination system has already been reported in many pathogenic organisms such as E.coli, *Salmonella* and *Yersinia* species [22–24]. More recently, the lambda red recombination based genome deletion was also reported in *Salomella* enterica serovar Pullorum in which the cigR deletion mutant was further tested for its virulence and biofilm formation [25]. The lambda red genes (i.e., gam, bet, and exo) from bacteriophage lambda are vital to support homologous recombination of linearized dsDNA [26]. Gam retards host endogenous nuclease activities of RecBCD and SbcCD; shielding the dsDNA for recombination apparently avoids the degradation of linear dsDNA template in the cell [27,28]. The Exo acts as an exonuclease which truncates linearized dsDNA in 5′→3′ direction from both ends, forming 3′ single-stranded tails for dsDNA [29]. Beta swiftly binds the ssDNA template ends to complementary strand to target DNA [30].

In this study, we have employed a simplified genome engineering strategy using CRISPR/Cas9 parallel with the conventional lambda red recombineering. This pairing increases the genome editing efficiency of the overall system with shorter time period required than previously reported methods [31]. An animal infection model was used to study the effect of *ssaU* mutant on the pathogenicity of this bacteria. This study demonstrates the versatility of CRISPR/Cas9 as a robust bioengineering tool for attenuated live vaccine development by targeting the different virulence-associated genes of pathogenic and resistant bacteria.

2. Materials and Methods

2.1. Strains, Culture Conditions, Plasmids, and Oligos

All the bacterial cultures along with the plasmids used in this study are mentioned in the Table 1. *E.* coli Top 10 cells used for transformation were cultured at 37 °C in Tryptic Soy Broth (TSB), supplemented with 100µg/mL of ampicillin or 50 µg/mL of kanamycin as required. *S.* Gallinarum (wild type) strain and transformants were grown aerobically at 37 °C in TSB supplemented with 50 µg/mL of kanamycin (kan), 100 µg/mL of ampicillin (amp) or chloramphenicol (cam) 25 µg/mL as per requirement. All the DNA oligonucleotide sequences used in this study are listed in Table 2. The linear DNA fragments designed for DSB repairing were amplified through Q5® High-Fidelity DNA Polymerase (NEB) after fusion using NEBuilder® HiFi DNA Assembly as described in the Supplementary Material.

Table 1. Plasmids and bacterial strains used in this study.

Sr. No.	Strain/Plasmid	Characteristics	Reference/Source
1.	E.coli Top 10	Tetracycline, hsdR, mcrA	Thermofisher Scientific
2.	*Salmonella* Gallinarum	Poultry origin	Addgene
3.	pCas9	pACYC184, Chloramphenicol, DH5alpha	Addgene (Plasmid #42876)
4.	pRed	Kanamycin, Lambda Red recombinase, DH5alpha, RepA101ts	This study
5.	pET22b (+)	pelB, His•Tag®, Ampicillin	Addgene (Plasmid # 69744-3)
6.	RP-18	pelB, His•Tag®, Ampicillin, 1.8 kb HAs	This study
7.	ssaU/G3	pACYC184 Chloramphenicol, gRNA3	This study
8.	ssaU/G4	Chloramphenicol, pACYC184, gRNA4	This study

Table 2. Oligonucleotide sequences designed for this study.

Oligonucleotide	Sequence (5′–3′)
HA1_ssaU_FP	aaaatcTCTAGAAGCGGTATCCTGTTGAATTATACC
HA1_ssaU_Rp	AATAACGTTTCAGGAATTTTATCTCTTCtttctgtagtctgttctgttttc
HA2_ssaU_Fp	gaaaacagaacagactacagaaaaGAAGAGATAAAATTCCTGAAACGTTATT
HA2_ssaU_Rp:	ttatgaCTCGAGTGCTGCTTGCTGCGGTTTACCAGA
gRNA3F_ ssaU	aaacTTATTTTTTGAAGTGGAACGg
gRNA3R_ ssaU	aaaacCGTTCCACTTCAAAAAATAA
gRNA-4F-ssaU	aaacACAGAAAAGAAATTACGTGAg
gRNA-4R- ssaU	aaaacTCACGTAATTTCTTTTCTGT
ssaU _screening_Fp:	ACGTCTATGCCGGTAGTGTTGGT
ssaU _screening_Rp:	CATTTGTATGGCTGTGGTTACCG
Cas9_R	ATAGTGACTGGCGATGCTGTC
Pcas-red-Fp	ATTCACTTTTTCTTCACAACCG
Pcas-red-Rp	TATCACCAGTGGGTTTACTTTC
lambda-Fp	AAGCAGACAGGACATGAGCGGATACATATTTG
lambda-Rp	CGCTCATGTCCTGTCTGCTTACATAAACAG
ssau_seq-Fp	GTATCAGCTTCTCTCTTCCT
ssau-seq-Rp	CACCTTTATCGTCAAGCACT

(Underline indicates restriction recognition sequence; lowercase indicates overhang regions).

2.2. DNA Editing Template Construction, gRNA Cloning and Induction Method

Homology arms ~1800 bp as DNA editing template were cloned in the pET22b (+) vector. Homology arms flanking up and downstream regions of the selected *ssaU* gene was amplified through overlap primers HA1_ssaU_Fp and HA1_ssaU_Rp for upstream while HA2_ssaU_Fp and HA2_ssaU_Rp for downstream region followed by fusion through NEB Hifi DNA® assembly kit (see Supplementary Material). This assembly mixture was later amplified by using primer HA1_ssaU _Fp and HA2_ssaU_Rp and cloned in pET22b (+) vector named as RP-18. For dsDNA repair template, PCR product of fused homology arm fragments was used after PCR purification. In order to introduce targeted DSB, RNA spacer (5′-CGTTCCACTTCAAAAAATAA-3′) and (5′-TCACGTAATTTCTTTTCTGT-3′) were cloned in pCas9 and named as ssaU/G3, ssaU/G4, respectively. A Golden Gate assembly protocol was used for cloning of space sequence in plasmid (see Supplementary Material) [32,33]. As off target effects of gRNA has been reported, which could lead to secondary DSB or mutations, we used online available software Cas-OFFinder [34] (Bae et al., 2014) to rule out any secondary target site possibilities. Successful cloning of spacer sequence in pCas9 was confirmed by using primers Cas9_R as reverse and gRNA3F_ssaU or gRNA4F_ssaU as forward primers for ssaU/G3 and ssaU/G4, respectively (Figure 1). Spacer cloning was confirmed by Sanger sequencing (Supplementary Material Figure S1).

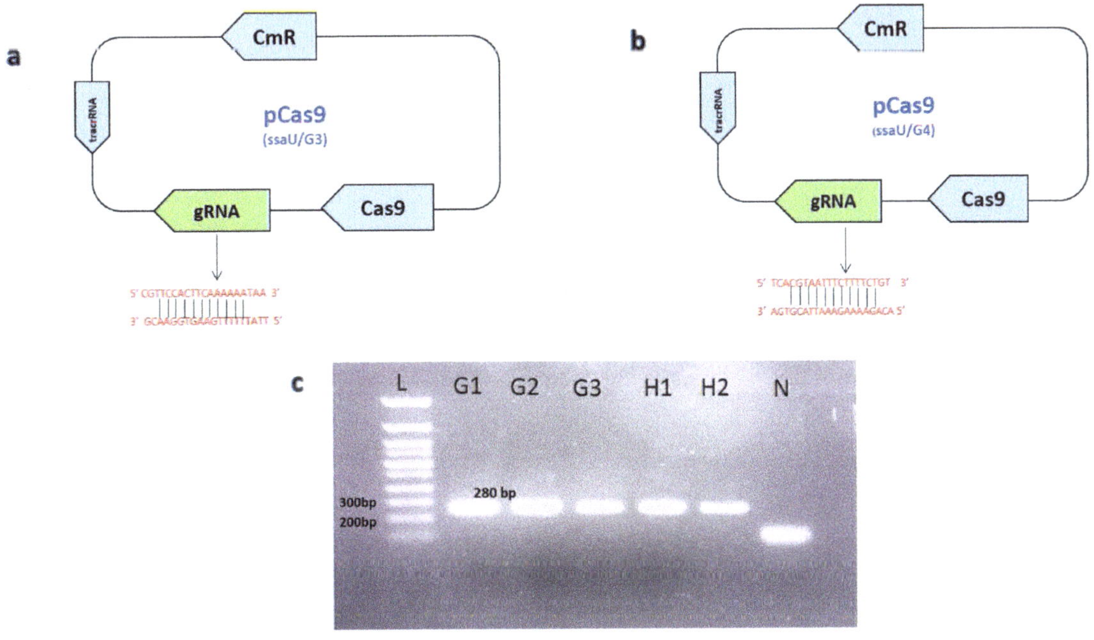

Figure 1. Confirmation of gRNA cloning in pCas9 through PCR. (**a**) Graphical representation of ssau/G3 plasmid showing spacer cloning using golden gate assembly. (**b**) The plasmid map of ssaU/G4 along with its spacer sequence. (**c**) PCR confirmation of gRNA cloning in pCas9. Lane (G1-G3) illustrating positive result with 280 bp amplified fragment, showing cloning of gRNA3 in ssau/G3 plasmid, while lane (H1 H2) represents the successful cloning of second gRNA named ssaU/G4 plasmid. Lane (N) represents negative control. Lane (L) shows 100 base pair plus DNA ladder (Bioron).

Mutants were identified using screening primers designed outside of the frame of homology arm sequence named ssaU _screening_Fp and ssaU _screening_Rp. Mutants were further confirmed by sanger sequencing.

In all of the experiments exhibiting recombineering, controls were kept in place to assess the electroporation of plasmids. At least three replicates were employed to assess the recombination of the plasmid/dsDNA template for each recombineering experiment performed. The formula for calculating transformation and electroporation efficiency is provided in the Supplementary Material. Statistical Analysis including arithmetic means with SD and *p*-value using two-way ANOVA was performed using the software tool Graphpad prism version 8.0.1 GraphPad Software, La Jolla, CA, USA, www.graphpad.com (accessed on 20 July 2022).

2.3. Mutant Strain Preparation

The red recombineering plasmid (pRed) was electroporated into the S. Gallinarum cells selected on 50 μg/mL kanamycin and confirmed through colony PCR using Pcas-red-Fp and Pcas-red-Rp as forward and reverse primers, respectively. Overnight cultures of the recombineering plasmid transformed S. Gallinarum cells (500 μL) were shifted to 30 mL TSB at 30 °C for ~1.5 h with shaking at 200 rpm. When $OD_{600\ nm}$ of cells reached 0.2, the cells were induced by 10 mM arabinose for expression of lambda recombineering components (Exo, Gam, Beta) and further incubated at 30 °C for ~1 h with shaking. These cells were then used for electrocompetent cells preparation.

2.4. Electroporation

A single colony of the wild type *S.* Gallinarum was inoculated in 3 mL TSB overnight. Overnight culture from this single colony was diluted (500 μL) in 50 mL of the TSB and incubated at 37 °C with shaking at 200 rpm. When OD_{600} of the cells reached 0.2–0.3, the cells were subjected to chilled ice for ~10 min and later centrifuged at 5000 rpm for 5 min at 4 °C. The supernatant was discarded and cells were washed with 20 mL sterile distilled water once and twice with sterile 10% glycerol solution at 4 °C. Finally, the cells were resuspended in 1 mL 10% glycerol solution and 100 uL aliquots were stored at −80 °C.

According to electroporation requirements, aliquots were thawed on ice for 10 min. Then the cells were mixed with 8 μg of the required plasmid/linearized DNA and transferred to pre-chilled 2 mm electroporation cuvette (Bio-Rad). Cells were then pulsed at 25 KV/cm, 1100 Ω, and 25 μF, with time constant of 5.0 milli-second. Immediately after the pulse, cells were mixed with the TSB (1 mL) and incubated at 37 °C for ~2 h, which were then plated on TSA plates containing corresponding antibiotics. Plates were incubated at 37 °C for 24–36 h or until the colonies were evident. Mutants were screened using cPCR.

2.5. Poultry Experimental Model to Evaluate Virulence

To infect broiler birds, wild type *S.* Gallinarum (WT-SG18) and modified (ΔssaU_SG18) were used. A total of 200 disease-free male broiler chickens of the Cobb 500 strain were used in this study which were procured from a commercial hatchery. The birds were kept in sterile pens in an environmentally-controlled room with a 20-h light cycle. The birds were distributed randomly into four groups: SG_WT (S. Gallinarum positive control), SG_Negative (not infected with any pathogens), complement strain (cSG18) and ΔssaU_SG18 mutant group, with five equal sized replicates containing 10 birds each on the first day. On day 7, cloacal swabs were used to randomly sample birds for the screening of *SG* infection. Throughout the duration of the experiment, the birds had unfettered access to antibiotic-free food and water. At 16 days of age, birds were administered 1 mL of normal saline containing approximately 1×10^8 colony-forming units (CFU) of each strain via oral gavage. Birds were fed orally with *SG* in order to mimic the natural route of infection. After administration, birds were routinely observed for gross signs of disease. Throughout the trial, birds were weighed each week to record any changes in weight gain. Gross examination and post-mortem investigations of both dead and critically ill birds were conducted 21 days post-infection.

3. Results

3.1. Lethality of Cas9 Induced Double Stranded Break and Efficiency of Repair Plasmids

We employed multiple techniques for optimizing CRISPR/Cas9-mediated genome editing in the SG, by targeting the selected *ssaU* gene encoding type three secretion system apparatus (TTSS), which is located on *Salmonella* pathogenicity island 2 (SPI-2) comprising 1059 nucleotides. Initially, two plasmid approach was employed in which the modified pCas9 plasmid [35] containing the constitutively expressing Cas9 nuclease for DSB and repair template plasmid RP-18 was used.

Co-electroporation of the modified pCas9 ssaU/G3 and RP-18 (DNA editing plasmid) into SG cells resulted in no clones. However, colonies consistently appeared on the control plate when electroporated with simple pCas9 plasmid (with no guide RNA) and RP-18, with transformation efficiency of ($8.2 \times 10^2 \pm 25$ cfu/µg $n = 3$), which rule out the toxic effect of empty Cas9 nuclease and suggest the lethality of Cas9 nuclease due to cloned gRNA causing genomic DSB [32]. Clones were not evident even after utilizing a higher amount (10 µg) of repair template plasmid or decreasing the ssaU/G3 (< 50 ng) concentration.

This suggested that the DSB induced by ssaU/G3 was not resolved by the homology-directed repair, meanwhile native non-homologous end joining (NHEJ) was also not efficient in the SG [36]. We further cloned a second spacer RNA in pCas9 (ssaU/G4) targeting another region of the *ssaU* gene (Figure 1), however, no colonies were observed after electroporation with the plasmid based homology templates RP-18. A two-step electroporation was also performed, where RP-18 was electroporated first into SG to allow the cross-over event to occur before the electroporation of ssaU/G3 or ssaU/G4. However, this strategy still yielded no clones.

3.2. Coupling CRISPR/Cas9 with Lambda Red Recombineering Results in Successful Gene Knock Out

To overcome the lower recombination capacity of the bacterial cells after DSB, we used the lambda red recombineering plasmid pRed available in our lab amplified from pCas in combination with pCas9 [37]. After the induction with 10 Mm arabinose and followed by the co-electroporation with ssaU/G3 (300 ng) and linearized dsDNA repair template (8 µg) clones were apparent on the selection plate. The colonies that appeared were screened through colony PCR. Surprisingly a maximum of ($70 \pm 2.5\%$ $n = 3$) clones showed successful deletion of the said gene. The PCR screens of the *ssaU* gene deletion is shown in (Figure 2), unraveling the high potential of this unified system for efficient gene deletion in *S. Gallinarum.*

Deletion of the *ssaU* gene was further confirmed through Sanger sequencing (Supplementary Material Figure S2), whereas no mutation was observed in the flanking upstream and downstream regions of the deleted gene confirming the precise genome modification.

3.3. Clearance of ssaU Mutant Strain from Experimental Birds

In order to determine the systemic infection, bacteria were isolated from visceral organs of the birds, primarily the liver and ileum, after 4, 7 and 14 days post-infection. The WT-SG18 strain recovered from the liver and intestine of infected birds (positive control group) showed 1.3×10^3, 5.0×10^5 and 1.6×10^8 CFU/g after 4, 7 and 14 days, respectively, post-infection indicating systemic bacteremia. Our results are in agreement with the previously reported study showing the highest number of viable counts that were isolated from the liver after 14 days post-infection with S. Gallinarum [38]. In contrary, ΔssaU_SG18 strain could not be recovered from any of the tissues, even following enrichment with selenite broth, although it was recovered from the caecum 2 days post-infection. The post-mortem examination of WT-SG18 group exhibited the typical signs of fowl typhoid pathology (Figure 3), including hepatosplenomegaly with necrotic foci on liver and spleen, bronze discoloration of liver and hemorrhages in the anterior small intestine. In contrast, the

ΔssaU_SG18 strain showed no signs and symptoms of disease at postmortem examination throughout the period of experiment (Figure 3).

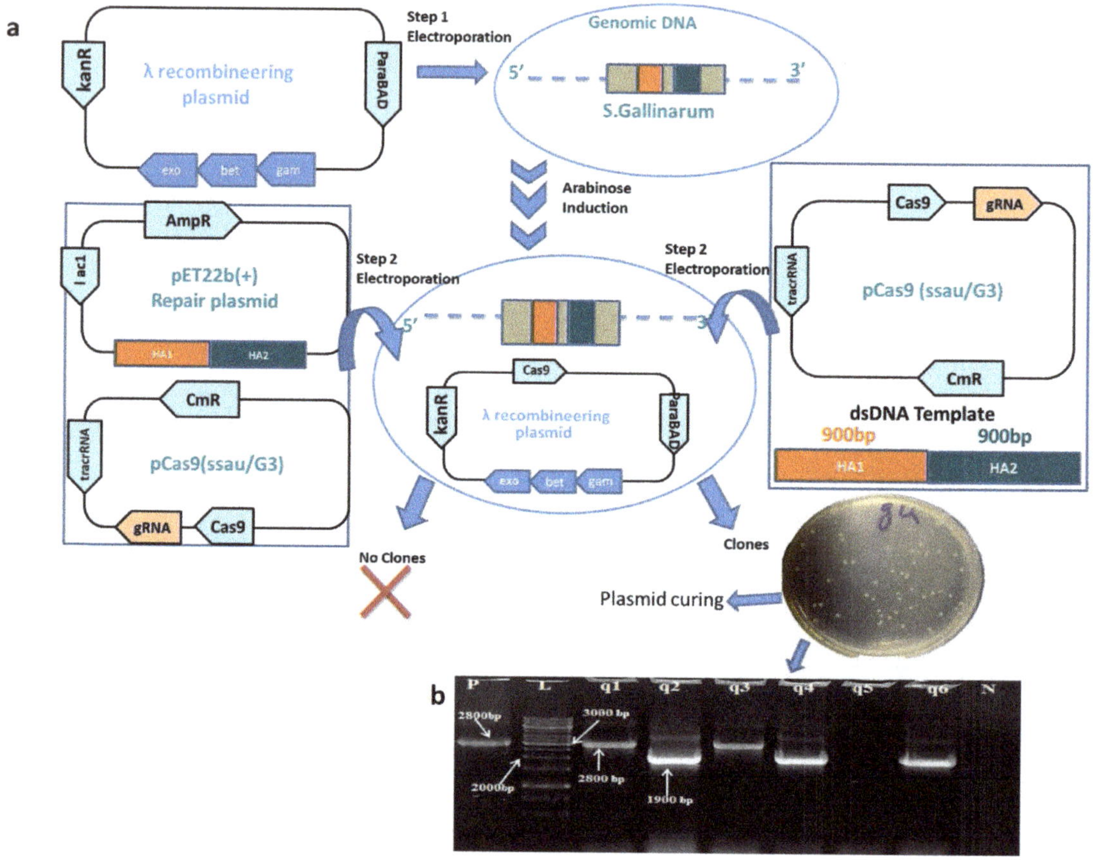

Figure 2. Coupling CRISPR/Cas9 machinery with lambda recombineering for efficient gene deletion from S. Gallinarum genome. (**a**) Stepwise strategy used for deletion of ssaU gene from SG genome using lambda recombineering and CRISPR/Cas9 system. Step (1) indicating the electroporation of pRed. Step (2) co-electroporation of (ssaU/G3) along with RP-18 DNA editing plasmid in SG does not yield any edited clones while co-electroporation of ssaU/G3 combined with dsDNA editing template ≃1800 bp yielded > 50 clones. (**b**) Colony PCR of clones obtained after co electroporation of ssaU/G3 along with double stranded DNA editing template. Lane (q2, q4, q6) indicated gene deletion with a band length of 1.9 kb. While lane (q1, q3) showed no gene deletion with fragment length of ≃2.8 kb. Lane (N) shows negative control reaction. Lane (P) indicate positive control reaction using SG wild type genome as DNA template amplifying a product of ≃2.8 kb. Lane (L) indicates a 1 Kb DNA ladder (Thermo fisher Scientific).

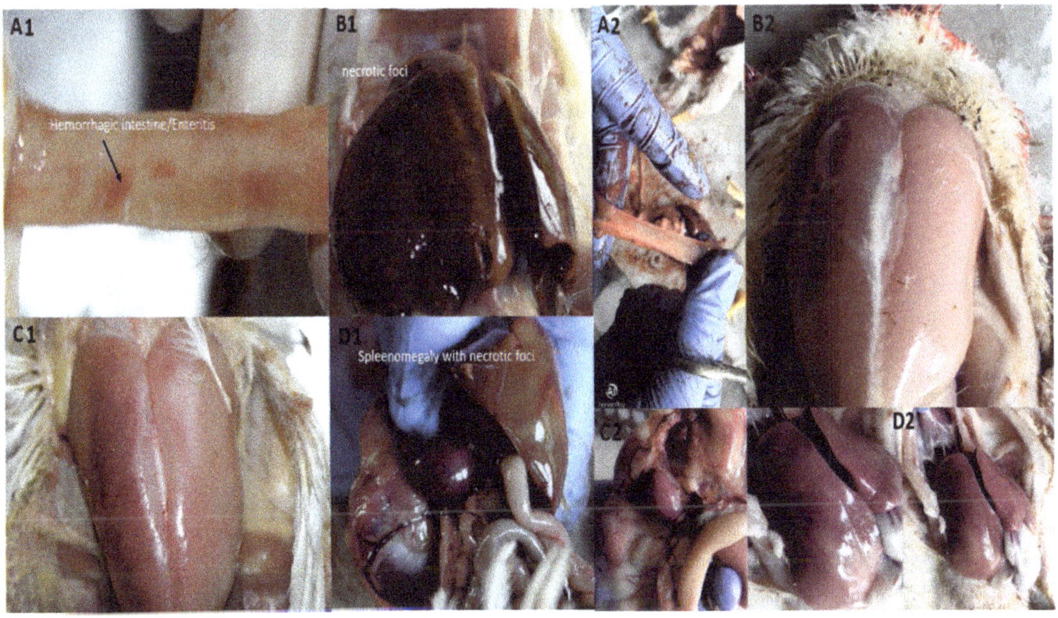

Figure 3. Gross pathological examination of dead experimental birds. Birds infected with WT-SG18 showed (**A1**) intestinal hemorrhages, (**B1**) necrotic foci on liver with discoloration, (**C1**) fever and inflammation (**D1**) splenomegaly. In contrast ΔssaU_SG18 infected birds (**A2**,**B2**,**C2**,**D2**) showed no gross pathological lesions of intestine, muscles, spleen and liver, respectively.

Histopathological analysis of the liver, spleen and intestine revealed lesions showing marked necrosis, hemorrhages, congestion and infiltration of heterophils (avian polymorphonuclear cells) characteristic of fowl typhoid (Figure 4) [39].

Our results suggest that the avirulent strain ΔssaU_SG18 was unable to invade from the intestine to the liver, which is consistent with a previously reported work where the avirulent *ssaU* mutant of *S.* Gallinarum was not recovered from tissues of birds, hence suggesting their rapid clearance after infection [40]. The particular loss of SPI-2 function in serovar Typhimurium leading to attenuation of typhoidal disease in mice models has been previously reported [41]. The complement strain (cSG18) which was constructed using pTZ19R to achieve continuous expression of *ssaU* showed virulence identical to that of WT-SG18, further confirming the role of this gene in the pathogenicity of bacteria. Weight gain and progression of mortality for both WT-SG18, negative control and ΔssaU_SG18 during the period of experiment is shown in (Figure 5). The final weight gain of ΔssaU_SG18 and negative control revealed no significant difference (p-value = 0.8166), whereas the WT-SG18 group exhibited higher significance (p-value < 0.0001) due to disease progression and systemic infection leading to abrupt loss in weight gain. The overall mortality data also showed higher significance (p-value < 0.0001), where no mortality was observed in the ΔssaU_SG18 group, while the WT-SG18 group demonstrated an overall mortality of 32 ± 4.4% starting at day 6 (Figure 5B).

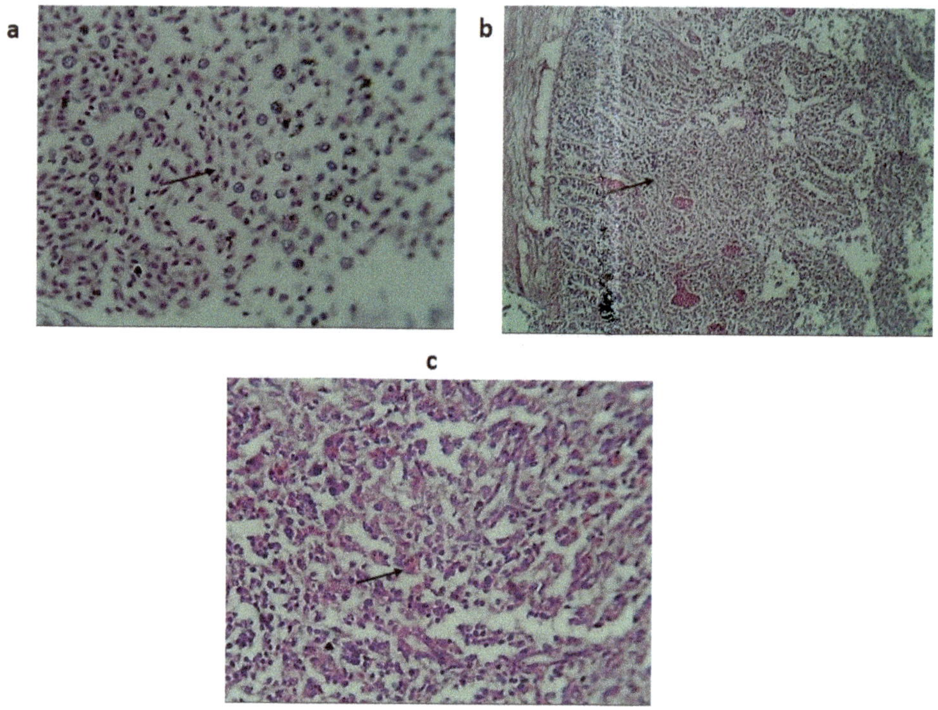

Figure 4. Histopathological findings of birds orally treated with WT-SG18. (**a**) Histological section of the liver showing hemorrhage, vacuolar degeneration of hepatocytes (black arrow) and Infiltration of mononuclear inflammatory cells. (**b**) Histological section of intestine showing coagulative necrosis (black arrow). (**c**) Section of spleen showing hemorrhage, infiltration of mononuclear inflammatory cells and necrosis (black arrow) (H and E stain).

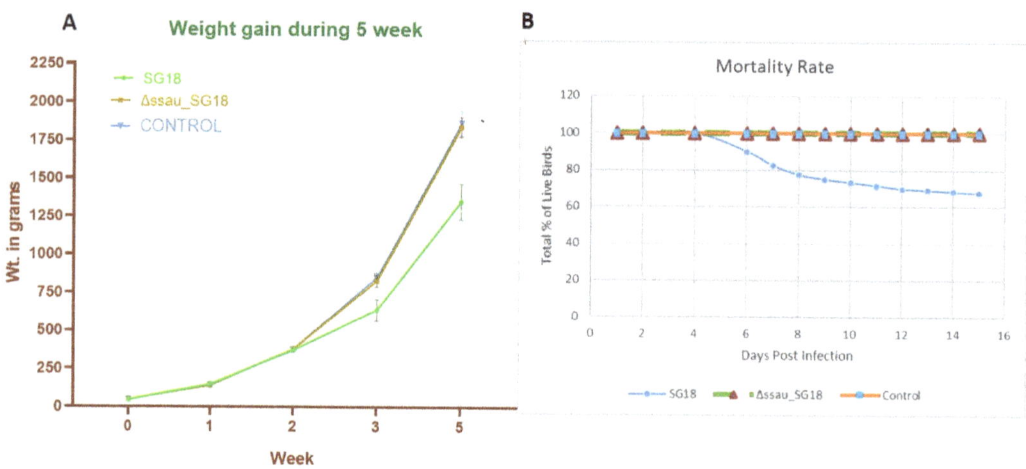

Figure 5. (**A**) Comparison of weight gain between WT-SG18, Δssau_SG18 mutant and control group at 5th week of experiment. (**B**) Progression of mortality in chickens recorded over 15 days post-infection with SG18 and ΔssaU_SG18 strains.

4. Discussion

For the past 20 years, genetic engineering using recombineering has been employed as a robust tool to modify the bacterial genome [42]. More recently, the CRISPR/Cas9 system has highlighted its potential also as a powerful tool for the genetic manipulations and hence revolutionizing genome engineering. Therefore, we have demonstrated the use of such a coupled system to modify the genome of SG which provides a streamlined approach for knock-out mutant generation. The first approach to genetically manipulating the SG genome using two plasmids (ssaU/G3 and RP18) did not produce any mutant clones. We hypothesized that the plasmid-based DNA editing template was less efficient. The lower efficiency of plasmid-based repair templates has already been reported in the previous studies [43].

Considering this, we moved on to the lambda red recombineering proteins which have been proven to be influential in enhancing the recombination capacity of bacteria [44,45]. By using the SG cells expressing lambda protein along with ssaU/G3 and dsDNA template. The recombination efficiency in case of gene knockout was 70 $\pm$ 2.5% using the CRISPR/Cas9 and recombineering system together. We were unable to obtain any clone using only lambda recombination alone and not CRISPR/Cas9 system. The plasmid-based homologous recombination using CRISPR/Cas9 causes the vector integration event (VIE) problems (homologous recombination by double crossover), which further needs screening steps for the selection of the required positive mutant [17,32]. Additionally, the lambda red recombination alone depended upon the integration of the resistant marker, hence this coupled methodology provides a straightforward design for S. Gallinarum genome editing that too with higher efficiency. As this approach does not require construction of HDR plasmid, it provides a reliable methodology for precise knock-out production.

A complete deletion of the *ssaU* gene from S. Gallinarum resulted in in vivo attenuation; however, further deletion of other virulent genes from the Δssau_SG18 strain will make this strain a suitable candidate for vaccine design. TTSS is required for triggering apoptosis in the epithelial cells [46]. In both apoptosis (enteric phase) and pyroptosis (delayed macrophage death), TTSS is instrumental in the translocation of effectors into the host cell during the early phase of pathogenesis [47], whereas the effector proteins (*SpvB, SseL, SlrP*), which are crucial for host cellular dysfunction and deterioration, are expressed in the later stages of infections [48,49]. Several TTSS components have been found to be crucial for *SpvB*-induced actin-depolymerization and cellular cytotoxicity, lending support to the fact that *SpvB* is translocated into the host cell via TTSS [50,51]. The role of *SpvB* as potential virulent effector protein has previously been identified in our lab where *SpvB* mutant of SG was also unable to produce disease symptoms in the poultry infection model [52]. Hence, it can be speculated that the TTSS act as a vital machinery to deliver the effector proteins inside the host cell which give rise to the induction of virulence. Here we have highlighted the significance of the SPI-2 TTSS in a naturally occurring route of systemic, typhoid-like Salmonella infection that has economical and veterinary significance in developing countries. Following this optimized CRISPR/Cas9-based manipulation, studies are underway in our lab to produce an indigenous vaccine strain by deletion of a larger pathogenicity island to control this disease which has been a major concern for Pakistan poultry industry.

5. Conclusions

This study shows the importance of the SPI-2 TTSS in the infection of typhoid-like salmonellosis that has significant economic and veterinary importance in under developed countries. The SPI-2 TTSS allows the serovar Gallinarum to cause the disease by facilitating the bacterial survival within the macrophages and reticuloendothelial system. This coupled CRISPR/Cas9 and lambda recombineering system can be formulated for the live vaccine development against indigenous S. Gallinarum by large-scale deletion of *Salmonella* pathogenicity islands or relative virulence genes [32].

Supplementary Materials: The following supporting information can be downloaded at: https://www.mdpi.com/article/10.3390/biomedicines10123028/s1, Figure S1: Confirmation of gRNA cloning by Sanger sequencing., Figure S2: Confirmation of precise gene knock-out by Sanger sequencing, Calculation of transformation and recombination efficiency, Detailed protocol for Genome engineering of *Salmonella* Gallinarum.

Author Contributions: H.T. (Hamza Tahir) and A.B. equally contributed as the first author in experimentation, designing and analysis of data. Along with drafting and revising the manuscript. H.T. (Hafsa Tariq), Z.H. (Zulquernain Haider) and A.U. helped in performing experiments and analysis. Z.H. (Zafar Hayat) helped in animal experimentation. S.U.R. contributed to designing research and final approval of the manuscript. All authors have read and agreed to the published version of the manuscript.

Funding: This research was funded by International Development Research Centre (IDRC) grant number 109051-002 and APC was funded by IDRC.

Institutional Review Board Statement: The study was conducted according to the guidelines of the Declaration of Helsinki, and approved by the Ethical review committee, Office of Research Innovation and Commercialization-University of Veterinary and Animal Sciences, Lahore, Pakistan, ORIC code no. DR/932 Dated 30 August 2019.

Data Availability Statement: Further details related to primer designing and plasmid used, including additional photographs of experiments are provided.

Acknowledgments: This work was carried out with financial support from UK government, Department of Health and Social Care (DHSC), Global AMR Innovation Fund (GAMRIF) and the International Development Research Centre Ottawa, Canada, (Grant No. 109051-002). The views expressed herein do not necessarily represent those of IDRC or its Board of Governors.

Conflicts of Interest: Authors claim no conflict of interest on any grounds.

Abbreviations

CRISPR, Clustered Regularly Interspaced Short Palindromic Repeats; Cas, CRISPR-associated; cPCR, colony PCR; DSBs, double-strand breakages NHEJ, non-homologous end-joining; ORF, open reading frame; PAM, protospacer adjacent motif; vector integration event; WT, wild type; PAM, protospacer adjacent motif; SP1-2, Salmonella pathogenicity island 2; TTSS, Type Three Secretion System.

References

1. Hosen, J.; Rahman, M.; Alam, J.; Das, Z.; Khan, M.; Haider, M. Pathology of Fowl Typhoid and Molecular Detection of its Pathogen. *Ann. Bangladesh Agric.* **2019**, *23*, 49–60. [CrossRef]
2. Maciorowski, K.G.; Jones, F.T.; Pillai, S.D.; Ricke, S.C. Incidence, sources, and control of food-borne Salmonella spp. in poultry feeds. *World's Poult. Sci. J.* **2007**, *60*, 446–457. [CrossRef]
3. Kaiser, P.; Rothwell, L.; Galyov, E.E.; Barrow, P.A.; Burnside, J.; Wigley, P. Differential cytokine expression in avian cells in response to invasion by Salmonella typhimurium, Salmonella enteritidis and Salmonella gallinarumThe GenBank accession numbers for the sequences reported in this paper are AI982185 for chicken IL-6 cDNA and AJ250838 for the partial chicken IL-6 genomic sequence, respectively. *Microbiology* **2000**, *146*, 3217–3226. [PubMed]
4. Silva, E.; Snoeyenbos, G.; Weinack, O.M.; Smyser, C. Studies on the use of 9R strain of Salmonella gallinarum as a vaccine in chickens. *Avian Dis.* **1981**, *25*, 38–52. [CrossRef] [PubMed]
5. USDA. *Animal Health Monitoring & Surveillance-National Animal Health Reporting System*; USDA: Washington, DC, USA, 2009.
6. Nwiyi, P.; Omadamiro, O. Seroprevalence and isolation of chicken infected with Salmonella: Haematological and Pathological Evaluation. *J. Anim. Feed. Res.* **2012**, *2*, 483–487.
7. Beaudette, F.R. Fowl typhoid and bacillary white diarrhoea. *Vet. J. (1900)* **1930**, *86*, 381–383. [CrossRef]
8. Ezema, W.; Onuoha, E.; Chah, K. Observations on an outbreak of fowl typhoid in commercial laying birds in Udi, South Eastern Nigeria. *Comp. Clin. Pathol.* **2009**, *18*, 395–398. [CrossRef]
9. Ramos-Morales, F. Impact of *Salmonella enterica* Type III Secretion System Effectors on the Eukaryotic Host Cell. *ISRN Cell Biol.* **2012**, *2012*, 787934. [CrossRef]
10. Hueck, C.J. Type III protein secretion systems in bacterial pathogens of animals and plants. *Microbiol. Mol. Biol. Rev.* **1998**, *62*, 379–433. [CrossRef]

11. Stévenin, V.; Chang, Y.Y.; Le Toquin, Y.; Duchateau, M.; Gianetto, Q.G.; Luk, C.H.; Salles, A.; Sohst, V.; Matondo, M.; Reiling, N.; et al. Dynamic Growth and Shrinkage of the Salmonella-Containing Vacuole Determines the Intracellular Pathogen Niche. *Cell Rep.* **2019**, *29*, 3958–3973.e3957. [CrossRef]
12. Ruiz-Albert, J.; Yu, X.J.; Beuzón, C.R.; Blakey, A.N.; Galyov, E.E.; Holden, D.W. Complementary activities of SseJ and SifA regulate dynamics of the Salmonella typhimurium vacuolar membrane. *Mol. Microbiol.* **2002**, *44*, 645–661. [CrossRef] [PubMed]
13. Shah, D.H.; Lee, M.-J.; Park, J.-H.; Lee, J.-H.; Eo, S.-K.; Kwon, J.-T.; Chae, J.-S. Identification of Salmonella gallinarum virulence genes in a chicken infection model using PCR-based signature-tagged mutagenesis. *Microbiology* **2005**, *151*, 3957–3968. [CrossRef] [PubMed]
14. Hensel, M. Salmonella pathogenicity island 2. *Mol. Microbiol.* **2000**, *36*, 1015–1023. [CrossRef] [PubMed]
15. Darwin, K.H.; Miller, V.L. Molecular basis of the interaction of Salmonella with the intestinal mucosa. *Clin. Microbiol. Rev.* **1999**, *12*, 405–428. [CrossRef]
16. Deltcheva, E.; Chylinski, K.; Sharma, C.M.; Gonzales, K.; Chao, Y.; Pirzada, Z.A.; Eckert, M.R.; Vogel, J.; Charpentier, E. CRISPR RNA maturation by trans-encoded small RNA and host factor RNase III. *Nature* **2011**, *471*, 602–607. [CrossRef]
17. Jinek, M.; Chylinski, K.; Fonfara, I.; Hauer, M.; Doudna, J.A.; Charpentier, E. A programmable dual-RNA–guided DNA endonuclease in adaptive bacterial immunity. *Science* **2012**, *337*, 816–821. [CrossRef]
18. Mojica, F.J.; Díez-Villaseñor, C.; García-Martínez, J.; Almendros, C. Short motif sequences determine the targets of the prokaryotic CRISPR defence system. *Microbiology* **2009**, *155*, 733–740. [CrossRef]
19. Jiang, W.; Bikard, D.; Cox, D.; Zhang, F.; Marraffini, L.A. RNA-guided editing of bacterial genomes using CRISPR-Cas systems. *Nat. Biotechnol.* **2013**, *31*, 233–239. [CrossRef]
20. Copeland, N.G.; Jenkins, N.A.; Court, D.L. Recombineering: A powerful new tool for mouse functional genomics. *Nat. Rev. Genet.* **2001**, *2*, 769–779. [CrossRef]
21. Court, D.L.; Sawitzke, J.A.; Thomason, L.C. Genetic engineering using homologous recombination. *Annu. Rev. Genet.* **2002**, *36*, 361–388. [CrossRef]
22. Derbise, A.; Lesic, B.; Dacheux, D.; Ghigo, J.M.; Carniel, E. A rapid and simple method for inactivating chromosomal genes in Yersinia. *FEMS Immunol. Med. Microbiol.* **2003**, *38*, 113–116. [CrossRef] [PubMed]
23. Hu, K.; Shi, Z.; Wang, H.; Feng, E.; Huang, L. Study on gene knockout using red system in Shigella flexneri. *Wei Sheng Wu Xue Bao Acta Microbiol. Sin.* **2003**, *43*, 740–746.
24. Karlinsey, J.E. λ-Red genetic engineering in Salmonella enterica serovar Typhimurium. *Methods Enzymol.* **2007**, *421*, 199–209. [PubMed]
25. Yin, J.; Xia, J.; Tao, M.; Xu, L.; Li, Q.; Geng, S.; Jiao, X. Construction and characterization of a cigR deletion mutant of Salmonella enterica serovar Pullorum. *Avian Pathol.* **2016**, *45*, 569–575. [CrossRef] [PubMed]
26. Yu, D.; Ellis, H.M.; Lee, E.-C.; Jenkins, N.A.; Copeland, N.G.; Court, D.L. An efficient recombination system for chromosome engineering in Escherichia coli. *Proc. Natl. Acad. Sci. USA* **2000**, *97*, 5978–5983. [CrossRef]
27. Karu, A.; Sakaki, Y.; Echols, H.; Linn, S. The gamma protein specified by bacteriophage gamma. Structure and inhibitory activity for the recBC enzyme of Escherichia coli. *J. Biol. Chem.* **1975**, *250*, 7377–7387. [CrossRef]
28. Murphy, K.C. Lambda Gam protein inhibits the helicase and chi-stimulated recombination activities of Escherichia coli RecBCD enzyme. *J. Bacteriol.* **1991**, *173*, 5808–5821. [CrossRef]
29. Little, J.W. An exonuclease induced by bacteriophage λ: II. Nature of the enzymatic reaction. *J. Biol. Chem.* **1967**, *242*, 679–686. [CrossRef]
30. Katashkina, J.I.; Hara, Y.; Golubeva, L.I.; Andreeva, I.G.; Kuvaeva, T.M.; Mashko, S.V. Use of the lambda Red-recombineering method for genetic engineering of Pantoea ananatis. *BMC Mol. Biol.* **2009**, *10*, 34. [CrossRef]
31. Pyne, M.E.; Moo-Young, M.; Chung, D.A.; Chou, C.P. Coupling the CRISPR/Cas9 system with lambda red recombineering enables simplified chromosomal gene replacement in Escherichia coli. *Appl. Environ. Microbiol.* **2015**, *81*, 5103. [CrossRef]
32. Wang, Y.; Zhang, Z.-T.; Seo, S.-O.; Lynn, P.; Lu, T.; Jin, Y.-S.; Blaschek, H.P. Bacterial genome editing with CRISPR-Cas9: Deletion, integration, single nucleotide modification, and desirable "clean" mutant selection in Clostridium beijerinckii as an example. *ACS Synth. Biol.* **2016**, *5*, 721–732. [CrossRef] [PubMed]
33. Chen, W.; Zhang, Y.; Yeo, W.-S.; Bae, T.; Ji, Q. Rapid and efficient genome editing in Staphylococcus aureus by using an engineered CRISPR/Cas9 system. *J. Am. Chem. Soc.* **2017**, *139*, 3790–3795. [CrossRef] [PubMed]
34. Bae, S.; Park, J.; Kim, J.-S. Cas-OFFinder: A fast and versatile algorithm that searches for potential off-target sites of Cas9 RNA-guided endonucleases. *Bioinformatics* **2014**, *30*, 1473–1475. [CrossRef] [PubMed]
35. Song, X.; Huang, H.; Xiong, Z.; Ai, L.; Yang, S. CRISPR-Cas9D10A nickase-assisted genome editing in Lactobacillus casei. *Appl. Environ. Microbiol.* **2017**, *83*, e01259-17. [CrossRef]
36. Arroyo-Olarte, R.D.; Bravo Rodríguez, R.; Morales-Ríos, E. Genome Editing in Bacteria: CRISPR-Cas and Beyond. *Microorganisms* **2021**, *9*, 844. [CrossRef]
37. Jiang, Y.; Chen, B.; Duan, C.; Sun, B.; Yang, J.; Yang, S. Multigene editing in the Escherichia coli genome via the CRISPR-Cas9 system. *Appl. Environ. Microbiol.* **2015**, *81*, 2506–2514. [CrossRef]
38. Batista, D.F.A.; Freitas Neto, O.C.; Barrow, P.A.; de Oliveira, M.T.; Almeida, A.M.; Ferraudo, A.S.; Berchieri, A., Jr. Identification and characterization of regions of difference between the Salmonella Gallinarum biovar Gallinarum and the Salmonella Gallinarum biovar Pullorum genomes. *Infect. Genet. Evol.* **2015**, *30*, 74–81. [CrossRef]

39. Shivaprasad, H. Fowl typhoid and pullorum disease. *Rev. Sci. Et Tech. (Int. Off. Epizoot.)* **2000**, *19*, 405–424. [CrossRef]
40. Jones, M.A.; Wigley, P.; Page, K.L.; Hulme, S.D.; Barrow, P.A. Salmonella enterica serovar Gallinarum requires the Salmonella pathogenicity island 2 type III secretion system but not the Salmonella pathogenicity island 1 type III secretion system for virulence in chickens. *Infect. Immun.* **2001**, *69*, 5471–5476. [CrossRef]
41. Coburn, B.; Li, Y.; Owen, D.; Vallance, B.A.; Finlay, B.B. Salmonella enterica serovar Typhimurium pathogenicity island 2 is necessary for complete virulence in a mouse model of infectious enterocolitis. *Infect. Immun.* **2005**, *73*, 3219–3227. [CrossRef]
42. Zhang, Y.; Buchholz, F.; Muyrers, J.P.; Stewart, A.F. A new logic for DNA engineering using recombination in Escherichia coli. *Nat. Genet.* **1998**, *20*, 123–128. [CrossRef] [PubMed]
43. Schwartz, M.L.; Davis, M.W.; Rich, M.S.; Jorgensen, E.M. High-efficiency CRISPR gene editing in C. elegans using Cas9 integrated into the genome. *PLoS Genet.* **2021**, *17*, e1009755. [CrossRef] [PubMed]
44. Yao, R.; Liu, D.; Jia, X.; Zheng, Y.; Liu, W.; Xiao, Y. CRISPR-Cas9/Cas12a biotechnology and application in bacteria. *Synth. Syst. Biotechnol.* **2018**, *3*, 135–149. [CrossRef] [PubMed]
45. Feng, X.; Zhao, D.; Zhang, X.; Ding, X.; Bi, C. CRISPR/Cas9 assisted multiplex genome editing technique in Escherichia coli. *Biotechnol. J.* **2018**, *13*, 1700604. [CrossRef] [PubMed]
46. Kim, J.M.; Eckmann, L.; Savidge, T.C.; Lowe, D.C.; Witthöft, T.; Kagnoff, M.F. Apoptosis of human intestinal epithelial cells after bacterial invasion. *J. Clin. Investig.* **1998**, *102*, 1815–1823. [CrossRef]
47. Paesold, G.; Guiney, D.G.; Eckmann, L.; Kagnoff, M.F. Genes in the Salmonella pathogenicity island 2 and the Salmonella virulence plasmid are essential for Salmonella-induced apoptosis in intestinal epithelial cells. *Cell. Microbiol.* **2002**, *4*, 771–781. [CrossRef]
48. Rytkönen, A.; Poh, J.; Garmendia, J.; Boyle, C.; Thompson, A.; Liu, M.; Freemont, P.; Hinton, J.C.; Holden, D.W. SseL, a Salmonella deubiquitinase required for macrophage killing and virulence. *Proc. Natl. Acad. Sci. USA* **2007**, *104*, 3502–3507. [CrossRef]
49. Libby, S.J.; Lesnick, M.; Hasegawa, P.; Weidenhammer, E.; Guiney, D.G. The Salmonella virulence plasmid spv genes are required for cytopathology in human monocyte-derived macrophages. *Cell. Microbiol.* **2000**, *2*, 49–58. [CrossRef]
50. Browne, S.H.; Lesnick, M.L.; Guiney, D.G. Genetic requirements for Salmonella-induced cytopathology in human monocyte-derived macrophages. *Infect. Immun.* **2002**, *70*, 7126–7135. [CrossRef]
51. Browne, S.H.; Hasegawa, P.; Okamoto, S.; Fierer, J.; Guiney, D.G. Identification of Salmonella SPI-2 secretion system components required for *SpvB*-mediated cytotoxicity in macrophages and virulence in mice. *FEMS Immunol. Med. Microbiol.* **2008**, *52*, 194–201. [CrossRef]
52. Basit, A.; Tahir, H.; Haider, Z.; Tariq, H.; Ullah, A.; Rehman, S.U. CRISPR/Cas9-Based Deletion of *SpvB* Gene From Salmonella gallinarum Leads to Loss of Virulence in Chicken. *Front. Bioeng. Biotechnol.* **2022**, *10*, 885227. [CrossRef] [PubMed]

biomedicines

MDPI

Article

Identification and Elimination of Antifungal Tolerance in *Candida auris*

Samira Rasouli Koohi [1,†], Shamanth A. Shankarnarayan [1,†], Clare Maristela Galon [1] and Daniel A. Charlebois [1,2,*]

1 Department of Physics, University of Alberta, Edmonton, AB T6G 2R3, Canada
2 Department of Biological Sciences, University of Alberta, Edmonton, AB T6G 2E9, Canada
* Correspondence: dcharleb@ualberta.ca
† These authors contributed equally to this work.

Abstract: Antimicrobial resistance is a global health crisis to which pathogenic fungi make a substantial contribution. The human fungal pathogen *C. auris* is of particular concern due to its rapid spread across the world and its evolution of multidrug resistance. Fluconazole failure in *C. auris* has been recently attributed to antifungal "tolerance". Tolerance is a phenomenon whereby a slow-growing subpopulation of tolerant cells, which are genetically identical to susceptible cells, emerges during drug treatment. We use microbroth dilution and disk diffusion assays, together with image analysis, to investigate antifungal tolerance in *C. auris* to all three classes of antifungal drugs used to treat invasive candidiasis. We find that (1) *C. auris* is tolerant to several common fungistatic and fungicidal drugs, which in some cases can be detected after 24 h, as well as after 48 h, of antifungal drug exposure; (2) the tolerant phenotype reverts to the susceptible phenotype in *C. auris*; and (3) combining azole, polyene, and echinocandin antifungal drugs with the adjuvant chloroquine in some cases reduces or eliminates tolerance and resistance in patient-derived *C. auris* isolates. These results suggest that tolerance contributes to treatment failure in *C. auris* infections for a broad range of antifungal drugs, and that antifungal adjuvants may improve treatment outcomes for patients infected with antifungal-tolerant or antifungal-resistant fungal pathogens.

Keywords: adjuvant; antifungal tolerance/resistance; broth microdilution assay; *Candida auris*; disk diffusion assay; diskImageR; human fungal pathogen

1. Introduction

Antimicrobial resistance (AMR) threatens the advances of modern medicine. Antifungal resistance contributes significantly to the AMR problem [1,2], especially among immunocompromised patients [3,4]. A multitude of biological, sociological, and economic factors result in hundreds of millions of serious fungal infections and between 1 and 1.5 million fungal infection-related deaths per year globally [5,6]. AMR, among fungi, is of particular concern due to the limited number of classes of drugs available to treat invasive fungal infections (i.e., fungistatic azoles as well as fungicidal polyenes and echinocandins) [7]. This threat is exacerbated by the fact that no new class of antifungal drugs has reached the market in over a decade [8,9]. Climate change is also predicted to increase the prevalence of fungal infections, as fungi adapt to warmer temperatures to increase their geographic range and overcome the thermal protection barrier of their warm-blooded hosts [10].

Candida species of yeast are the most common causes of fungal infections [11]. One *Candida* species that is increasingly of concern is *Candida auris* [12], due to its resistance to antifungal drugs and healthcare-associated outbreaks [13]. *C. auris* has now been reported on all inhabited continents and in over 47 countries [14,15]. Particularly concerning, is that *C. auris* is multidrug resistant (i.e., non-susceptible to at least one agent in three or more classes of antimicrobials) [16–18], and, in some cases, it has been found to be pandrug-resistant (i.e., non-susceptible to all agents in all antimicrobial classes) [18,19]. *C. auris* has mortality rates of up to 45% among patients with bloodstream infections [20].

"Tolerance" is a phenomenon whereby a slow-growing subpopulation of cells, which are thought to be genetically identical to susceptible cells, emerges during antifungal drug treatment [21]. Antifungal tolerance is distinct from antifungal resistance, in that resistance is the result of heritable genetic changes and resistant cells grow above the minimum inhibitory concentration (MIC) in a concentration-dependent manner (i.e., MIC increases in resistance, but it does not increase in tolerance). In contrast, tolerance is a reversible phenomenon whereby cells grow slowly above MIC (i.e., they exhibit growth at "supra-MIC"). Tolerance manifests from the phenotypic heterogeneity intrinsic to a given fungal isolate, such that any cell within an isogenic population can reproduce the fractions of susceptible and tolerant cells present prior to the initiation of antifungal treatment. Cross tolerance has been observed in *C. albicans*, whereby strains tolerant to posaconazole also exhibit tolerance to other azole drugs [22]. Though the molecular mechanisms underlying tolerance in *Candida* species are still largely unknown, preliminary studies have shown that tolerance is associated with multiple genetic components that differ between isolates, including Hsp90-faciliated azole tolerance in *C. auris* [23]. Aneuploidy has also been shown to alter antifungal tolerance in *C. albicans* [24,25]. It is unknown if *C. auris* is tolerant to non-azole classes of antifungal drugs.

Clinical assays have not been designed to detect antifungal tolerance [26,27]. Quantitatively measuring tolerances of infecting isolates may provide prognostic insights concerning the success of mono- and combination-antifungal therapies [28]. Broth microdilution assays and disk diffusion assays, coupled with the image analysis software diskimageR, have been successfully used to quantify antifungal tolerance in research laboratories [29]. Most clinical diagnostic tests are performed on cultures grown for 24 h and therefore cannot detect drug-tolerant cells, which are typically visually evident after 48 h of growth [21]. Tolerance, along with host factors, immune status, and pharmacological issues [30], may explain why some patients do not respond to drug therapy despite being infected with fungi that have been determined, by traditional antimicrobial susceptibility testing methods, to be susceptible to a particular drug (i.e., cells that do not grow above MIC at 24 h, the standard endpoint MIC measurement for *Candida* species) [21,28]. "Trailing growth" (the clinical term for tolerance) leads to poor response to fluconazole in *C. tropicalis* in wax moth larvae [31] and mouse models [32], and high levels of tolerance are associated with *C. albicans* infections in patients treated with fluconazole [33].

Adjuvant drugs have the potential to sustain the vital functions of antimicrobial drugs [21]. Non-antifungal agents have been shown to enhance the effectiveness of azole drugs against resistant *Candida* species and other pathogenic fungi, including *Aspergillus fumigatus*, *Cryptococcus neoformans*, and the dimorphic fungus *Histoplasma capsulatum* [34–36]. Specifically, the antimalarial drug chloroquine, in combination with fluconazole, exhibited enhanced antifungal activity against *C. albicans*, *C. tropicalis*, *C. glabrata*, *C. parapsilosis*, and *C. krusei* (teleomorph is known as *Issatchenkia orientalis* and *Pichia kudriavzevii* [11]) isolates in vitro [37]. Whether or not tolerance and resistance to azoles or to other classes of antifungal drugs can be eliminated in *C. auris* using adjuvant antifungal therapies, remains to be investigated. Another study explored the activity of doxycycline, pyrvinium pamoate, along with chloroquine, as adjuvants in combination with fluconazole in clinical *C. albicans* isolates, and found increased antifungal activity [29]. Chloroquine is a member of the quinoline family and is used to treat diseases including malaria, amebiasis, rheumatoid arthritis, discoid, and systemic lupus erythematosus [38–40]. Chloroquine causes iron depletion, leading to a decrease in membrane sterol availability and downregulates the *ERG11* gene [41]. We hypothesize that the combining chloroquine with common antifungal drugs will eliminate antifungal tolerance in *C. auris*.

The main aims of our study are to use broth microdilution and disk diffusion assays, together with diskImageR, to investigate if tolerance to all three classes of antifungal drugs occurs in *C. auris*, and if this tolerance can be eliminated by adjuvant antifungal therapy. We find that *C. auris* is tolerant to several fungistatic and fungicidal drugs: fluconazole, itraconazole, posaconazole, voriconazole, amphotericin B, and caspofungin. We demonstrate that antifungal tolerance is detectable at 24 h, as well as at 48 h, and that tolerance is a reversible phenomenon. Finally, we are reporting for the first time that in some isolates combining antifungal drugs with the adjuvant chloroquine eliminates tolerance and resistance in *C. auris*.

2. Materials and Methods

2.1. Strains, Media, and Growth Conditions

C. auris isolates were obtained from clinical samples from the Alberta Precision Laboratories (APL)—Public Health Laboratory (ProvLab).

All strains and isolates (Table S1) were preserved in 25% glycerol at −80 °C until further use. The strains and isolates were revived by culturing from frozen stock on YPD agar plates (yeast extract: Sigma Aldrich, #8013-01-2; bacto peptone: Difco, #9295043) and incubated at 35 °C for 48 h. Fresh subcultures were made on YPD agar plates and incubated at 35 °C for 24 h prior to conducting microbroth dilution and disk diffusion assays (Section 2.3).

2.2. DNA Extractions, PCR, and Sequencing

The initial identification of all *C. auris* isolates was performed using matrix-assisted laser desorption ionization—time of flight (MALDI-TOF) mass spectrometry [42,43] by the APL—ProvLab. The molecular identity of these isolates was confirmed by amplifying and sequencing the Internal Transcribed Spacer (ITS) region of ribosomal DNA. The primers ITS-5 (5′-GGAAGTAAAAGTCGTAACAAGG-3′) and ITS-4 (5′-TCCTCCGCTTATTGATATGC-3′) were used to amplify the ITS region (Integrated DNA Technologies). Genomic DNA was extracted using manual phenol–chloroform–isoamyl alcohol method [44]. The concentration of the extracted DNA was measured using a microvolume µDrop Plate (Thermo Fisher Scientific, Mississauga, ON, Canada, #N12391). The template and the primers were mixed in concentrations of 7.5 ng/µL and 0.25 µM, respectively, to a final volume of 10 µL. Sanger sequencing was then performed using a 3730 Genetic Analyzer (Thermo Fisher Scientific, Mississauga, ON, Canada, #A41046) at the Molecular Biology Services Unit at the University of Alberta. The resulting sequences were subjected to nucleotide BLAST analysis [45], which revealed 100% similarly to the standard strains. The *C. auris* isolates' ITS sequences were submitted to NCBI with the accession number OP984814-OP984818.

2.3. Broth Microdilution and Disk Diffusion Assays

The MIC for each isolate was first determined via broth microdilution assays following CLSI M27 guidelines [46]. All the isolates were tested in 96-well U-bottom microwell plates (Thermo Fisher Scientific, Mississauga, ON Canada, #163320) against fluconazole (Sigma-Aldrich, Oakville, ON, Canada, #F8929) (0.12–64 µg/mL), amphotericin B (Sigma-Aldrich, Canada, #A9528) (0.03–16 µg/mL), itraconazole (Sigma-Aldrich, Oakville, ON, Canada, #16657) (0.03–16 µg/mL), posaconazole (Sigma-Aldrich, Oakville, ON, Canada, #SML2287) (0.03–16 µg/mL), voriconazole (Sigma Aldrich, Oakville, ON Canada, #P20005) (0.03–16 µg/mL), micafungin (Sigma-Aldrich, Oakville, ON, Canada, #208538) (0.015–8 µg/mL), caspofungin (Sigma-Aldrich, Oakville, ON, Canada, #179463-17-3) (0.015–8 µg/mL), and anidulafungin (Sigma-Aldrich, Oakville, ON, Canada, #166663-25-8) (0.03–16 µg/mL). These antifungals were dissolved in DMSO (fluconazole, itraconazole, voriconazole, posaconazole, anidulafungin, and amphotericin B) or water (caspofungin and micafungin); the concentration of the antifungal microwell plates were twice the final concentration tested with the inoculum added. Freshly cultured *Candida* species (*C. auris*, *C. parapsilosis* (ATCC 22019), and *I. orientalis* (ATCC 6258)) at 24 h of incubation at 35 °C were used as inoculum. Inoculum of 100 µL consisting of 2–5 × 10^3 cells were used to inoculate the antifungal microwell plates. After inoculation, the microwell plates were incubated at 35 °C and evaluated after 24 h and 48 h to determine the MICs.

Disk diffusion assays (DDAs) were carried out as per CLSI M44-A2 guidelines [47] against fluconazole (25 μg), itraconazole (50 μg), posaconazole (5 μg), voriconazole (1 μg), amphotericin B (20 μg), and caspofungin (5 μg). MHA medium with 2% dextrose (Sigma Aldrich, Oakville, ON, Canada, #50-99-7) and 0.5 μg/mL methylene blue dye (Sigma Aldrich, Oakville, ON, Canada, #03978) was used to perform the disk diffusion assays. After 24 h of growth, 5–10 colonies were picked and liquid suspensions of *C. auris* were made by reconstituting colonies in 2 mL of normal saline (Sigma Aldrich, Oakville, ON, Canada, #S8776). The optical density (OD) was measured using a Varioskan LUX microplate reader (Thermo Fisher Scientific, Mississauga, ON, Canada, #N16044) at 530 nm, and adjusted to an OD of 0.09–0.13, which corresponded to 1–5 $\times$ 10^6 cells/mL. The adjusted solution was utilized to swab on the Muller–Hinton agar (MHA) using sterile cotton swabs (Fisher Scientific, Saint-Laurent, Quebec, Canada, #22-029-683). An antifungal disk was placed on each plate after inoculating and drying the agar plates. The plates were then incubated for 24 to 48 h at 35 °C. All experiments were performed in triplicate.

2.4. Photography and Image Preprocessing

Photographs of each disk diffusion plate were taken after 24 h and 48 h at the maximum possible resolution (6000 by 4000 pixels with an aspect ratio of 3:2) using a Canon EOS Rebel SL3 camera with a Canon EF-S 35 mm f/2.8 Macro IS STM macro lens. The camera settings were as follows: ISO 800, white balance, picture type "neutral", time 1/100 s, center focused against a plain black background from a fixed distance. The photos were taken and then the size of each photograph was standardized by cropping the edges and bringing all images to the same resolution.

2.5. Quantifying Tolerance via Supra-MIC Growth and Fraction of Growth

Tolerant subpopulations grow slowly in drug concentrations above MIC [21]. We used established methods to quantify tolerance, namely, supra-MIC growth from microbroth dilution assays and the fraction of growth (FoG) in the zone of inhibition (ZOI) from disk diffusion assays (Section 2.3).

The MIC for each isolate was determined using CLSI supplement M60 guidelines [48]. The MIC readings were recorded at 24 h and 48 h post inoculation. Tentative breakpoints provided by the Centers for Disease Control and Prevention for *C. auris* were considered to differentiate them as susceptible or resistant [49]. *I. orientalis* and *C. parapsilosis* were used as reference strains to ensure that the antifungal MIC range in each experiment was within CLSI guidelines.

Supra-MIC growth (*SMG*) was determined by subjecting the antifungal microwell plates used for measuring MICs to spectrophotometric reading at 630 nm after 24 h and 48 h of incubation at 35 °C. *SMG* was calculated as an average growth per well above MIC-normalized to total growth without antifungals [28]:

$$SMG = \frac{average\ growth\ per\ well\ above\ MIC}{growth\ without\ antifungal} \quad (1)$$

The software program diskImageR [29] analyses photographs of disk diffusions assays. diskImageR utilizes the image processing program ImageJ [50] and the programming language *R* [51]. We used diskImageR to measure the tolerance and resistance of *C. auris* isolates to fungistatic and fungicidal drugs from photographs of the disk diffusion assay plates (Section 2.4; Figures S1 and S2). All disk diffusion experiments were repeated in triplicate using antifungal disks placed in the center of MHA plates incubated at 35 °C for 24 and 48 h (Figure S3). After the photographs were imported by diskImageR into ImageJ, the size of each photograph was standardized and the "find particles" macro was used to find the center of the antimicrobial diffusion disk. The radius of the ZOI (RAD) and the FoG in the ZOI were measured where 20%, 50%, and 80% of growth was inhibited (RAD_{20},

RAD_{50}, and RAD_{80}, and FoG_{20}, FoG_{50}, and FoG_{80}, respectively). The RAD measures the degree of susceptibility/resistance, and FoG measures the degree of tolerance. The RAD for all disk diffusions assay plates were also measured manually (using a ruler), and the FoGs were also analyzed using ImageJ [52]. ImageJ analysis for estimating pixel intensity to obtain FoG was carried out by importing photographs to ImageJ software and setting "on" the measurements such as "mean grey value", minimum and maximum grey "area", and fixing the "area" for ZOI. The "measure" macro was then used to measure the pixel intensity. For photographs of 48 h DDA plates, the same parameters were restored to their 24 h counterparts, and the pixel intensity was measured within ZOI. When there are colonies at border of the ZOI (e.g., Figure S3B), diskImageR considers it as the area outside of the ZOI, and the measured RAD is smaller than the manually measured RAD; consequently, the FoG_{20} measured by diskImageR is also inaccurate. Therefore, in these cases, the RAD was obtained by manually measuring the RAD and by measuring the FoG using ImageJ (Figure S4) [50]. When isolates were highly tolerant, resulting in many colonies in the ZOI (Figure S5B) or complete confluence in the ZOI (Figure S5D) after 48 h, diskImageR reported RAD and FoG as "NA" (Not Applicable).

2.6. Experiments to Determine Effectiveness Adjuvant-Antifungal Treatment

The synergies among antifungals (fluconazole, itraconazole, posaconazole, voriconazole, amphotericin B, and caspofungin) and adjuvant (chloroquine) against *C. auris*, *C. parapsilosis*, and *I. orientalis* were evaluated using DDAs (Section 2.3) and broth microdilution methods with minor modifications. For DDAs, a syringe-filtered chloroquine diphosphate salt (Sigma-Aldrich, #C6628) solution was added to MHA media after autoclaving to a final concentration of 1031.8 μg/mL. After inoculation of *C. auris* and the control strains, the MHA plates containing chloroquine were incubated in the dark as chloroquine light sensitive. These plates were read and photographed at 24 h and 48 h. *C. auris* isolates and control strains were lawn cultured (i.e., the entire surface of the agar plate was covered by swabs dipped in the liquid culture) on the MHA plates containing chloroquine with and without antifungal disks, to respectively determine the effect of antifungal chloroquine and chloroquine alone on *C. auris*. Whereas for the broth microdilution method, the concentration for different antifungal drugs were as mentioned in Section 2.3 and the chloroquine concentration ranged from 8 to 512 μg/mL. Synergistic activity of chloroquine with different antifungals was tested using the checkerboard method as previously described [37]. Both antifungal drugs (50 μL) and chloroquine (50 μL) were dispensed to sterile 96 well U bottom microtiter plates and prepared inoculum (100 μL) as per Section 2.3 was inoculated. Plates were then incubated at 35 °C. MIC and SMG results were read at 24 h and 48 h.

3. Results

3.1. Identification of Resistance in C. auris from Broth Microdilution Assays

To determine if the *C. auris* isolates were resistant to the antifungal drugs used in our study, we performed antifungal susceptibility testing at 24 and 48 h using the broth microdilution method (Section 2.3). The MICs for the *C. auris* isolates indicated that three isolates were susceptible to the fungicidal and fungistatic drug tested, whereas *C. auris* isolate 2 was not susceptible to fluconazole, and *C. auris* isolate 5 was not susceptible to fluconazole, voriconazole, caspofungin, and amphotericin B (Figure 1 and Table S2). The quality control strains *C. parapsilosis* and *I. orientalis* were within the recommended ranges. No change in MIC was observed at 24 and 48 h except for *C. auris* isolates 1 and 2 against amphotericin B.

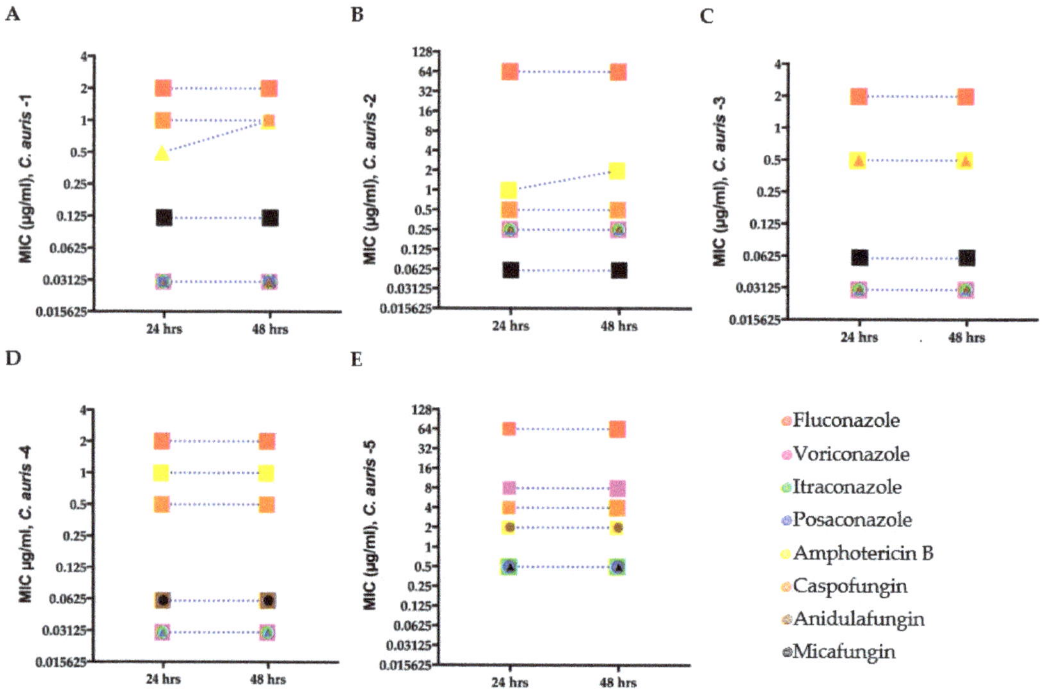

Figure 1. Minimum inhibitory concentration (MIC) for clinical *C. auris* isolates (**A–E**) growing in antifungal microwell plates to determine susceptibility/resistance to antifungal drugs. Mean MICs of five clinical *C. auris* isolates measured after 24 and 48 h for four fungistatic drugs (fluconazole, itraconazole, posaconazole, and voriconazole) and two fungicidal drugs (amphotericin B and caspofungin). Different symbols denote *C. auris* isolates with the same MIC.

3.2. Identification of Resistance in C. auris from Disk Diffusion Assays

To confirm the resistance of the *C. auris* isolates determined by the broth microdilution assays (Section 3.1), we performed the corresponding disk diffusion assays. In agreement with the microbroth dilution method, resistance was noted in *C. auris* isolate 2 for fluconazole and *C. auris* isolate 5 for fluconazole and voriconazole (RAD = 0 mm in all three instances; Figure 2B,E). However, *C. auris* isolate 5 exhibited a ZOI to amphotericin B (RAD = 7 mm) and caspofungin (RAD = 6 mm) at 24 h (Figure 2E). As expected, and in agreement with previous work [28], there was an inverse correlation between RAD and MIC (Pearson test, $r = -0.58$, $p = 0.007$).

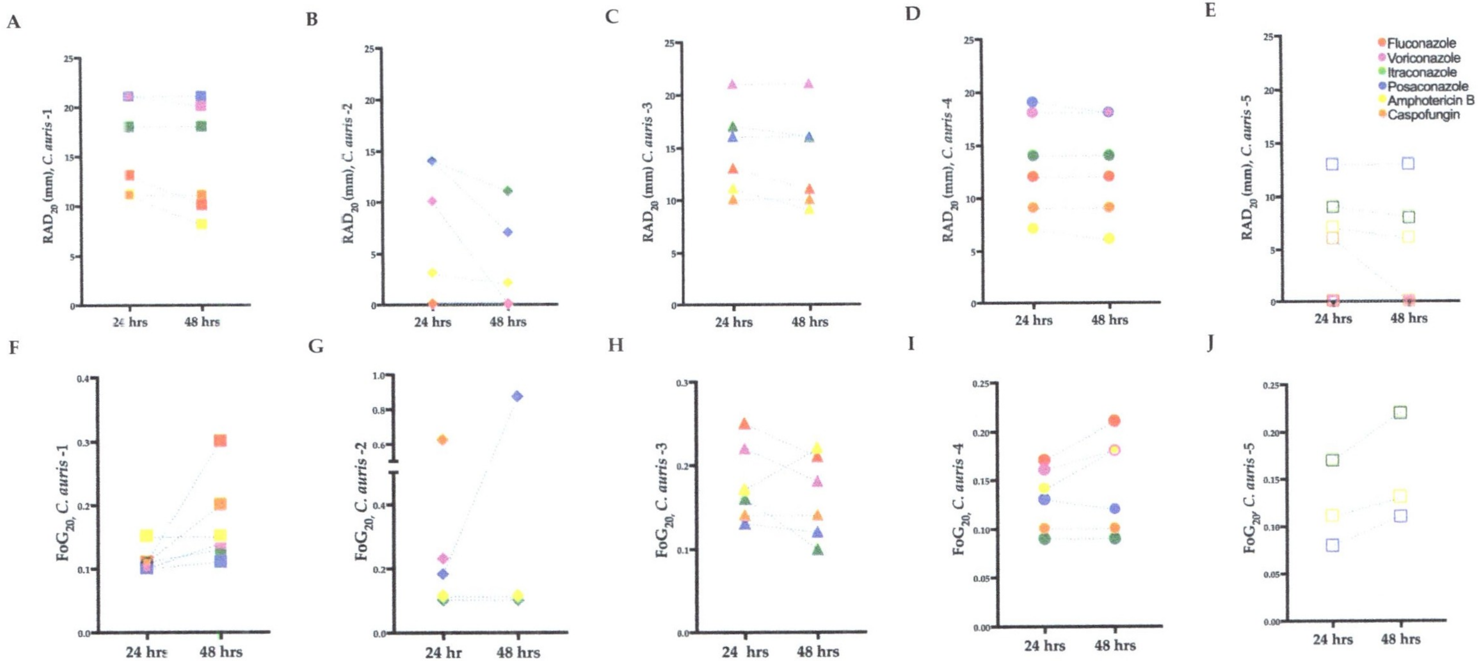

Figure 2. Radius of the zone of inhibition (RAD) (**A–E**) and fraction of growth in the zone of inhibition (FoG_{20}) (**F–J**) for *C. auris* isolates treated with antifungal drugs. Mean RAD where 20% of growth is inhibited (RAD_{20}) at 24 and 48 h. (**B**) Mean FoG where 20% of growth is inhibited (FoG_{20}) 24 and 48 h. *C. auris* isolate 2 treated with caspofungin at 48 h is not plotted in (**B**), as it exhibited FoG in the entire ZOI (i.e., a "NA" data point was generated by diskImageR [29]); the reduction in RAD and FoG_{20} for *C. auris* isolate 3 in (**C**) and (**H**), respectively, is due to the exclusion of FoG_{20} within the ZOI by diskImageR (see Section 2.5 for details). *C. auris* isolate 5 exhibited resistances to fluconazole and caspofungin.

3.3. Identification of Tolerance in *C. auris* from Broth Microdilution Assays

To determine if tolerant subpopulations existed within the non-resistant *C. auris* isolates, we carried out an SMG analysis (Section 2.5). A statistically significant increase in SMG was observed after 48 h for *C. auris* isolate 1 to fluconazole and itraconazole (Independent *t*-test, $p = 0.009$ and $p = 0.001$, respectively), *C. auris* isolates 2 and 4 to voriconazole (Independent *t*-test, $p = 0.0009$ and $p = 0.0014$, respectively), and *C. auris* isolate 2 to caspofungin (Independent *t*-test, $p = 0.006$), indicating the presence of tolerance (Figure 3B,D). There was also a non-significant increase in SMG at 48 h for *C. auris* isolates 3 and 4 to fluconazole, *C. auris* isolates 2, 3, and 5 to itraconazole, *C. auris* isolates 1, 2, 3, 4, and 5 to posaconazole, *C. auris* isolates 1, 3, and 5 to voriconazole, and *C. auris* isolate 1 and 5 to amphotericin B. No tolerance was observed for *C. auris* isolate 4 to itraconazole and caspofungin, and *C. auris* isolates 2, 3, and 4 to amphotericin B (Figure 3F–I). There was a decrease in SMG for *C. auris* isolate 1 against amphotericin B. This occurred because the growth of isolates in wells without antifungals increased over 48 h, which in turn reduced the SMG (as described in Equation (1)).

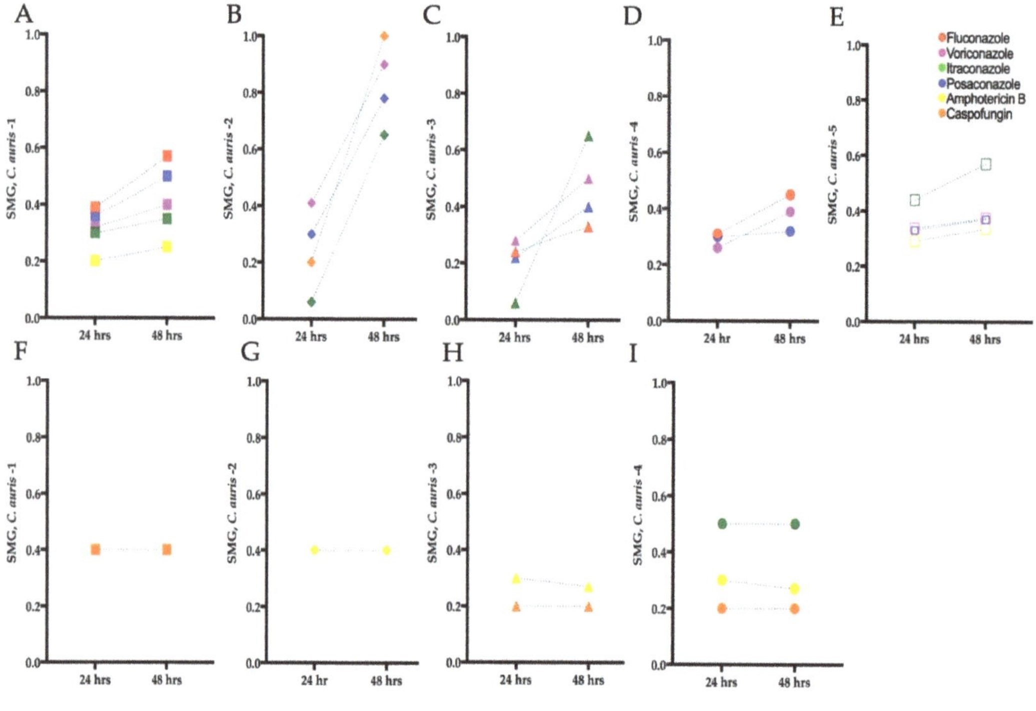

Figure 3. Tolerance from supra-MIC growth (SMG) for clinical *C. auris* isolates grown in antifungal microwell plates. (**A–E**) Mean SMG of tolerant isolates after 24 and 48 h. (**F–I**) Mean SMG of non-tolerant isolates after 24 and 48 h. *C. auris* 5 was resistant to fluconazole and caspofungin hence tolerance/non-tolerance could not be determined for these isolate–antifungal combinations.

3.4. Identification of Tolerance in *C. auris* from Disk Diffusion Assays

To confirm the tolerance of the *C. auris* isolates determined by the broth microdilution assays (Section 3.3), we performed the corresponding DDAs. All the *C. auris* isolates with higher SMG exhibited higher FoG_{20} at 48 h (Figure 2F–J). The FoG_{20} within the ZOI ranged between 0.08 and 0.62 and 0.09 and 0.87 at 24 h and 48 h, respectively (Figure 2). *C. auris* isolate 2 exhibited the highest FoG_{20} against caspofungin at 24 h (0.62) and against posaconazole at 48 h (0.87). Similarly, at 24 h the highest pixel intensity occurred for *C. auris* isolate 3 against posaconazole (195, Figure 4A) and the highest SMG occurred for *C. auris* isolate 2 against caspofungin (1.0, Figure 4A). At 48 h, the highest pixel intensity and SMG were measured for *C. auris* isolate 2 against voriconazole (197 and 0.90, respectively; Figure 4B).

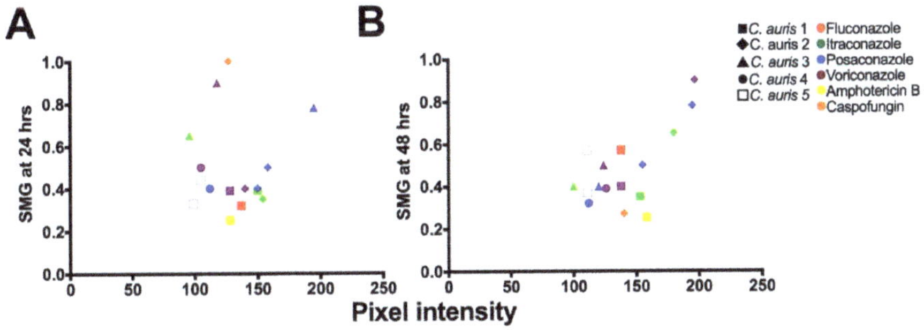

Figure 4. Correlation analysis for mean supra-MIC growth (SMG) and mean pixel intensity measured by ImageJ [51] to determine tolerance. (**A**) Analysis performed after 24 h of growth ($R^2 = 0.3128$; Pearson correlation test, $p = 0.0469$). (**B**) Analysis performed after 48 h of growth ($R^2 = 0.2862$; Pearson correlation test, $p = 0.0085$).

There was no correlation between FoG_{20} and RAD levels (Pearson test, $r = -0.25$, $p = 0.28$), as expected based on previous work which established that the FoG_{20} and RAD measure different drug responses [28,29]. The was significant correlation between SMG measured by diskImageR and pixel intensity measured by *ImageJ* (Figure 4), which occurred as both SMG and pixel intensity increase when tolerant subpopulations are present.

Overall, there was no significant difference between diskImageR and manual readings of the RAD (Independent *t*-test, $p = 0.5634$ and $p = 0.8453$ for readings at 24 h and 48 h, respectively; Figure S4). There was also no significant difference for FoG_{20} readings using diskImageR and ImageJ at 24 h (Unpaired *t*-test, $p = 0.35$). However, there was a statistically significant difference for FoG_{20} reading using diskImageR and ImageJ at 48 h (Unpaired *t*-test, $p = 0.022$). The difference in these FoG_{20} readings resulted from the fact that diskImageR was unable to distinguish the border of the ZOI among tolerant isolates, which was obscured by tolerant colonies at 48 h.

Among reference strains, only *C. parapsilosis* exhibited tolerance to fluconazole and voriconazole (Figure S6). The FoG_{20} and SMG for fluconazole and voriconazole is presented in Table S3. No tolerance was observed for the other antifungal drugs considered in this study against *C. parapsilosis*. *I. orientalis* did not exhibit tolerance to any of the antifungal agents tested.

3.5. Tolerance in *C. auris* Is a Reversable Phenomenon

Next, we investigated if the antifungal tolerance that we discovered in *C. auris* was a reversible phenomenon. To investigate this, we sub-cultured colonies growing inside and outside of the ZOI and repeated the microbroth dilution and disk diffusion experiments (Figure S7). There was no difference between the MICs of original colonies and colonies from inside and outside ZOI at both 24 and 48 h (Table S4). RAD, FoG_{20}, and SMG, obtained from *C. auris* colonies isolated from inside and outside the ZOI, also did not show any statistically significant differences. These results indicate that the antifungal tolerant colonies in our experiments could reversibly generate antifungal-susceptible colonies.

3.6. Elimination of Tolerance and Resistance in *C. auris* via Adjuvant-Antifungal Treatment

To eliminate the tolerance observed in our clinical *C. auris* isolates (Sections 3.3 and 3.4), a previously known adjuvant chloroquine [37] was combined with the antifungal drugs fluconazole, itraconazole, posaconazole, voriconazole, amphotericin B, and caspofungin. Chloroquine-antifungal disk diffusion assays and broth microdilution assays were performed on all five clinical *C. auris* isolates, as well as on the *C. parapsilosis* and *I. orientalis* reference strains (Table S1). Chloroquine alone did not have any antifungal effect on either *C. auris* isolates or the reference strains (Figure S9).

Tolerance and resistance were reduced or eliminated in some of our clinical *C. auris* isolates by combing chloroquine with antifungal drugs. *C. auris* isolate 1 showed an increase in RAD for fluconazole, posaconazole, amphotericin B, and caspofungin in presence of chloroquine compared to the RAD measured with these antifungal drugs alone at 48 h (Figure 5A–J). Similar results were found for: *C. auris* isolate 2 for posaconazole, voriconazole, amphotericin B, and caspofungin; *C. auris* 3 for fluconazole, posaconazole, voriconazole, amphotericin B, and caspofungin; *C. auris* isolate 4 for itraconazaole and amphotericin B; and *C. auris* isolate 5 for itraconazole and caspofungin (elimination of resistance for caspofungin), which all displayed an increase in RAD when these antifungal drugs were combined with chloroquine. Correspondingly, the FoG_{20} was reduced in presence of chloroquine for *C. auris* isolate 1 when combined with posaconazole, amphotericin B, and caspofungin (Figure 5K–T). However, no effect was observed when chloroquine was combined with fluconazole, itraconazole, or voriconazole. Similar adjuvant antifungal FoG_{20} results were obtained for *C. auris* isolate 2 against posaconazole, voriconazole, amphotericin B, and caspofungin; *C. auris* isolate 3 against fluconazole, itraconazole, posaconazole, voriconazole, amphotericin B, and caspofungin; *C. auris* isolate 4 against amphotericin B; and *C. auris* isolate 5 against itraconazole and caspofungin. No effect of chloroquine was observed for *C. auris* isolate 4 against fluconazole and posaconazole, nor for *C. auris* isolate 5 against amphotericin B. The FoG_{20} for *C. auris* isolate 2 for voriconazole and caspofungin and *C. auris* isolate 5 for caspofungin could not be measured at 48 h without chloroquine as there was no ZOI. However, we were able to measure the ZOI in some of these isolates in the presence of chloroquine, indicating an adjuvant effect of chloroquine on tolerance as well as on resistance. The reference strain *I. orientalis* (resistant to fluconazole) exhibited a ZOI against fluconazole when supplemented with chloroquine (Table S5; Figure S8). However, *C. parapsilosis* was not significantly affected by the presence of chloroquine (Table S5; Figure S8).

Similar effects on antifungal tolerance were obtained in adjuvant antifungal broth microdilution assays (Figure 6). Tolerance decreased for all chloroquine–antifungal drug combinations in the following isolates: *C. auris* isolate 1 (except for itraconazole), *C. auris* isolate 2 (except for voriconazole and itraconazole), *C. auris* isolate 3, *C. auris* isolate 4 (except for fluconazole), and *C. auris* isolate 5 (except for posaconazole and amphotericin B) all exhibited reduced SMG with chloroquine–antifungal drug at 48 h compared to SMG at 24 h with chloroquine–antifungal drug. As *C. auris* isolate 2 was resistant to fluconazole, SMG was not calculated. However, chloroquine did not show any effect on *C. auris* isolate-2 against itraconazole and voriconazole at 48 h compared to 24 h which is in concordance with disk diffusion assay. Whereas the SMG for *C. auris* isolate 5 was reduced against itraconazole. However, SMG could not be calculated to fluconazole and caspofungin, due to the growth at highest concentration. Similar to the disk diffusion assays, chloroquine did not show any effect on *C. auris* isolate 1 against itraconazole, *C. auris* isolate 2 against voriconazole and itraconazole, *C. auris* isolate 4 against fluconazole, and *C. auris* isolate-5 against posaconazole and amphotericin B. The MICs of all the *C. auris* isolates and control strains remained unchanged in presence of chloroquine.

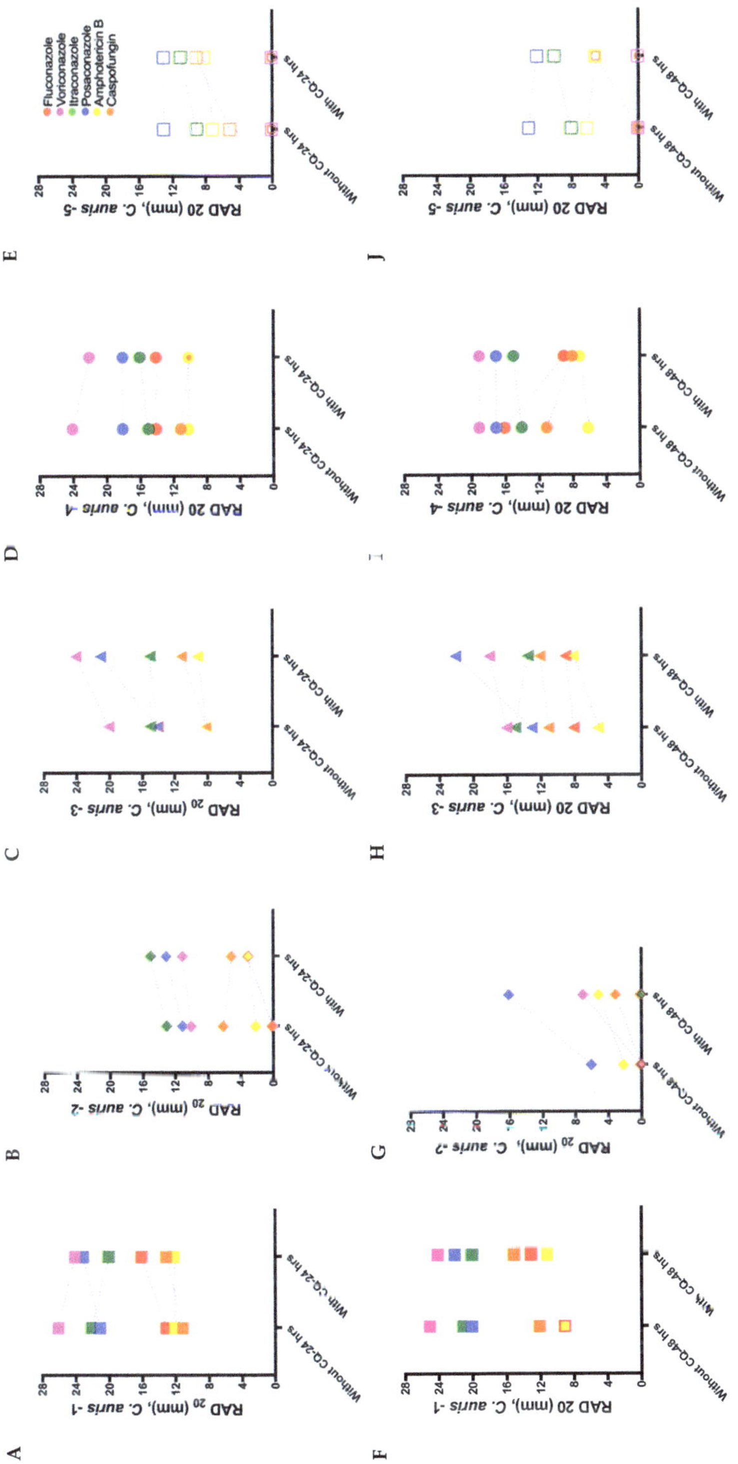

Figure 5. Cont.

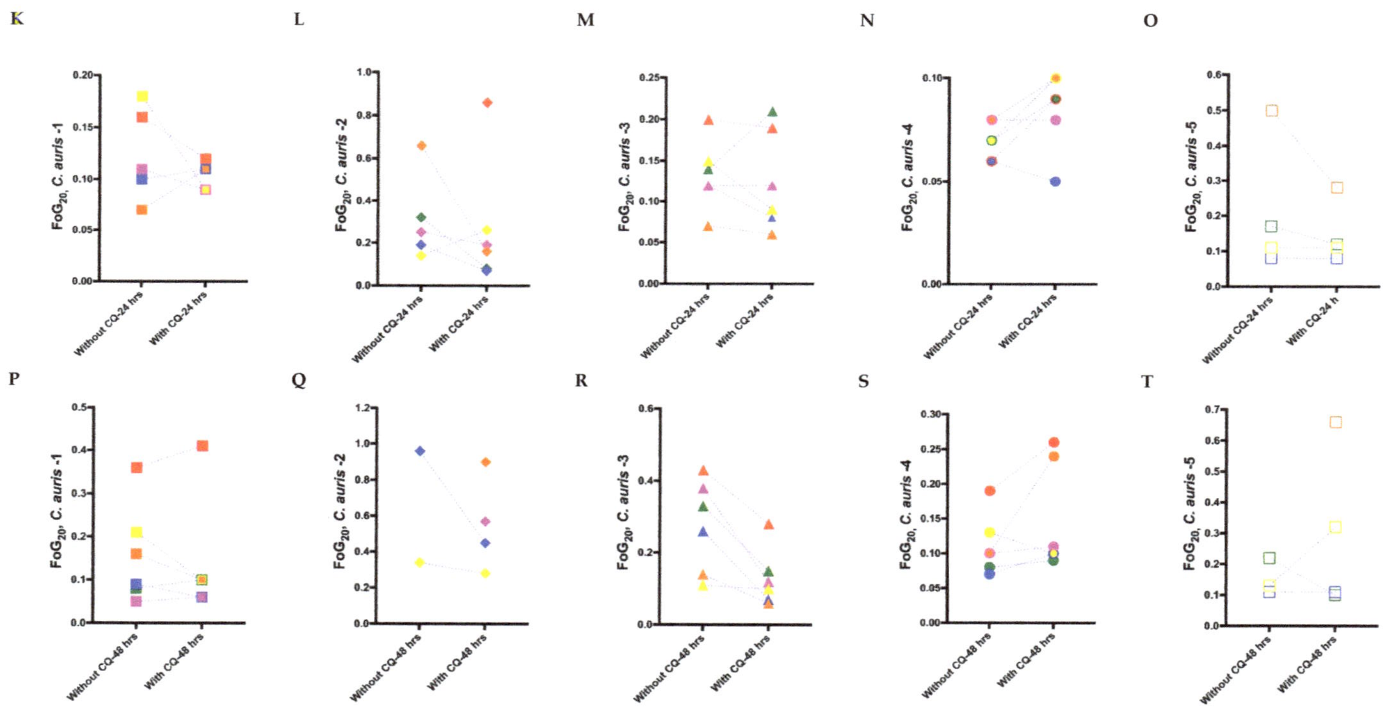

Figure 5. Radius of the zone of inhibition (RAD) and fraction of growth in the zone of inhibition (FoG_{20}) measurements for *C. auris* isolates for adjuvant antifungal disk diffusion assays. (**A–E**) Mean RAD measured for the *C. auris* isolates at 24 h against antifungal drugs with and without the adjuvant chloroquine. (**F–J**) Mean RAD measured for the *C. auris* isolate at 48 h against antifungal drugs with and without chloroquine. (**K–O**) Mean FoG_{20} measured using diskImageR [29] for all *C. auris* isolates at 24 h against antifungal drugs with and without chloroquine. (**P–T**) Mean FoG_{20} measured using diskImageR for the *C. auris* isolates at 48 h against antifungal drugs with and without chloroquine. Note that the single data points in (**L**), (**Q**), and (**T**) at 48 h are due to the mitigation of resistance in presence of chloroquine, as FoG_{20} could not be measured for these isolates at 24 h because of their resistance to the corresponding antifungal drugs.

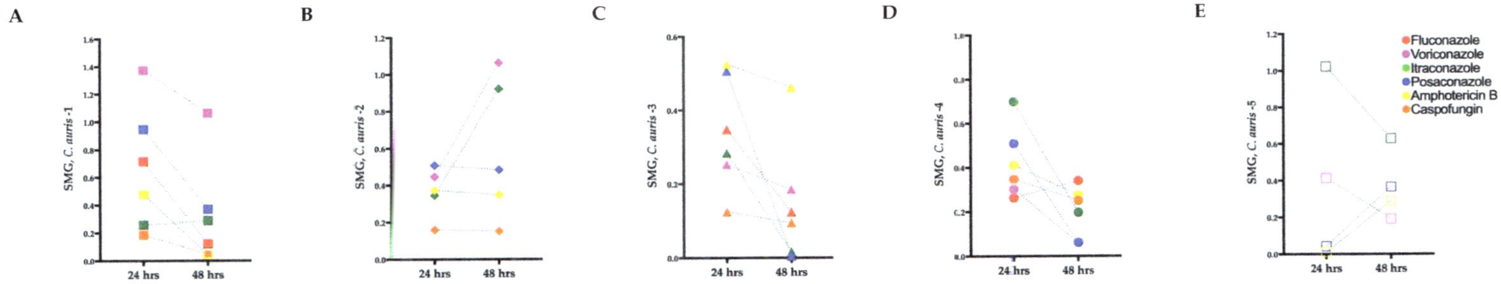

Figure 5. Supra-MIC growth (SMG) of *C. auris* isolates 1 to 5 (**A**–**E**) for adjuvant antifungal broth microdilution assays. Chloroquine did not show any adjuvant effect on C. auris isolate 1 when combined with itraconazole, nor for *C. auris* isolate 2 when combined with itraconazole or voriconazole at 48 h compared to 24 h. Similarly, no adjuvant effect was noted for *C. auris* isolate 4 against fluconazole nor for *C. auris* isolate 5 against posaconazole and amphotericin B. Since *C. auris* isolate 2 is resistant to fluconazole and *C. auris* isolate 5 is resistant to fluconazole and caspofungin, the SMGs were not calculated for these isolate–adjuvant–antifungal combinations.

4. Discussion

We report for the first time that some clinical *C. auris* isolates are tolerant to fungistatic drugs (fluconazole, voriconazole, itraconazole, and posaconazole) and to fungicidal drugs (amphotericin B and caspofungin). We also found azole tolerance in *C. parapsilosis* (fluconazole and voriconazole), but not in *I. orientalis* which was intrinsically resistant to fluconazole. We were able to detect tolerance after 24 h, as well as after 48 h by FoG_{20}, of antifungal treatment using diskImageR [29] and ImageJ [52]. These findings suggest that a distinct subpopulation among *C. auris* was able to survive and grow slowly in the presence of different antifungal drugs. Since *C. auris* is a multidrug-resistant pathogen, the presence of tolerance further narrows treatment options. Previous reports suggest that tolerant subpopulations among infecting *Candida* species are strongly associated with mortality among candidemia patients [53]. Therefore, clinical diagnostic laboratories should also test for antifungal tolerance along with standard antifungal susceptibility/resistance tests to increase the efficacy of antifungal treatment. Furthermore, existing tolerance quantification methods could be adapted to detect tolerance after 24 h and 48 h to broaden the scope of standard antimicrobial susceptibility testing in medical diagnostic laboratories. The fluconazole tolerance that we observed in *C. auris* was in agreement with previous studies on *C. albicans* [28] and *C. auris* [23], as well as with related clinical studies on "trailing growth" (reduced but persistent visible growth of *Candida* species in fluconazole concentrations above MIC [32,54,55]).

The tolerance to fungistatic and fungicidal drugs observed in some of the clinical *C. auris* isolates in our study appears to be a reversible phenomenon, as previously described for clinical *C. albicans* isolates [56]. The tolerant cells growing inside ZOI upon subculture are indistinguishable from the parental population, suggesting the presence of phenotypic heterogeneity instead of genetic variation. *C. auris* isolates cultured from inside and outside the ZOI did not show any significant changes in the average RAD, MIC, or SMG levels. This reversible tolerance that we observed in *C. auris* may result from stochastic phenotype switching or an induced response activated by the presence of antifungal drugs inside of the cell. The general mechanism underlying tolerance in *C. auris* remains to be elucidated in future work, to be aided, for instance, by mathematical modeling and synthetic biology [57], tracking single cell growth and gene expression trajectories in microfluidic devices [58,59], as well as genetic sequence and aneuploidy analyses [60].

The tolerance in some of our *C. auris* isolates was reduced or eliminated in vitro by combining azole, polyene, and echinocandin antifungal drugs with the antimalarial drug chloroquine. Chloroquine reduced tolerance for some *C. auris* isolate–antifungal combinations, while chloroquine did not have an adjuvant effect for other combinations. The mechanism underlying this strain-dependent phenomenon remains to be elucidated. Combining chloroquine with antifungal drugs had a partial effect on resistance in some of the *C. auris* isolates investigated in this study. Specifically, *C. auris* isolate 5, which was resistant to caspofungin and voriconazole (RAD = 0 mm), had a small increase in the ZOI (RAD < 12 mm) in presence of chloroquine. Correspondingly, *C. auris* isolate 2 had no ZOI for caspofungin, but had a small ZOI (RAD = 6 mm) in presence of chloroquine. The RADs for these cases were smaller than those for the sensitive *C. auris* isolates in our experiments. Chloroquine did not affect the MICs of the *C. auris* isolates in our study. Chloroquine also affected fluconazole resistance in *I. orientalis* (Table S5), though tolerant subpopulations in *C. parapsilosis* were unaffected by chloroquine. Altogether, these results suggest that combining chloroquine with antifungal drugs may have a partial mitigation effect on resistance in *C. auris*. Though the mechanism of action is unknown, it is likely related to iron depletion caused by chloroquine and its downregulation of the *ERG11* gene [35,41]. Iron depletion is known to decrease membrane sterols and increase membrane fluidity, leading to increased uptake of antifungal drugs into the cell [61]. The downregulation of *ERG11* gene, which synthesizes lanosterol alpha demethylase enzyme, is also known to be an important rate-limiting enzyme for the synthesis of ergosterol [62].

Due to the limited number of *C. auris* isolates that we were able to acquire, the results presented in this study serve as a proof of concept that *C. auris* is tolerant to fungistatic and fungicidal drugs, and that this antifungal tolerance can be mitigated by using chloroquine as an adjuvant. Further in vitro validation of these results in additional *C. auris* isolates, as well as subsequent investigations using in vivo model systems, will be pursued in future research. Another limitation of our study is that we did not have access to patient details and antifungal treatment history due to privacy regulations.

Overall, this study advances our understanding of antifungal treatment failure in *C. auris* and identifies opportunities for the clinical detection of antifungal tolerance as well as the development of targeted adjuvant antifungal therapies against tolerant and resistant invasive candidiasis.

Supplementary Materials: The following supporting information can be downloaded at: https://www.mdpi.com/article/10.3390/biomedicines11030898/s1, Table S1: *Candida* isolates and strains; Table S2: minimum inhibitory concentrations (MICs) of *C. auris* isolates; Table S3: mean MIC, SMG, FoG_{20}, and RAD for reference strains *Issatchenkia orientalis* and *C. parapsilosis* measured at 24 and 48 h for different antifungal drugs. Table S4: reversibility of tolerance phenotype in *Candida auris*. Table S5: effect of chloroquine (CLQ) on reference strains *Issatchenkia orientalis* and *Candida parapsilosis*. Figure S1: quantification of antifungal tolerance in a disk diffusion assay using the image analysis program diskImageR [29]. Figure S2: detecting tolerance in *Candida auris* from disk diffusion assays (DDAs) using diskImageR. Figure S3: representative disk diffusion assays (DDA) images of fluconazole (FLU) tolerance in *Candida auris* and *Candida parapsilosis*. Figure S4: comparison between diskImageR and manual radius of the zone of inhibition (RAD) measurements. Figure S5: azole tolerance in *Candida auris*. Figure S6: azole tolerance in *Candida parapsilosis* reference strain. Figure S7: reversibility of tolerance in a representative *Candida auris* isolate 2 against voriconazole. Figure S8: disk diffusion assays (DDAs) of antifungal adjuvant treatment in *Candida auris* isolates and *Issatchenkia orientalis* and *Candida parapsilosis* reference strains. Figure S9: *Candida auris* isolates and *Candida parapsilosis* and *Issatchenkia orientalis* reference strains growing on Mueller–Hinton agar (MHA) media with chloroquine.

Author Contributions: Conceptualization, D.A.C.; methodology, S.A.S., S.R.K. and D.A.C.; experimental investigation, S.R.K., S.A.S. and C.M.G.; formal analysis and visualization, S.R.K. and S.A.S.; data curation, S.A.S. and S.R.K.; writing, original draft preparation, review, and editing, D.A.C., S.A.S., S.R.K. and C.M.G.; supervision, D.A.C. and S.A.S.; funding acquisition, D.A.C. All authors have read and agreed to the published version of the manuscript.

Funding: This research was funded by grants to D.C. by the Government of Canada's New Frontiers in Research Fund—Exploration grant program (NFRFE-2019-01208), the Canadian Foundation for Innovation's John R. Evans Leaders Fund (CFI-40558), the Government of Alberta's Research Capacity Program (RCP-21-008-SEG), and an Audrey and Randy Lomnes Early Career Endowment Award. C.M.G. was funded by the DOST-SEI-UAlberta S&T Graduate Scholarship Program.

Institutional Review Board Statement: Not applicable.

Informed Consent Statement: Not applicable.

Data Availability Statement: The data presented in this study are openly available in Mendeley Data at doi: 10.17632/69zp86jxpz.1.

Acknowledgments: We thank Tanis Dingle and Alberta Precision Laboratories—Public Health Laboratory for providing the *C. auris* isolates. We acknowledge the Molecular Biology Services Unit at the University of Alberta for assistance with genetic sequencing. Finally, we thank the late André T. Charlebois (1953–2020) for guidance on the camera setup and for advice on the macro photography of yeasts. Finally, we are grateful to the anonymous Reviewer 2 for their thorough review and helpful feedback on the manuscript.

Conflicts of Interest: The authors declare no conflict of interest. The funders had no role in the design of the study; in the collection, analyses, or interpretation of data; in the writing of the manuscript; or in the decision to publish the results.

References

1. Perlin, D.S.; Rautemaa-Richardson, R.; Alastruey-Izquierdo, A. The Global Problem of Antifungal Resistance: Prevalence, Mechanisms, and Management. *Lancet Infect. Dis.* **2017**, *17*, e383–e392. [CrossRef] [PubMed]
2. Gow, N.A.R.; Johnson, C.; Berman, J.; Coste, A.T.; Cuomo, C.A.; Perlin, D.S.; Bicanic, T.; Harrison, T.S.; Wiederhold, N.; Bromley, M.; et al. The Importance of Antimicrobial Resistance in Medical Mycology. *Nat. Commun.* **2022**, *13*, 5352. [CrossRef] [PubMed]
3. Alangaden, G.J. Nosocomial Fungal Infections: Epidemiology, Infection Control, and Prevention. *Infect. Dis. Clin. North Am.* **2011**, *25*, 201–225. [CrossRef] [PubMed]
4. Andes, D.R.; Safdar, N.; Baddley, J.W.; Alexander, B.; Brumble, L.; Freifeld, A.; Hadley, S.; Herwaldt, L.; Kauffman, C.; Lyon, G.M.; et al. The Epidemiology and Outcomes of Invasive Candida Infections among Organ Transplant Recipients in the United States: Results of the Transplant-Associated Infection Surveillance Network (TRANSNET). *Transpl. Infect. Dis.* **2016**, *18*, 921–931. [CrossRef] [PubMed]
5. O'Neill, J. Tackling Drug-Resistant Infections Globally: Final Report and Recommendations: The Review on Antimicrobial Resistanc. 2016. Available online: https://Amr-Review.Org (accessed on 10 November 2022).
6. Woolhouse, M.; Farrar, J. Policy: An Intergovernmental Panel on Antimicrobial Resistance. *Nature* **2014**, *509*, 555–557. [CrossRef]
7. Krysan, D.J. The Unmet Clinical Need of Novel Antifungal Drugs. *Virulence* **2017**, *8*, 135–137. [CrossRef]
8. Wall, G.; Lopez-Ribot, J.L. Current Antimycotics, New Prospects, and Future Approaches to Antifungal Therapy. *Antibiotics* **2020**, *9*, 445. [CrossRef]
9. McCarty, T.P.; Pappas, P.G. Antifungal Pipeline. *Front. Cell. Infect. Microbiol.* **2021**, *11*, 1–11. [CrossRef]
10. Nnadi, N.E.; Carter, D.A. Climate Change and the Emergence of Fungal Pathogens. *PLoS Pathog.* **2021**, *17*, e1009503. [CrossRef]
11. Turner, S.A.; Butler, G. The Candida Pathogenic Species Complex. *Cold Spring Harb. Perspect. Med.* **2014**, *4*, a019778. [CrossRef]
12. Cendejas-Bueno, E.; Kolecka, A.; Alastruey-Izquierdo, A.; Theelen, B.; Groenewald, M.; Kostrzewa, M.; Cuenca-Estrella, M.; Gómez-López, A.; Boekhout, T. Reclassification of the Candida Haemulonii Complex as Candida Haemulonii (C. Haemulonii Group I), C. Duobushaemulonii Sp. Nov. (C. Haemulonii Group II), and C. Haemulonii Var. Vulnera Var. Nov.: Three Multiresistant Human Pathogenic Yeasts. *J. Clin. Microbiol.* **2012**, *50*, 3641–3651. [CrossRef] [PubMed]
13. Forsberg, K.; Woodworth, K.; Walters, M.; Berkow, E.L.; Jackson, B.; Chiller, T.; Vallabhaneni, S. Candida Auris: The Recent Emergence of a Multidrug-Resistant Fungal Pathogen. *Med. Mycol.* **2019**, *57*, 1–12. [CrossRef] [PubMed]
14. Chow, N.A.; Gade, L.; Tsay, S.V.; Forsberg, K.; Greenko, J.A.; Southwick, K.L.; Barrett, P.M.; Kerins, J.L.; Lockhart, S.R.; Chiller, T.M.; et al. Multiple Introductions and Subsequent Transmission of Multidrug-Resistant Candida Auris in the {USA}: A Molecular Epidemiological Survey. *Lancet Infect. Dis.* **2018**, *18*, 1377–1384. [CrossRef]
15. Centers for Disease Control and Prevention Tracking Candida Auris: Candida Auris Fungal Diseases CDC. Available online: https://www.cdc.gov/fungal/candida-auris/tracking-c-auris.html (accessed on 1 April 2021).
16. Arendrup, M.C.; Patterson, T.F. Multidrug-Resistant Candida: Epidemiology, Molecular Mechanisms, and Treatment. *J. Infect. Dis.* **2017**, *216*, S445–S451. [CrossRef] [PubMed]
17. Rhodes, J.; Fisher, M.C. Global Epidemiology of Emerging Candida Auris. *Curr. Opin. Microbiol.* **2019**, *52*, 84–89. [CrossRef] [PubMed]
18. Magiorakos, A.-P.; Srinivasan, A.; Carey, R.B.; Carmeli, Y.; Falagas, M.E.; Giske, C.G.; Harbarth, S.; Hindler, J.F.; Kahlmeter, G.; Olsson-Liljequist, B.; et al. Multidrug-Resistant, Extensively Drug-Resistant and Pandrug-Resistant Bacteria: An International Expert Proposal for Interim Standard Definitions for Acquired Resistance. *Clin. Microbiol. Infect.* **2012**, *18*, 268–281. [CrossRef]
19. Lyman, M.; Forsberg, K.; Reuben, J.; Dang, T.; Free, R.; Seagle, E.E.; Sexton, D.J.; Soda, E.; Jones, H.; Hawkins, D.; et al. Notes from the Field: Transmission of Pan-Resistant and Echinocandin-Resistant Candida Auris in Health Care Facilities — Texas and the District of Columbia, January–April 2021. *MMWR. Morb. Mortal. Wkly. Rep.* **2021**, *70*, 1022–1023. [CrossRef] [PubMed]
20. Chen, J.; Tian, S.; Han, X.; Chu, Y.; Wang, Q.; Zhou, B.; Shang, H. Is the Superbug Fungus Really so Scary? A Systematic Review and Meta-Analysis of Global Epidemiology and Mortality of Candida Auris. *BMC Infect. Dis.* **2020**, *20*, 827. [CrossRef]
21. Berman, J.; Krysan, D.J. Drug Resistance and Tolerance in Fungi. *Nat. Rev. Microbiol.* **2020**, *18*, 319–331. [CrossRef]
22. Kukurudz, R.J.; Chapel, M.; Wonitowy, Q.; Adamu Bukari, A.-R.; Sidney, B.; Sierhuis, R.; Gerstein, A.C. Acquisition of Cross-Azole Tolerance and Aneuploidy in Candida Albicans Strains Evolved to Posaconazole. *G3 Genes | Genomes | Genetics* **2022**, *12*, jkac156. [CrossRef]
23. Kim, S.H.; Iyer, K.R.; Pardeshi, L.; Muñoz, J.F.; Robbins, N.; Cuomo, C.A.; Wong, K.H.; Cowen, L.E. Genetic Analysis of Candida Auris Implicates Hsp90 in Morphogenesis and Azole Tolerance and Cdr1 in Azole Resistance. *MBio* **2019**, *10*, e02529-18. [CrossRef]
24. Healey, K.R.; Kordalewska, M.; Ortigosa, C.J.; Singh, A.; Berrío, I.; Chowdhary, A.; Perlin, D.S. Limited ERG11 Mutations Identified in Isolates of Candida Auris Directly Contribute to Reduced Azole Susceptibility. *Antimicrob. Agents Chemother.* **2018**, *62*, e01427-18. [CrossRef]
25. Mount, H.O.; Revie, N.M.; Todd, R.T.; Anstett, K.; Collins, C.; Costanzo, M.; Boone, C.; Robbins, N.; Selmecki, A.; Cowen, L.E. Global Analysis of Genetic Circuitry and Adaptive Mechanisms Enabling Resistance to the Azole Antifungal Drugs. *PLoS Genet.* **2018**, *14*, e1007319. [CrossRef]
26. Coenye, T.; De Vos, M.; Vandenbosch, D.; Nelis, H. Factors Influencing the Trailing Endpoint Observed in Candida Albicans Susceptibility Testing Using the CLSI Procedure. *Clin. Microbiol. Infect.* **2008**, *14*, 495–497. [CrossRef] [PubMed]

27. Arendrup, M.C.; Cuenca-Estrella, M.; Lass-Flörl, C.; Hope, W.; Arendrup, M.C.; Hope, W.W.; Flörl, C.; Cuenca-Estrella, M.; Arikan, S.; Barchiesi, F.; et al. EUCAST Technical Note on the EUCAST Definitive Document EDef 7.2: Method for the Determination of Broth Dilution Minimum Inhibitory Concentrations of Antifungal Agents for Yeasts EDef 7.2 (EUCAST-AFST). *Clin. Microbiol. Infect.* **2012**, *18*, E246–E247. [CrossRef]
28. Rosenberg, A.; Ene, I.V.; Bibi, M.; Zakin, S.; Segal, E.S.; Ziv, N.; Dahan, A.M.; Colombo, A.L.; Bennett, R.J.; Berman, J. Antifungal Tolerance Is a Subpopulation Effect Distinct from Resistance and Is Associated with Persistent Candidemia. *Nat. Commun.* **2018**, *9*, 2470. [CrossRef]
29. Gerstein, A.C.; Rosenberg, A.; Hecht, I.; Berman, J. DiskimageR: Quantification of Resistance and Tolerance to Antimicrobial Drugs Using Disk Diffusion Assays. *Microbiology* **2016**, *162*, 1059–1068. [CrossRef] [PubMed]
30. Lepak, A.J.; Andes, D.R. Antifungal Pharmacokinetics and Pharmacodynamics. *Cold Spring Harb. Perspect. Med.* **2014**, *5*, a019653. [CrossRef] [PubMed]
31. Bhattacharya, S.; Holowka, T.; Orner, E.P.; Fries, B.C. Gene Duplication Associated with Increased Fluconazole Tolerance in Candida Auris Cells of Advanced Generational Age. *Sci. Rep.* **2019**, *9*, 5052. [CrossRef]
32. Astvad, K.M.T.; Sanglard, D.; Delarze, E.; Hare, R.K.; Arendrup, M.C. Implications of the EUCAST Trailing Phenomenon in Candida Tropicalis for the In Vivo Susceptibility in Invertebrate and Murine Models. *Antimicrob. Agents Chemother.* **2018**, *62*, e01624-18. [CrossRef]
33. Colombo, A.L.; Guimarães, T.; Sukienik, T.; Pasqualotto, A.C.; Andreotti, R.; Queiroz-Telles, F.; Nouér, S.A.; Nucci, M. Prognostic Factors and Historical Trends in the Epidemiology of Candidemia in Critically Ill Patients: An Analysis of Five Multicenter Studies Sequentially Conducted over a 9-Year Period. *Intensive Care Med.* **2014**, *40*, 1489–1498. [CrossRef] [PubMed]
34. Islahudin, F.; Khozoie, C.; Bates, S.; Ting, K.-N.; Pleass, R.J.; Avery, S. V Cell Wall Perturbation Sensitizes Fungi to the Antimalarial Drug Chloroquine. *Antimicrob. Agents Chemother.* **2013**, *57*, 3889–3896. [CrossRef]
35. Levitz, S.M.; Harrison, T.S.; Tabuni, A.; Liu, X. Chloroquine Induces Human Mononuclear Phagocytes to Inhibit and Kill Cryptococcus Neoformans by a Mechanism Independent of Iron Deprivation. *J. Clin. Investig.* **1997**, *100*, 1640–1646. [CrossRef] [PubMed]
36. Newman, S.L.; Gootee, L.; Brunner, G.; Deepe, G.S. Chloroquine Induces Human Macrophage Killing of Histoplasma Capsulatum by Limiting the Availability of Intracellular Iron and Is Therapeutic in a Murine Model of Histoplasmosis. *J. Clin. Investig.* **1994**, *93*, 1422–1429. [CrossRef] [PubMed]
37. Li, Y.; Wan, Z.; Liu, W.; Li, R. Synergistic Activity of Chloroquine with Fluconazole against Fluconazole-Resistant Isolates of Candida Species. *Antimicrob. Agents Chemother.* **2015**, *59*, 1365–1369. [CrossRef]
38. Bjelle, A.; Björnham, A.; Larsen, A.; Mjörndal, T. Chloroquine in Long-Term Treatment of Rheumatoid Arthritis. *Clin. Rheumatol.* **1983**, *2*, 393–399. [CrossRef]
39. Powell, S.J. Therapy of Amebiasis. *Bull. N. Y. Acad. Med.* **1971**, *47*, 469–477.
40. Dima, A.; Jurcut, C.; Chasset, F.; Felten, R.; Arnaud, L. Hydroxychloroquine in Systemic Lupus Erythematosus: Overview of Current Knowledge. *Ther. Adv. Musculoskelet. Dis.* **2022**, *14*, 1759720X211073001. [CrossRef]
41. Lan, C.-Y.; Rodarte, G.; Murillo, L.A.; Jones, T.; Davis, R.W.; Dungan, J.; Newport, G.; Agabian, N. Regulatory Networks Affected by Iron Availability in Candida Albicans. *Mol. Microbiol.* **2004**, *53*, 1451–1469. [CrossRef]
42. Singhal, N.; Kumar, M.; Kanaujia, P.K.; Virdi, J.S. MALDI-TOF Mass Spectrometry: An Emerging Technology for Microbial Identification and Diagnosis. *Front. Microbiol.* **2015**, *6*, 791. [CrossRef]
43. Dingle, T.C.; Butler-Wu, S.M. Maldi-Tof Mass Spectrometry for Microorganism Identification. *Clin. Lab. Med.* **2013**, *33*, 589–609. [CrossRef]
44. Turenne, C.Y.; Sanche, S.E.; Hoban, D.J.; Karlowsky, J.A.; Kabani, A.M. Rapid Identification of Fungi by Using the ITS2 Genetic Region and an Automated Fluorescent Capillary Electrophoresis System. *J. Clin. Microbiol.* **1999**, *37*, 1846–1851. [CrossRef] [PubMed]
45. National Library of Medicine. Available online: https://blast.ncbi.nlm.nih.gov/Blast.cgi?PROGRAM=blastn&BLAST_SPEC=GeoBlast&PAGE_TYPE=BlastSearch (accessed on 12 October 2022).
46. CLSI. *Reference Method for Broth Dilution Antifungal Susceptibility Testing of Yeasts*; Clinical and Laboratory Standards Institute: Wayne, PA, USA, 2017.
47. CLSI. *Method for Antifungal Disk Diffusion Susceptibility Testing of Yeasts: Approved Guideline—Second Edition*, 2nd ed.; Clinical and Laboratory Standards Institute: Wayne, PA, USA, 2008; ISBN 1562385321.
48. CLSI. *Performance Standards for Antifungal Susceptibility Testing of Yeasts*, 1st ed.; CLS.; Clinical and Laboratory Standards Institute: Wayne, PA, USA, 2017.
49. CDC 2019 Antifungal Susceptibility Testing and Interpretation | Candida Auris | Fungal Diseases | CDC. Available online: https://www.cdc.gov/fungal/candida-auris/c-auris-antifungal.html (accessed on 24 December 2019).
50. ImageJ. Available online: https://imagej.nih.gov/ij/ (accessed on 28 July 2022).
51. R Core Team. *R: A Language and Environment for Statistical Computing*; R Foundation for Statistical Computing: Vienna, Austria, 2018.
52. Abràmoff, M.D.; Magalhães, P.J.; Ram, S.J. Image Processing with ImageJ. *Biophotonics Int.* **2004**, *11*, 36–41. [CrossRef]
53. Levinson, T.; Dahan, A.; Novikov, A.; Paran, Y.; Berman, J.; Ben-Ami, R. Impact of Tolerance to Fluconazole on Treatment Response in Candida Albicans Bloodstream Infection. *Mycoses* **2021**, *64*, 78–85. [CrossRef] [PubMed]

54. Marcos-Zambrano, L.J.; Escribano, P.; Bouza, E.; Guinea, J. Susceptibility of Candida Albicans Biofilms to Caspofungin and Anidulafungin Is Not Affected by Metabolic Activity or Biomass Production. *Med. Mycol.* **2016**, *54*, 155–161. [CrossRef]
55. Delarze, E.; Sanglard, D. Defining the Frontiers between Antifungal Resistance, Tolerance and the Concept of Persistence. *Drug Resist. Updat.* **2015**, *23*, 12–19. [CrossRef]
56. Farquhar, K.S.; Rasouli Koohi, S.; Charlebois, D.A. Does Transcriptional Heterogeneity Facilitate the Development of Genetic Drug Resistance? *BioEssays* **2021**, *43*, 2100043. [CrossRef]
57. Farquhar, K.S.; Flohr, H.; Charlebois, D.A. Advancing Antimicrobial Resistance Research Through Quantitative Modeling and Synthetic Biology. *Front. Bioeng. Biotechnol.* **2020**, *8*, 583415. [CrossRef]
58. Jo, M.C.; Liu, W.; Gu, L.; Dang, W.; Qin, L. High-Throughput Analysis of Yeast Replicative Aging Using a Microfluidic System. *Proc. Natl. Acad. Sci. USA* **2015**, *112*, 9364–9369. [CrossRef]
59. Charlebois, D.A.; Hauser, K.; Marshall, S.; Balázsi, G. Multiscale Effects of Heating and Cooling on Genes and Gene Networks. *Proc. Natl. Acad. Sci. USA* **2018**, *115*, E10797–E10806. [CrossRef]
60. Yang, F.; Lu, H.; Wu, H.; Fang, T.; Berman, J.; Jiang, Y. Aneuploidy Underlies Tolerance and Cross-Tolerance to Drugs in Candida Parapsilosis. *Microbiol. Spectr.* **2021**, *9*, e00508-21. [CrossRef] [PubMed]
61. Prasad, T.; Chandra, A.; Mukhopadhyay, C.K.; Prasad, R. Unexpected Link between Iron and Drug Resistance of Candida Spp.: Iron Depletion Enhances Membrane Fluidity and Drug Diffusion, Leading to Drug-Susceptible Cells. *Antimicrob. Agents Chemother.* **2006**, *50*, 3597–3606. [CrossRef] [PubMed]
62. Ostrowsky, B.; Greenko, J.; Adams, E.; Quinn, M.; O'Brien, B.; Chaturvedi, V.; Berkow, E.; Vallabhaneni, S.; Forsberg, K.; Chaturvedi, S.; et al. Candida Auris Isolates Resistant to Three Classes of Antifungal Medications—New York, 2019. *MMWR Morb. Mortal. Wkly. Rep.* **2020**, *69*, 6–9. [CrossRef] [PubMed]

biomedicines

MDPI

Article

The Antifungal and Antibiofilm Activities of Caffeine against *Candida albicans* on Polymethyl Methacrylate Denture Base Material

Doaa M. AlEraky [1,*], Hatem M. Abuohashish [1], Mohammed M. Gad [2], Muneer H. Alshuyukh [3], Amr S. Bugshan [1], Khalid S. Almulhim [4] and Maha M. Mahmoud [1]

[1] Department of Biomedical Dental Sciences, College of Dentistry, Imam Abdulrahman Bin Faisal University, P.O. Box 1982, Dammam 31441, Saudi Arabia
[2] Department of Substitutive Dental Sciences, College of Dentistry, Imam Abdulrahman Bin Faisal University, P.O. Box 1982, Dammam 31441, Saudi Arabia
[3] College of Dentistry, Imam Abdulrahman Bin Faisal University, P.O. Box 1982, Dammam 31441, Saudi Arabia
[4] Department of Restorative Dental Sciences, College of Dentistry, Imam Abdulrahman Bin Faisal University, P.O. Box 1982, Dammam 31441, Saudi Arabia
* Correspondence: dmaleraky@iau.edu.sa

Abstract: Background: In this study, the effect of pure caffeine was established against *Candida albicans* (*C. albicans*) using different microbiological techniques. Methods: Broth microdilution and colony forming units (CFUs) assays were used to detect the minimum inhibitory concentration (MIC) and minimum fungicidal concentration (MFC). The Live/Dead fluorescent dyes were implemented to determine the yeast viability. Polymethyl methacrylate acrylic resin (PMMA) discs were prepared to evaluate caffeine's effects against adherent *C. albicans* using microplate reader, CFUs, and scanning electron microscope (SEM). Results: caffeine's MIC was detected around 30 mg/mL, while the MFC was considered at 60 mg/mL. In an agar-well diffusion test, the inhibition zones were wider in caffeine groups. The Live/Dead viability test verified caffeine's antifungal effects. The optical density of the adherent *C. albicans* on PMMA discs were lower at 620 nm or 410 nm in caffeine groups. CFU count was also reduced by caffeine treatments. SEM revealed the lower adherent *C. albicans* count in caffeine groups. The effect of caffeine was dose-dependent at which the 60 mg/mL dose demonstrated the most prominent effect. Conclusion: The study reinforced caffeine's antifungal and antibiofilm properties and suggested it as an additive, or even an alternative, disinfectant solution for fungal biofilms on denture surfaces.

Keywords: caffeine; *Candida albicans*; polymethyl methacrylate acrylic resin (PMMA); denture; stomatitis

1. Introduction

Candida albicans (C. albicans) is the most frequent fungus that asymptomatically colonizes half of the population worldwide. Yet, the opportunistic mode of this pathogen has been established and associated with several severe infections and a high rate of morbidity and mortality [1]. The virulence factors of *C. albicans* are responsible for the advanced fungal infections in the oral cavity and digestive tract, in addition to mucocutaneous, systemic, and invasive infections [2].

Biofilm formation is one of the significant virulence factors of *C. albicans;* the pathogenesis of *C. albicans* is established on the irreversible attachment and construction of the biofilm biomass on the living tissues of the host or on prosthetic devices and implants [3,4]. Adhesion is the initial step of biofilm formation, then morphological changes develop cumulatively by the number of fungus cells and produce extracellular polymeric substances that impact the final biofilm architecture [1,5]. A single species can form biofilm, yet in vivo, combined different species generally underly the formation of biofilm. Previous studies have observed the biofilm maturation in an animal model at 24 h; however, in vitro, it was detected from 38 to 72 h [6,7].

Antifungal agents such as nystatin and miconazole are used to treat mild to moderate candidiasis in the oral cavity for a period of one to two weeks. Oral or intravenous administration of fluconazole is commonly used in case of severe infections [8]. Recently, the prolonged use of fluconazole and its fungistatic, not fungicidal, role has contributed to the growing resistance of *C. albicans* [9]. Studies have shown that the biofilm formation might provoke the candida resistance to antifungals agents [10]. The remarkable ability of *C. albicans* to develop resistance is a public health burden with the limited arsenal of treatment options [11,12] and it is crucial to direct our research studies to explore new antifungal modalities [13]. Notably, the effect of natural compounds on the biofilm activity of drug-resistant *C. albicans* was demonstrated in previous studies [14].

Denture stomatitis is an oral disease that arises due to poor denture hygiene and is generally caused by *C. albicans*. Chlorhexidine (CHX) has been used as a disinfectant agent to diminish the adhesion of *C. albicans* on the denture-based resins [15]. CHX is a biguanide disinfectant [16] that is frequently used in dentistry as an antiseptic oral rinse, mouthwash. CHX has shown a high fungicidal effect and can be used as a topical antifungal agent. However, the proper use of CHX for clinical and dental purposes is important and should be specifically used [15,17].

Caffeine is a natural classical purine alkaloid belonging to the methylxanthines family (1,3,7-trimethixanthine). It is produced in more than eighty herbal species and considered the most common behavior-affecting substance globally [18]. Since the chemical compound has been isolated, pharmaceutical studies have been applied to explore the function of caffeine and the effect of caffeine on the immune and nervous systems has been established [18,19]. Few research studies have reported the antifungal activities of caffeine and recommended further investigations to describe the potential inhibitory effect of caffeine on the growth of *C. albicans* and biofilm formation [20,21]. The aim of our study was to evaluate the antifungal activity of caffeine on *C. albicans* and to assess its antibiofilm activity on polymethyl methacrylate resin (PMMA) denture discs using different microbiological methods. The null hypothesis of the present work is that the antifungal effects of caffeine would not differ from the control.

2. Materials and Methods

2.1. Chemicals and Materials

Caffeine (Cat. # C080100) was provided from Toronto Research Chemicals Inc. (2 Brisbane Road Toronto, ON, Canada). Live/Dead® yeast viability kit (Cat. # L7009) was purchased from Thermo Fisher Scientific (Waltham, MA, USA). Sabouraud Dextrose Agar (SDA) (Cat. # MM-M-024) and Sabouraud Dextrose Broth (Cat. # MM-M-205) were provided from Molequle-on® (Auckland, New Zealand). Mueller–Hinton Agar (Cat. # CM0337) was provided from Oxoid (Basingstoke, UK), and the 96-well plates were purchased from CELLTREAT (Pepperell, MA, USA). Heat-polymerized acrylic denture base resin (Major.Base.20, Major Prodotti Dentari, Moncalieri, Italy) was used for resin specimens' preparation.

2.2. Antifungal Susceptibility Testing

2.2.1. Broth Microdilution Assay

The antifungal susceptibility testing of caffeine at 15, 30, and 60 mg/mL concentrations was tested against *C. albicans* reference strain (ATCC 10231). *C. albicans* from a glycerol stock was streaked onto SDA plates 2–5 days before the exterminate and the plates were incubated at 37 ± 1 °C for two days. Isolated colonies were inoculated in 20 mL broth media and incubated at 30 ± 1 °C overnight with shaking. The yeast suspension was adjusted at 0.5 McFarland to produce approximately 1×10^6 CFU/mL. The broth microdilution method was performed according to the guidelines from the Clinical and Laboratory Standards Institute (CLSI) M27 protocol for *Candida species* [22] with slight modification [23].

In a 96-well plate, multiple concentrations of caffeine solutions were prepared in sterile phosphate-buffered saline (PBS) by two-fold dilution, 100 μL of each concentration was prepared in triplicate. Then, 100 μL diluted *Candida* (2.5×10^3 CFU/mL) in liquid media was added to each well. CHX (Avalon Pharma, Riyadh, Saudi Arabia) at a concentration of 0.2% was used as a standard antifungal agent also, and positive and negative growth wells were performed as control samples. The microdilution plates were incubated at 37 ± 1 °C, the optical density (OD) of the plate was determined at 620 nm after one and two days using a microplate reader (Bio-Rad xMark TM Microplate Spectrophotometer; Hercules, CA, USA). The minimum inhibitory concentration (MIC) was determined as the lowest concentration at which no turbidity of the yeast growth was observed [24]. To detect the Minimal Fungicidal Concentration (MFC), 10 μL from each well was inoculated on SDA plates and incubated for two days at 37 ± 1 °C and MFC value that was determined at the lowest concentration that showed equal or less than 2 colony-forming units.

2.2.2. Agar-Well Diffusion

The surface of three SDA plates were inoculated by spreading 100 μL of C. albicans over the entire surface. After this, a 6 mm hole was punched in the agar under aseptic conditions using a microtip. A volume of 50 μL from different concentrations of caffeine, as well as CHX, PBS (control), and distilled water (blank), were added to the wells. Then, the treated plates were incubated overnight at 37 ± 1 °C. After one day, the inhibition zones of caffeine-inoculated wells were visualized and compared with positive- and negative-control wells. The diameter of each inhibition zone was measured (mm) and divided by 2 after subtraction of the diameter of the hole.

2.2.3. Determination of Yeast Viability

The yeast viability was evaluated using the LIVE/DEAD® Yeast Viability Kit (Cat. # L7009) that was purchased from Thermo Fisher Scientific. After one day incubation at 37 ± 1 °C, the yeast suspension with different caffeine concentrations was centrifuged for 5 min at 10,000× *g*. After decanting the supernatant, the remaining pellets were resuspended in another 1 mL of the wash buffer in 1.5 mL tubes. For the cellular staining, 1 μL of FUN® 1 cell stain (Component-A) and 5 μL of Calcofluor® White M2R (Component-B) were added to each yeast suspension sample. Samples were then incubated in the dark for 30 min at 30 °C. Hereafter, 5 μL from each yeast suspension-stained sample was added on a glass slide and covered with a coverslip. The slides were inspected over an inverted fluorescent microscope (Nikon Eclipse Ts2R, Nikon Instruments Inc., Melville, NY, USA). From each sample, four different fields were visualized. Images were taken at excitation/emission wavelength of 488/530 nm and 365/435 nm for the FUN® 1 cell stain and Calcofluor® White M2R, respectively.

2.3. *Polymethyl Methacrylate (PMMA) Resin Preparation*

A total of 60 specimens of polymethyl methacrylate heat polymerized acrylic resin discs with a thickness of 2 mm, a diameter of 10 mm diameter, and polished surfaces were prepared according to the manufacturer's guidelines. A digital caliper was used for specimens' dimensions confirmation, and approved specimens were stored in distilled water for two days at 37 ± 1 °C before testing procedures. The PMMA discs composed five groups (n = 10), as follows: Group 1: control group containing PBS; Group 2: the standard (CHX); Group 3: caffeine 60 mg/mL (CAF-60); Group 4: caffeine 30 mg/mL (CAF-30); and Group 5: caffeine 15 mg/mL (CAF-15).

2.4. *Antibiofilm Activity Assay*

Under aseptic conditions in the laminar flow safety cabinet, each disc was sterilized with 95% ethanol on each side then subjected to 1 h ultraviolet illumination. The discs were placed in 12-well plates according the 5 groups with 1 mL of artificial saliva and incubated for 2 h. The overnight culture of *C. albicans* was adjusted to 0.5 McFarland, and 1 mL of crobial effects of caffeine are well-documented. On the fungal level, a wound dressing that delivers caffeine alone or in combination with ascorbic acid showed a clear antifungal activity against *C. albicans* using the disc diffusion method [32]. In one study, caffeine enhanced the antifungal activity of chloroquine through induction of the cell wall perturbation [33]. Caffeine and its salts caused alterations in the structure of *C. albicans* in the Mittag study [34]. Sabie and Gadd reported that caffeine and other phosphodiesterase inhibitors might induce yeast–mycelium transition of *C. albicans* [35]. Different concentrations of caffeine triggered gene segregations of *C. albicans* and suppressed its cell replication [36]. The reported direct antifungal effects of caffeine in the current study were confirmed by CFU assay and the agar well-diffusion method, where caffeine, particularly the 60 mg/mL concentration, suppressed the number of *C. albicans* colonies and generated a fungal growth inhibition zone wider than CHX. Therefore, the highest concentration of caffeine was chosen to innovatively verify its antifungal effects using the florescent yeast Live/Dead viability assay. In this test, the reduced count of the viable *Candida* indicated that caffeine could disturb the integrity of the fungal plasma membrane as well as induce alterations in the metabolic function of the yeast.

Multiple mechanistic pathways have been suggested to explain the antimicrobial effects of caffeine. For instance, caffeine might inhibit microbial DNA synthesis by suppressing thymidine kinase and preventing the inclusion of thymidine or adenine bases, which might additionally potentiate the effects of other antimicrobial agents [37,38]. Furthermore, the antimicrobial effects of caffeine might be indirectly augmented by its combination with other antimicrobial agents that target microbial cell-wall synthesis, which facilitates the influx of caffeine into microorganisms [39]. The antifungal effects of caffeine might be attributed to the manipulation of the N-terminal domain and specific protein phosphatase Z1 (CaPpz1) of *C. albicans* [40]. Thus, CaPpz1 gene deletion improved *C. albicans* sensitivity to caffeine, which supports this mode of action [41]. Studies have also suggested the role of CaTip41, CaSCH9, CaDOA1, and CDR1-yEGFP3 genes in regulating caffeine antifungal properties [42–45]. Stichternoth et al. reported that Tor1 and Sch9 kinases are essential for caffeine antifungal effects through suppression of hyphal morphogenesis [46]. Sanglard explored the calcineurin pathway as an essential tool for tolerance to *C. albicans* metabolic inhibitors such as caffeine [47].

One of the novelty aspects of the present work is that we tested the influence of caffeine alone against the adherent *C. albicans* on the surface of denture-based PMMA acrylic resin. One recent study conducted by Alfaifi et al. found that caffeine can suppress the metabolic activity and biofilm formation of *C. albicans* in the presence of nicotine on PMMA acrylic denture resin [20]. Nicotine was used to replicate the smoking condition as a causative factor for denture stomatitis. In the Alfaifi et al. study, the effect of caffeine alone against PMMA adherent *C. albicans* was not evaluated. Moreover, the range of caffeine concentrations in the Alfaifi et al. study was similar to our assessed caffeine concentrations. In the present study, caffeine markedly and dose-dependently reduced adherent *Candida* count as indicated by reduced OD at both 620 nm and 410 nm wavelengths. Here, we anticipated that the adherent *Candida* cell count would be low. Given the fact that the OD of *Candida* cell suspension could be recognized at different wavelengths [48,49], we inspected the OD of the adherent *Candida* at 620 nm and 410 nm wavelengths. Moreover, CFU assay results of the adherent *Candida* revealed the marked effects of caffeine against *Candida* biofilm formation at all tested concentrations. Treatment of the PMMA discs with different concentrations of caffeine lowered the number of surviving adherent *C. albicans*, which lowered OD values and CFU count of caffeine groups, particularly the higher (60 mg/mL) dose. SEM examination confirmed these findings. The SEM images were comparable to other studies, where *C. albicans* showed similar adhesion and biofilm formation without treatment on the PMMA discs. Indeed, the pattern of the reduced *Candida* adhesion by caffeine on SEM images resembled previous studies in its structure and arrangements [50,51].

yeast suspension was added to each well with different concentration of caffeine, except for the control samples. The plates were incubated at 37 ± 1 °C for two days to allow the biofilm formation phase of *C. albicans*. Nonadherent cells were removed by washing the discs twice with PBS and placed in new, sterile, 12-well plastic plates. Finally, the biofilms were scraped and vortexed for 2 min at 3000 rpm speed to dislodge the adherent cells from the denture [25,26]. The adherent fungal cells were evaluated by microplate reader (Bio-Rad xMark TM Microplate Spectrophotometer, Hercules, CA, USA) at 620 and 410 nm.

2.5. Colony Forming Unit Assay

To assess the antibiofilm activity, Colony Forming Unit Assay was also performed. A volume of 100 µL of each well was inoculated on SDA plates and incubated at 37 ± 1 °C; the counting of colonies was performed within one day to easily distinguish the colonies before the overgrow and after two days to allow scoring of any slow grow isolates. After two days of incubation at 37 ± 1 °C, the colonies were then captured and counted using an Image J 1.37b image analysis system (National Institutes of Health, Bethesda, MD, USA). The experiment was independently performed by two microbiologists using three replicates' plates to ensure the reproducibility.

2.6. Scanning Electron Microscope

One randomly selected disc per group was used for the scanning electron microscope (SEM) procedure. The PMMA discs were initially fixed at room temperature in 2.5% glutaraldehyde solution. The discs were then dehydrated in ascending ethanol concentrations. Afterwards, the specimens were gold sputter-coated after mounting on metallic stubs (Quorum, Q150R ES, Lewes, UK). Samples were inspected under SEM instrument (Emcrafts, Gyeonggi-do, Korea) at 10kV at ×2000 magnification.

2.7. Statistical Analysis

Graph Pad Prism V-5 was employed as statistical software (GraphPad Software, Inc., La Jolla, CA, USA). All numerical values were presented as arithmetic means (±) their standard deviations (SD). After testing the normality with the Shapiro–Wilk test, one-way analysis of variance (ANOVA) followed by Turkey's post hoc test were used for data with normal distribution (microdilution analysis). Data without normal distribution (CFU and Agar-well diffusion) were analyzed using Kruskal–Wallis test and Dunn's post hoc test. The statistical significance for all analyses was considered when $p \leq 0.05$.

3. Results

Results of the microdilution test showed that caffeine at the 30 mg/mL concentration showed no visible turbidity of the yeast growth (MIC). Moreover, the results of the optical density using plate reader are presented in Figure 1. The growth of *C. albicans* was significantly ($p \leq 0.01$) reduced in CHX and caffeine (60 mg/mL) groups as compared to the control group following 24 h of incubation with growth media (Figure 1,A,B). The 30 and 15 mg/mL concentrations of caffeine did not demonstrate a statistically significant decrease in the percentage of *C. albicans* growth compared to the control group (Figure 1A,B).

In Figure 2, agar plates of the CFU test to enumerate the fungal colonies for further confirmation showed overgrowth of *C. albicans* in the control PBS group with no treatments (Figure 2A). This *Candida* growth was reduced, as shown in SDA plates from the CHX group (Figure 2B). Caffeine groups in 60, 30, and 15 mg/mL concentrations demonstrated marked and dose-dependent decreased *C. albicans* colony count on the SDA plates (Figure 2C–E). The CFU counts in CHX and caffeine (60 mg/mL) groups were statistically significantly ($p \leq 0.05$) low as compared to the control group (Figure 2F). Other concentrations of caffeine (30 and 15 mg/mL) showed a non-statistically significant lower CFU as compared to the control group (Figure 2F). Hence, caffeine at the 60 mg/mL concentration was considered as the MFC value as this concentration demonstrated less than 2 CFUs of *C. albicans*.

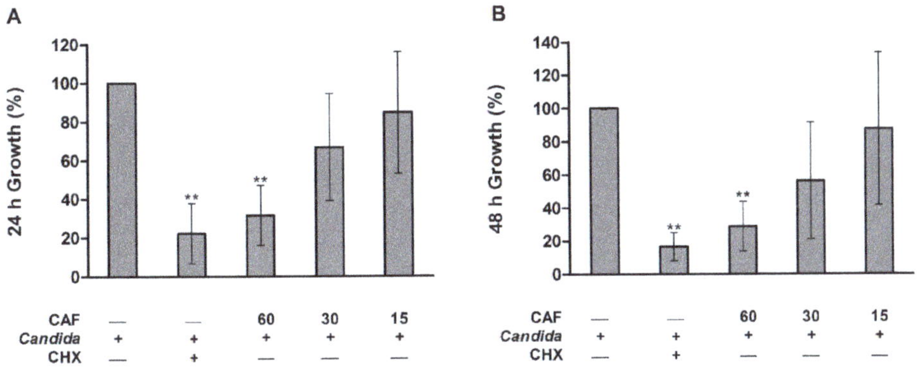

Figure 1. Microdilution test for caffeine (CAF) effects against *C. albicans* growth showing the percentage (%) growth calculated from the measured optical density (OD) of different groups measured at 620 nm. (**A**) After 24 h incubation at 37 °C with broth media. (**B**) After 48 h incubation at 37 °C with broth media. Data are expressed as mean ± SD (n = 6) and statistically analyzed using one-way ANOVA followed by Tukey's post hoc test. The significance was considered when ** $p \leq 0.01$ as compared with control.

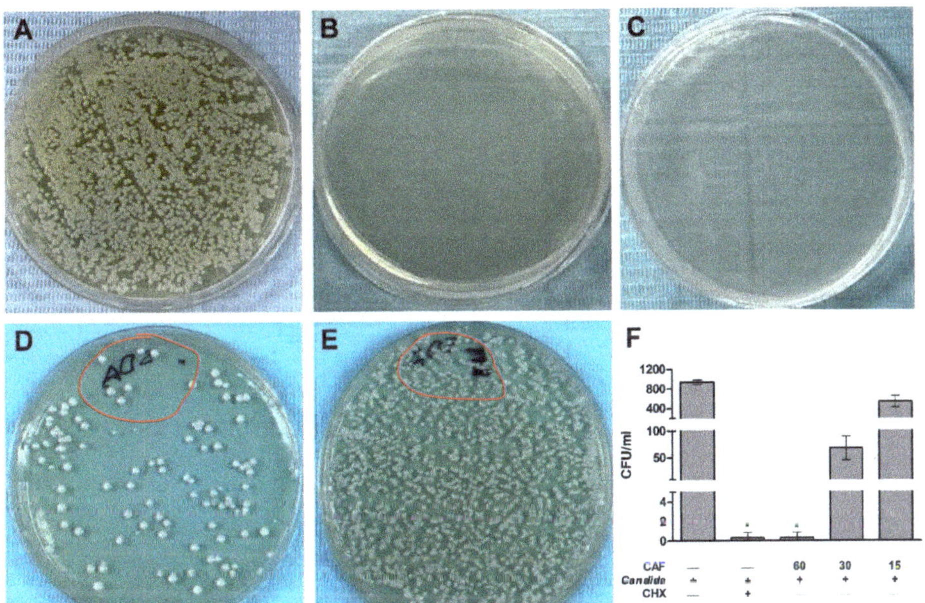

Figure 2. Antifungal Susceptibility Assay of caffeine (CAF) against *C. albicans* growth. (**A**) Fungal cell overgrowth in control group (PBS) containing no treatments. (**B**) Complete inhibition of the fungal cell growth in chlorhexidine (CHX) group. (**C**–**E**) Different concentrations of CAF including 60, 30, and 15 mg/mL, respectively, showing the dose-dependent effects of CAF and the gradual growth inhibition of fungal cells. (**F**) CFU count of different SDA plates. Data are expressed as mean ± SD (n = 3) and statistically analyzed using Kruskal–Wallis test and Dunn's post hoc test. The significance was considered when * $p \leq 0.05$ as compared with control.

The agar-well diffusion test on SDA plates demonstrated no growth inhibition zones in control (PBS), blank (distilled water), and caffeine (15 mg/mL) groups. On the other hand, CHX, caffeine (30 mg/mL) group, and caffeine (60 mg/mL) showed clear zones of *C. albicans* growth inhibition (Figure 3A). The inhibition zones on SDA plates presented as mm, and were significantly ($p \leq 0.05$) higher in CHX and caffeine (60 mg/mL) groups as compared to the control group (Figure 3B). The 30 mg/mL caffeine concentration showed a non-statistically significant increase in the growth inhibition zones as compared to the control group (Figure 3B).

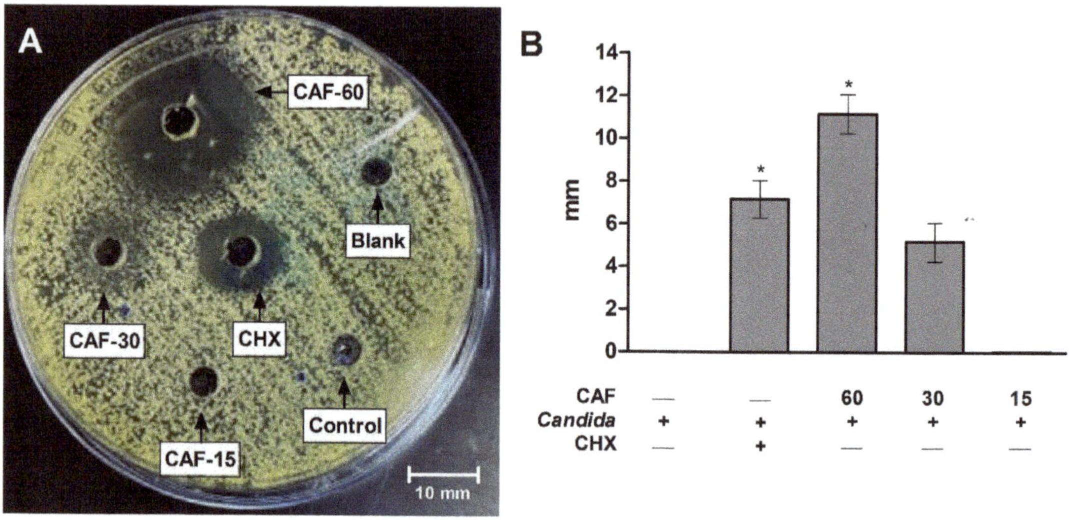

Figure 3. Agar-well diffusion method for caffeine (CAF) effects against *C. albicans* growth. (**A**) Different concentrations of CAF including 15 mg/mL, 30 mg/mL, and 60 mg/mL were tested. The higher two concentrations of CAF (30 and 60 mg/mL) showed the most prominent inhibition of the fungal growth as compared with control containing PBS alone, blank containing distilled water, and standard therapy containing chlorhexidine (CHX). (**B**) Measurements of zone of inhibition (mm). Data are expressed as mean ± SD (n = 3) and statistically analyzed using Kruskal–Wallis test and Dunn's post hoc test. The significance was considered when * $p \leq 0.05$ as compared with control.

The Live/Dead yeast viability test revealed the prominent effects of caffeine against *C. albicans* growth. The fluorescent microscope images showed a low number of fungal cells stained in light green with FUN-1 (Figure 4A), whereas a high number of cells stained red with Calcofluor White M2R were present (Figure 4D) in the control group without treatments. CHX group demonstrated higher dead fungal cells stained in green (Figure 4B) with markedly lower live red-stained cells (Figure 4E) when compared to the control group. The higher dose of caffeine (60 mg/mL) showed increased number of dead green-stained fungal cells (Figure 4C) and reduced count of the red-stained live cells (Figure 4F).

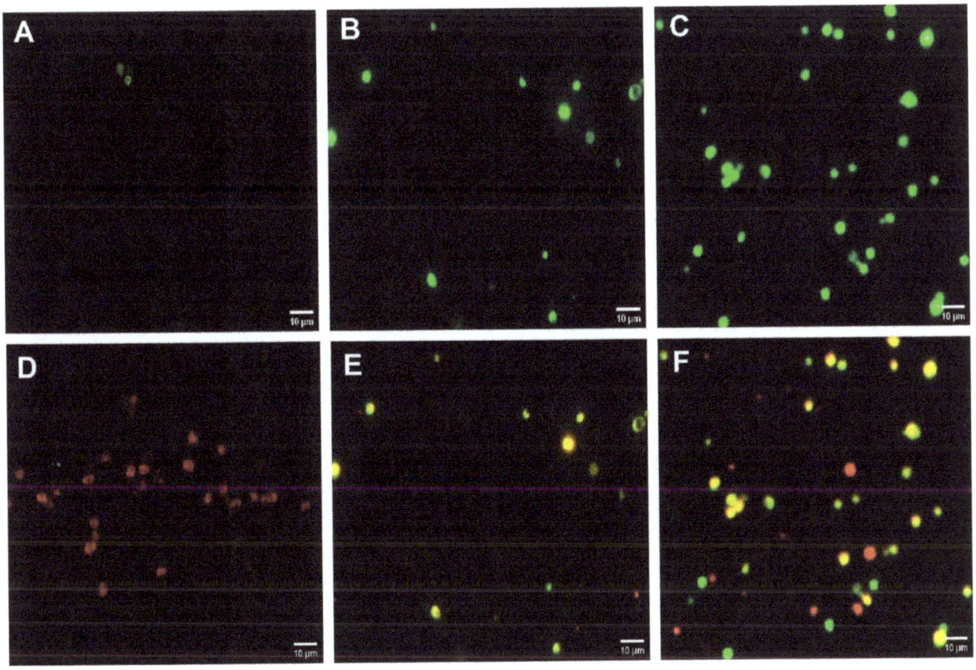

Figure 4. Live/Dead yeast viability method for caffeine (CAF) effects against *C. albicans* growth. The highest concentration of CAF (60 mg/mL) was tested. (**A–C**) Image showing dead fungal cells stained in green without treatments (**A**), with CHX treatment (**C**), and with CAF (60 mg/mL) treatment. (**D–F**) Image showing merged dead/live fungal cells stained in yellow-green or red-orange, respectively, after no treatments (**D**), after treatment with CHX (**E**), and after treatment with CAF (60 mg/mL) (**F**). Full plasma membrane integrity and metabolic function of yeast are needed to convert the yellow-green fluorescent into red-orange color.

Caffeine's effect was assessed against the adherent *C. albicans* to the PMMA discs using the microdilution test (Figure 5). The optical density (OD) measured at 620 nm or 410 nm revealed a significant ($p \leq 0.01$) reduction in adherent *C. albicans* count in the CHX group as compared to the control group (Figure 5). Similarly, the higher dose of caffeine (60 mg/mL) showed a significant ($p \leq 0.05$) decrease in adherent *C. albicans* count as compared to the control group at both weave lengths, whereas the 30 mg/mL dose significantly ($p \leq 0.05$) reduced the attached fungal count compared to the control group when measured at 620 nm (Figure 5).

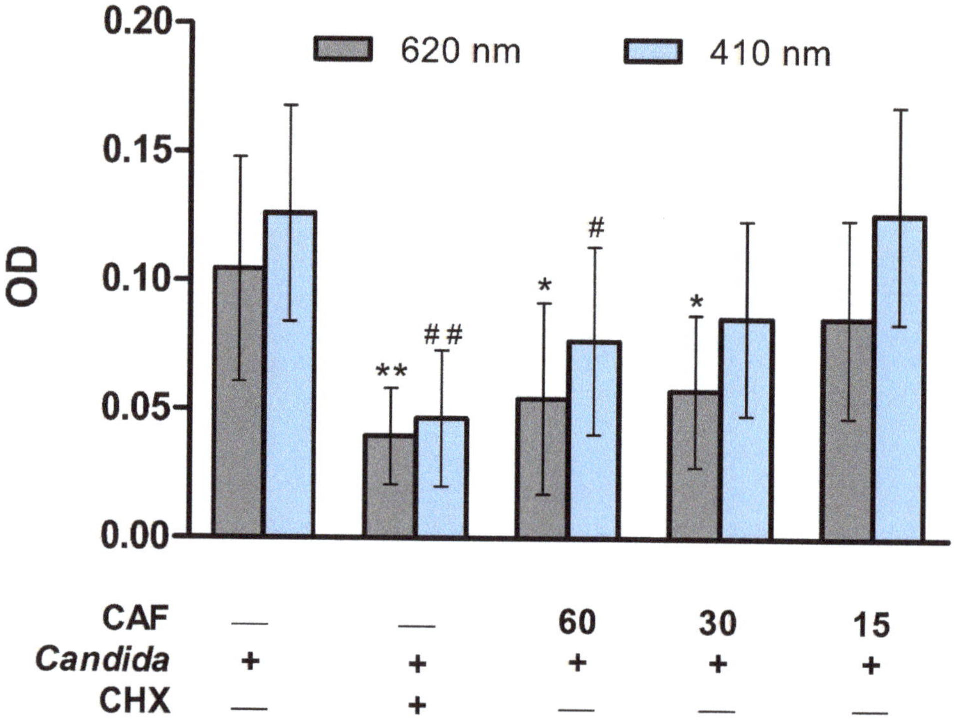

Figure 5. Microdilution test for caffeine (CAF) effects against the growth of *C. albicans* adherent to the PMMA discs showing the optical density (OD) of different groups measured at 620 nm and 410 nm. Data are expressed as mean ± SD (n = 10) and statistically analyzed using one-way ANOVA followed by Tukey's post hoc test. The significance was considered when * $p \leq 0.05$ and ** $p \leq 0.01$ as compared with control at 620 nm or # $p \leq 0.05$ and ## $p \leq 0.01$ as compared with control at 410 nm.

SDA plates of the CFU test are presented in Figure 6. The control group with no treatments showed the highest adherent *C. albicans* colony count (Figure 6A), whereas other plates from CHX (Figure 6B) and caffeine groups (Figure 6C–E) demonstrated a markedly lower colony count. The adherent *C. albicans* colony count on the PMMA discs, presented as CFU/mL, was significantly ($p \leq 0.05$) reduced by CHX and caffeine (60 mg/mL) treatments as compared to the control group (Figure 6F). Other concentrations of caffeine (30 and 15 mg/mL) showed a non-statistically significant reduction of the adherent *C. albicans* count as compared to the control group (Figure 6F).

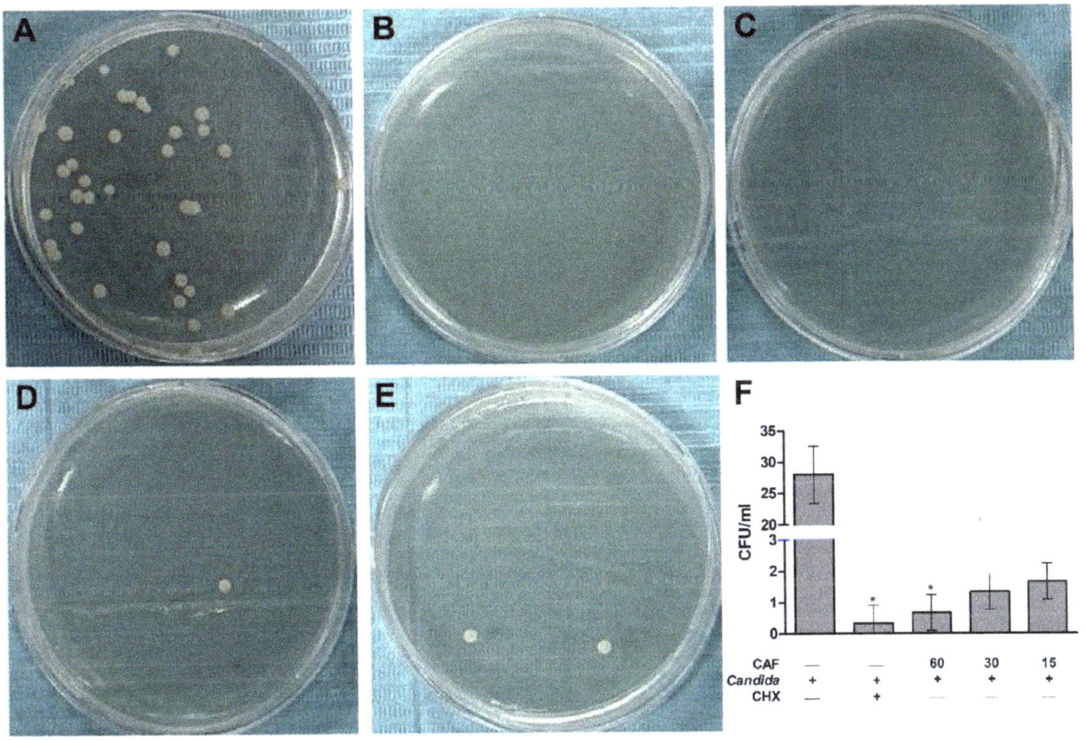

Figure 6. Colony-forming unit (CFU) method for caffeine (CAF) effects against the growth of *C. albicans* adherent to the PMMA discs. (**A**) Fungal cell growth in control group (PBS) containing no treatments. (**B**) Complete inhibition of the fungal cell growth in chlorhexidine (CHX) group. (**C–E**) Different descending concentrations of CAF including 60, 30, and 15 mg/mL, respectively, showing the dose-dependent effects of CAF and the gradual growth inhibition of fungal cells. (**F**) CFU count of different SDA plates. Data are expressed as mean ± SD (n = 3) and statistically analyzed using Kruskal–Wallis test and Dunn's post hoc test. The significance was considered when * $p \leq 0.05$ as compared with control.

Results of the SEM revealed adherent *C. albicans* cells to the surface of PMMA discs in the control group (Figure 7A), whereas this number was markedly reduced in the standard therapy group by CHX (Figure 7B). Images of the SEM also showed a lower count for the *C. albicans* cells attached to the surface of PMMA discs in caffeine groups. The effect of caffeine was dose-dependent, at which the 60 mg/mL dose demonstrated the most prominent reduction in adherent fungal cell count (Figure 7C), which was equivalent to the CHX group compared to the control group. Other concentrations of caffeine (30 and 15 mg/mL) also reduced the adherent *C. albicans* count as compared to the control group (Figure 7D,E).

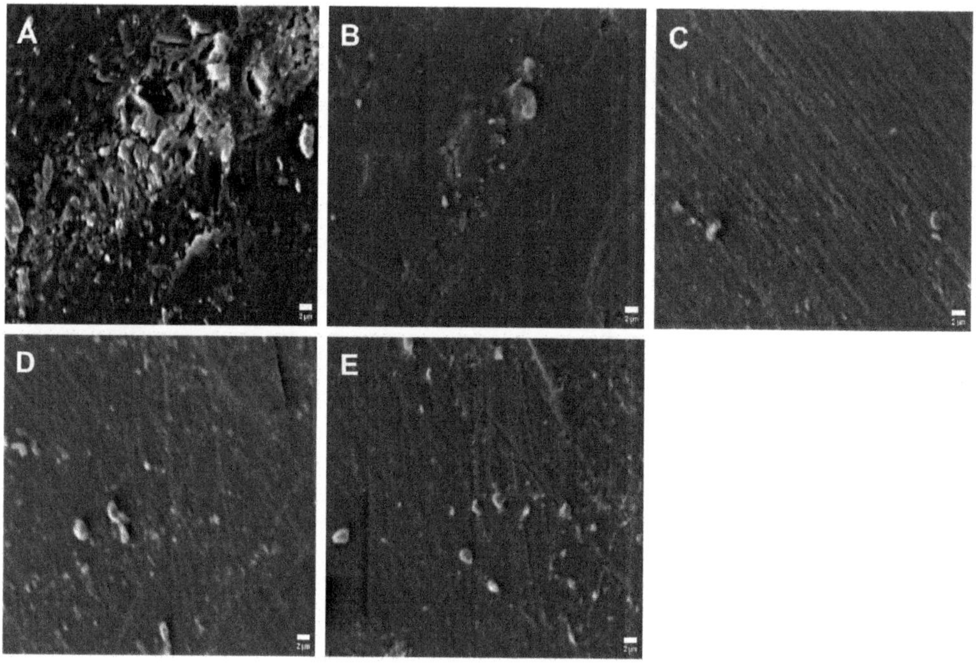

Figure 7. Scanning electron microscope (SEM) images showing the effects of caffeine (CAF) against the growth of *C. albicans* adherent to the PMMA discs at X5000. (**A**) Adherent fungal cells in control group (PBS) containing no treatments. (**B**) Limited number of the adherent fungal cells in chlorhexidine (CHX) group. (**C–E**) Different descending concentrations of CAF including 60, 30, and 15 mg/mL, respectively, showing the dose-dependent effects of CAF and the gradual reduction of the adherent fungal cells on the PMMA discs.

4. Discussion

C. albicans can be found in abundance on the surface of acrylic denture resin. Among the >700 microbial species that colonize the oral cavity, this fungus is considered the most common contributing factor in the pathogenesis of denture stomatitis [27]. Studies have reported that denture stomatitis is prevalent in almost 70% of denture wearers, which requires strict adherence to oral health instruction [28]. Accordingly, it could be anticipated that the daily consumed beverages might influence denture stomatitis and its caustic factors. Caffeine is a major component of coffee. It is also an active constituent in cocoa beverages, tea leaves, soft drinks, and chocolate-related products [29]. An average daily intake of 2 coffee cups is almost equivalent to 180 mg/d of caffeine [30]. The present study reinforced, using multiple techniques, that caffeine has a prominent direct antifungal property against *C. albicans*. Additionally, our findings demonstrated for the first time the influence of caffeine alone on *C. albicans* adherent to the PMMA acrylic denture resin discs, which was not documented in previous studies. PMMA is a globally utilized material in modern dentistry and is commonly used to fabricate dental prosthetics (such as dentures), artificial teeth, and orthodontic appliances. Therefore, PMMA discs represent the most suitable in vitro model for denture stomatitis [31]. Caffeine had antifungal activities and decreased *C. albicans* adhesion to PMMA denture base resin. Therefore, the null hypothesis was rejected.

In the present study, caffeine antifungal effects, particularly at the 60 mg/mL concentration, were comparable with the standard CHX therapy, indicating its prominent effects. Caffeine inhibited the growth of *C. albicans* after 24 and 48 h of incubation. The antimi-

The present work considered only the artificial saliva during the adherent *Candida* biofilm assay. The impact of salivary composition as well as pH was not studied. Additionally, the surface properties of the PMMA discs were not evaluated. Other surface properties or modified resin structures might influence the caffeine effects on the adherent *Candida* biofilm. The effect of pure caffeine was assessed in this study, which does not exactly mimic the condition case of daily beverages. Other excipients in the daily beverages should be also considered. These limitations might be considered in future research. Further studies are crucial to investigate the antifungal and antibiofilm effect of caffeine on different types of dentures used by edentulous patients and could be a promising solution to decrease the number of oral candidiasis cases. Moreover, in terms of denture base properties, the effects of caffeine with different concentrations on the strength of denture base resins are required.

5. Conclusions

The conscious use of a disinfectant is a fundamental element to prevent oral candidiasis infection. Taken together, the findings of the present in vitro study reinforced the antifungal properties of different caffeine concentrations against *C. albicans*. The 60 mg/mL concentration showed the most prominent effects that were comparable with standard CHX therapy. Caffeine also could reduce the antibiofilm activity of *C. albicans* and decrease the adhesion of the fungus on the PMMA acrylic denture base, which supports its potential use as an alternative disinfectant solution for fungal biofilms on denture surfaces.

Author Contributions: D.M.A.: methodology; validation; investigation; writing—original draft; supervision. H.M.A.: conceptualization; formal analysis; writing—original draft; visualization. M.M.G.: resources; writing—review and editing. M.H.A.: investigation; funding acquisition. A.S.B. and K.S.A.: resources; writing—review and editing. M.M.M.: writing—review and editing; supervision. All authors have read and agreed to the published version of the manuscript.

Funding: This study was funded by a grant from the Deanship of Scientific Research, Imam Abdulrahman Bin Faisal University, Saudi Arabia (Project no. 2020-035-Dent).

Institutional Review Board Statement: Not applicable.

Informed Consent Statement: Not applicable.

Data Availability Statement: All data supporting the findings of this study are available within the manuscript.

Acknowledgments: The authors are thankful to Omar Omar for his valuable assistance during the SEM imaging step and to Badr Saqr for his technical assistance during the coating step in the SEM procedure.

Conflicts of Interest: The authors declare no conflict of interest.

References

1. Lohse, M.B.; Gulati, M.; Johnson, A.D.; Nobile, C.J. Development and regulation of single- and multi-species *Candida albicans* biofilms. *Nat. Rev. Microbiol.* **2018**, *16*, 19–31. [CrossRef] [PubMed]
2. Talapko, J.; Juzbašić, M.; Matijević, T.; Pustijanac, E.; Bekić, S.; Kotris, I.; Škrlec, I. *Candida albicans*-The Virulence Factors and Clinical Manifestations of Infection. *J. Fungi* **2021**, *7*, 79. [CrossRef] [PubMed]
3. Gulati, M.; Nobile, C.J. *Candida albicans* biofilms: Development, regulation, and molecular mechanisms. *Microbes Infect.* **2016**, *18*, 310–321. [CrossRef]
4. Tsui, C.; Kong, E.F.; Jabra-Rizk, M.A. Pathogenesis of *Candida albicans* biofilm. *Pathog. Dis.* **2016**, *74*, ftw018. [CrossRef] [PubMed]
5. Cavalheiro, M.; Teixeira, M.C. Candida Biofilms: Threats, Challenges, and Promising Strategies. *Front. Med.* **2018**, *5*, 28. [CrossRef]
6. Chandra, J.; Mukherjee, P.K. Candida Biofilms: Development, Architecture, and Resistance. *Microbiol. Spectr.* **2015**, *3*, 3–4. [CrossRef]
7. Kucharíková, S.; Neirinck, B.; Sharma, N.; Vleugels, J.; Lagrou, K.; Van Dijck, P. In vivo *Candida glabrata* biofilm development on foreign bodies in a rat subcutaneous model. *J. Antimicrob. Chemother.* **2015**, *70*, 846–856. [CrossRef]
8. CDC. Candida Infections of the Mouth, Throat, and Esophagus. Available online: https://www.cdc.gov/fungal/diseases/candidiasis/thrush/index.html (accessed on 14 May 2022).
9. Zahra, N.; Hossein, Y.M. Antifungal Activity of Caffeine in Combination with Fluconazole Againstcandida Albicans. *Infect. Epidemiol. Microbiol. (Infect. Epidemiol. Med.)* **2016**, *2*, 18–21. [CrossRef]

10. Alikhani, T.; Daie Ghazvini, R.; Mirzaii, M.; Hashemi, S.J.; Fazli, M.; Rafat, Z.; Roostaei, D.; Ardi, P.; Kamali Sarvestani, H.; Zareei, M. Drug Resistance and Biofilm Formation in Candida Species of Vaginal Origin. *Iran. J. Public Health* **2022**, *51*, 913–918. [CrossRef]
11. Revie, N.M.; Iyer, K.R.; Robbins, N.; Cowen, L.E. Antifungal drug resistance: Evolution, mechanisms and impact. *Curr. Opin. Microbiol.* **2018**, *45*, 70–76. [CrossRef]
12. Cowen, L.E.; Sanglard, D.; Howard, S.J.; Rogers, P.D.; Perlin, D.S. Mechanisms of Antifungal Drug Resistance. *Cold Spring Harb. Perspect. Med.* **2014**, *5*, a019752. [CrossRef] [PubMed]
13. Fisher, M.C.; Alastruey-Izquierdo, A.; Berman, J.; Bicanic, T.; Bignell, E.M.; Bowyer, P.; Bromley, M.; Brüggemann, R.; Garber, G.; Cornely, O.A.; et al. Tackling the emerging threat of antifungal resistance to human health. *Nat. Rev. Microbiol.* **2022**, *20*, 557–571. [CrossRef] [PubMed]
14. Khan, A.; Azam, M.; Allemailem, K.S.; Alrumaihi, F.; Almatroudi, A.; Alhumaydhi, F.A.; Ahmad, H.I.; Khan, M.U.; Khan, M.A. Coadministration of Ginger Extract and Fluconazole Shows a Synergistic Effect in the Treatment of Drug-Resistant Vulvovaginal Candidiasis. *Infect. Drug Resist.* **2021**, *14*, 1585–1599. [CrossRef] [PubMed]
15. Brookes, Z.L.S.; Bescos, R.; Belfield, L.A.; Ali, K.; Roberts, A. Current uses of chlorhexidine for management of oral disease: A narrative review. *J. Dent.* **2020**, *103*, 103497. [CrossRef]
16. Poppolo, D.F.; Ouanounou, A. Chlorhexidine in Dentistry: Pharmacology, Uses, and Adverse Effects. *Int. Dent. J.* **2022**, *72*, 269–277. [CrossRef]
17. Shrestha, A.; Rimal, J.; Rao, A.; Sequeira, P.S.; Doshi, D.; Bhat, G.K. In vitro antifungal effect of mouth rinses containing chlorhexidine and thymol. *J. Dent. Sci.* **2011**, *6*, 1–5. [CrossRef]
18. Al Reef, T.; Ghanem, E. Caffeine: Well-known as psychotropic substance, but little as immunomodulator. *Immunobiology* **2018**, *223*, 818–825. [CrossRef]
19. Açıkalın, B.; Sanlier, N. Coffee and its effects on the immune system. *Trends Food Sci. Technol.* **2021**, *114*, 625–632. [CrossRef]
20. Alfaifi, A.A.; Lin, W.S.; Aldhaian, B.A.; Levon, J.A.; Gregory, R.L. Impact of caffeine on metabolic activity and biofilm formation of *Candida albicans* on acrylic denture resin in the presence of nicotine. *J. Prosthet. Dent.* **2020**, *123*, 875–879. [CrossRef]
21. Raut, J.S.; Chauhan, N.M.; Shinde, R.B.; Karuppayil, S.M. Inhibition of planktonic and biofilm growth of *Candida albicans* reveals novel antifungal activity of caffeine. *J. Med. Plants Res.* **2013**, *13*, 777–782. [CrossRef]
22. CLSI. Chapter 3. Antifungal broth dilutions susceptibility testing process for yeasts. In *M27: Reference Method for Broth Dilution Antifungal Susceptibility Testing of Yeasts*, 4th ed.; Alexander, B.D., Ed.; Clinical Laboratory Standards Institute: Wayne, PA, USA, 2017.
23. Balouiri, M.; Sadiki, M.; Ibnsouda, S.K. Methods for in vitro evaluating antimicrobial activity: A review. *J. Pharm. Anal.* **2016**, *6*, 71–79. [CrossRef] [PubMed]
24. Cantón, E.; Espinel-Ingroff, A.; Pemán, J. Trends in antifungal susceptibility testing using CLSI reference and commercial methods. *Expert Rev. Anti-Infect. Ther.* **2009**, *7*, 107–119. [CrossRef] [PubMed]
25. Fouda, S.M.; Gad, M.M.; Ellakany, P.; Al Ghamdi, M.A.; Khan, S.Q.; Akhtar, S.; Al Eraky, D.M.; Al-Harbi, F.A. Effect of Low Nanodiamond Concentrations and Polymerization Techniques on Physical Properties and Antifungal Activities of Denture Base Resin. *Polymers* **2021**, *13*, 4331. [CrossRef] [PubMed]
26. Mota, S.; Alves, R.; Carneiro, C.; Silva, S.; Brown, A.J.; Istel, F.; Kuchler, K.; Sampaio, P.; Casal, M.; Henriques, M.; et al. Candida glabrata susceptibility to antifungals and phagocytosis is modulated by acetate. *Front. Microbiol.* **2015**, *6*, 919. [CrossRef]
27. Wu, T.; Shi, W.; Loewy, Z.G.; He, X. Managing denture biofilm related diseases. *Dent. Open J.* **2015**, *2*, 80–86. [CrossRef]
28. Akpan, A.; Morgan, R. Oral candidiasis. *Postgrad. Med. J.* **2002**, *78*, 455–459. [CrossRef]
29. Daly, J.W. Caffeine analogs: Biomedical impact. *Cell Mol. Life Sci.* **2007**, *64*, 2153–2169. [CrossRef]
30. Temple, J.L.; Bernard, C.; Lipshultz, S.E.; Czachor, J.D.; Westphal, J.A.; Mestre, M.A. The Safety of Ingested Caffeine: A Comprehensive Review. *Front. Psychiatry* **2017**, *8*, 80. [CrossRef]
31. Lee, M.J.; Kim, M.J.; Oh, S.H.; Kwon, J.S. Novel Dental Poly (Methyl Methacrylate) Containing Phytoncide for Antifungal Effect and Inhibition of Oral Multispecies Biofilm. *Materials* **2020**, *13*, 371. [CrossRef]
32. Avizheh, L.; Peirouvi, T.; Diba, K.; Fathi-Azarbayjani, A. Electrospun wound dressing as a promising tool for the therapeutic delivery of ascorbic acid and caffeine. *Ther. Deliv.* **2019**, *10*, 757–767. [CrossRef]
33. Islahudin, F.; Khozoie, C.; Bates, S.; Ting, K.N.; Pleass, R.J.; Avery, S.V. Cell wall perturbation sensitizes fungi to the antimalarial drug chloroquine. *Antimicrob. Agents Chemother.* **2013**, *57*, 3889–3896. [CrossRef] [PubMed]
34. Mittag, H. Structural alterations in *Candida albicans* by caffeine and caffeine salts. *Mycoses* **1994**, *37*, 337–341. [CrossRef] [PubMed]
35. Sabie, F.T.; Gadd, G.M. Effect of nucleosides and nucleotides and the relationship between cellular adenosine 3′:5′-cyclic monophosphate (cyclic AMP) and germ tube formation in *Candida albicans*. *Mycopathologia* **1992**, *119*, 147–156. [CrossRef] [PubMed]
36. Sarachek, A.; Henderson, L.A. Recombinagenicity of caffeine for *Candida albicans*. *Mycopathologia* **1990**, *110*, 63–76. [CrossRef]
37. Bazzaz, B.S.F.; Fakori, M.; Khameneh, B.; Hosseinzadeh, H. Effects of Omeprazole and Caffeine Alone and in Combination with Gentamicin and Ciprofloxacin Against Antibiotic Resistant *Staphylococcus aureus* and *Escherichia coli* Strains. *J. Pharmacopunct.* **2019**, *22*, 49–54. [CrossRef]

38. Kang, T.M.; Yuan, J.; Nguyen, A.; Becket, E.; Yang, H.; Miller, J.H. The aminoglycoside antibiotic kanamycin damages DNA bases in Escherichia coli: Caffeine potentiates the DNA-damaging effects of kanamycin while suppressing cell killing by ciprofloxacin in Escherichia coli and Bacillus anthracis. *Antimicrob. Agents Chemother.* **2012**, *56*, 3216–3223. [CrossRef]
39. Cui, W.Q.; Wang, S.T.; Pan, D.; Chang, B.; Sang, L.X. Caffeine and its main targets of colorectal cancer. *World J. Gastrointest. Oncol.* **2020**, *12*, 149–172. [CrossRef]
40. Szabó, K.; Kónya, Z.; Erdődi, F.; Farkas, I.; Dombrádi, V. Dissection of the regulatory role for the N-terminal domain in *Candida albicans* protein phosphatase Z1. *PLoS ONE* **2019**, *14*, e0211426. [CrossRef]
41. Ádám, C.; Erdei, É.; Casado, C.; Kovács, L.; González, A.; Majoros, L.; Petrényi, K.; Bagossi, P.; Farkas, I.; Molnar, M.; et al. Protein phosphatase CaPpz1 is involved in cation homeostasis, cell wall integrity and virulence of *Candida albicans*. *Microbiology (Reading)* **2012**, *158*, 1258–1267. [CrossRef]
42. Feng, J.; Duan, Y.; Sun, W.; Qin, Y.; Zhuang, Z.; Zhu, D.; Sun, X.; Jiang, L. CaTip41 regulates protein phosphatase 2A activity, CaRad53 deactivation and the recovery of DNA damage-induced filamentation to yeast form in *Candida albicans*. *FEMS Yeast Res.* **2016**, *16*, fow009. [CrossRef]
43. Liu, W.; Zhao, J.; Li, X.; Li, Y.; Jiang, L. The protein kinase CaSch9p is required for the cell growth, filamentation and virulence in the human fungal pathogen *Candida albicans*. *FEMS Yeast Res.* **2010**, *10*, 462–470. [CrossRef] [PubMed]
44. Kunze, D.; MacCallum, D.; Odds, F.C.; Hube, B. Multiple functions of DOA1 in *Candida albicans*. *Microbiology (Reading)* **2007**, *153*, 1026–1041. [CrossRef] [PubMed]
45. Hernáez, M.L.; Gil, C.; Pla, J.; Nombela, C. Induced expression of the *Candida albicans* multidrug resistance gene CDR1 in response to fluconazole and other antifungals. *Yeast* **1998**, *14*, 517–526. [CrossRef]
46. Stichternoth, C.; Fraund, A.; Setiadi, E.; Giasson, L.; Vecchiarelli, A.; Ernst, J.F. Sch9 kinase integrates hypoxia and CO2 sensing to suppress hyphal morphogenesis in Candida albicans. *Eukaryot. Cell* **2011**, *10*, 502–511. [CrossRef]
47. Sanglard, D.; Ischer, F.; Marchetti, O.; Entenza, J.; Bille, J. Calcineurin A of *Candida albicans*: Involvement in antifungal tolerance, cell morphogenesis and virulence. *Mol. Microbiol.* **2003**, *48*, 959–976. [CrossRef]
48. Maver-Biscanin, M.; Mravak-Stipetic, M.; Jerolimov, V. Effect of low-level laser therapy on *Candida albicans* growth in patients with denture stomatitis. *Photomed. Laser Surg.* **2005**, *23*, 328–332. [CrossRef]
49. Kathwate, G.H.; Karuppayil, S.M. Antifungal properties of the anti-hypertensive drug: Aliskiren. *Arch. Oral Biol.* **2013**, *58*, 1109–1115. [CrossRef]
50. Mendonça e Bertolini, M.; Cavalcanti, Y.W.; Bordin, D.; Silva, W.J.; Cury, A.A. Candida albicans biofilms and MMA surface treatment influence the adhesion of soft denture liners to PMMA resin. *Braz. Oral Res.* **2014**, *28*, 61–66. [CrossRef]
51. Meirowitz, A.; Rahmanov, A.; Shlomo, E.; Zelikman, H.; Dolev, E.; Sterer, N. Effect of Denture Base Fabrication Technique on *Candida albicans* Adhesion In Vitro. *Materials* **2021**, *14*, 221. [CrossRef]

biomedicines

MDPI

Article

Expression of Claudin-9 (CLDN9) in Breast Cancer, the Clinical Significance in Connection with Its Subcoat Anchorage Proteins ZO-1 and ZO-3 and Impact on Drug Resistance

Xinguo Zhuang [1,2], Tracey A. Martin [1], Fiona Ruge [1], Jianyuan (Jimmy) Zeng [1], Xinyu (Amber) Li [1], Elyas Khan [3], Qingping Dou [1,3], Eleri Davies [4] and Wen G. Jiang [1,*]

1 School of Medicine, Cardiff University, Cardiff CF14 4XN, UK; zhuangx5@cardiff.ac.uk (X.Z.); martinta1@cardiff.ac.uk (T.A.M.); ruge@cardiff.ac.uk (F.R.); zengj8@cardiff.ac.uk (J.Z.); lix193@cardiff.ac.uk (X.L.); douq1@cardiff.ac.uk or doup@karmanos.org (Q.D.)
2 Department of Clinical Laboratory, The First Affiliated Hospital of Xiamen University, School of Medicine, Xiamen University, Xiamen 361003, China
3 Karmanos Cancer Institute, Department of Oncology, School of Medicine, Wayne State University, Detroit, MI 48201, USA; elyaskhan882@gmail.com
4 Wales Breast Centre, University Llandough Hospital, Cardiff and Vales University Health Board, Cardiff CF64 2XX, UK; eleri.davies8@wales.nhs.uk
* Correspondence: jiangw@cardiff.ac.uk

Abstract: (1) Introduction: Claudin-9 (CLDN9) is a member of the claudin protein family, a critical transmembrane protein family for tight junctions that are implemented in the progression of numerous cancer types. The present study investigated the role that CLDN9, along with the subcoat proteins, Zonula Occludens (ZOs), plays in clinical breast cancer and subsequent impact on drug response of patients. (2) Methods: CLDN9 protein and *CLDN9* transcript were determined and correlated with clinical and pathological indicators, together with the status of hormonal receptors. The levels of *CLDN9* transcript were also assessed against the therapeutic responses of the patients to chemotherapies by using a dataset from the TCGA database. Breast cancer cell models, representing different molecular subtypes of breast cancer, with differential expression of CLDN9 were created and used to assess the biological impact and response to chemotherapeutic drugs. (3) Results: Breast cancer tissues expressed significantly higher levels of the *CLDN9*, with the high levels being associated with shorter survival. CLDN9 was significantly correlated with its anchorage proteins ZO-1 and ZO-3. Integrated expression of CLDN9, ZO-1 and ZO-3 formed a signature that was significantly linked to overall survival (OS) ($p = 0.013$) and relapse-free survival (RFS) ($p = 0.024$) in an independent matter. *CLDN9* transcript was significantly higher in patients who were resistant to chemotherapies ($p < 0.000001$). CLDN9 connection to chemoresistance was particularly prominent in patients of ER-positive (ER(+)), Her-2-negative((Her-2(−)), ER(+)/Her-2(−) and triple-negative breast cancers (TNBCs), but not in patients with HER-2-positive tumors. In Her-2-negative MCF7 and MDA-MB-231 cancer cells, loss of CLDN9 significantly increased sensitivity to several chemotherapeutic drugs including paclitaxel, gemcitabine and methotrexate, which was not seen in Her-2(+) SKBR3 cells. However, suppressing Her-2 using neratinib, a permanent Her-2 inhibitor, sensitized cellular response to these chemodrugs in cells with CLDN9 knockdown. (4) Conclusions: CLDN9 is an important prognostic indicator for patients with breast cancer and also a pivotal factor in assessing patient responses to chemotherapies. Her-2 is a negating factor for the treatment response prediction value by CLDN9 and negating Her-2 and CLDN9 may enhance breast cancer cellular response to chemotherapeutic drugs.

Keywords: claudin; CLDN9; ZO-1; ZO-3; breast cancer; prognosis; chemotherapies; chemoresistance; cytotoxicity; Her-2; neratinib; estrogen receptor

1. Introduction

Claudins (CLDNs) are a family of small transmembrane proteins that are key elements for tight junctions. This protein family has twenty-four members each sharing a similar protein structure that has four membrane-spanning domains with both the C- and N-terminus located in the cytoplasmic region. Intracellularly, the N-terminus anchors the claudin protein to the cytoskeleton via interaction with proteins such as the Zonula Occludens (ZOs). Both termini are also involved in mediating cell signaling. Claudins, by interacting with the same or different claudins on the other cells via the extracellular domains, form a key part of the control mechanism of paracellular permeability and hence the function of tight junctions. The role of claudins in pathological conditions particularly in cancers including breast cancer have been explored in recent years. For example, CLDN1, CLDN3, CLDN7, CLDN16 and CLDN20 have been found to be downregulated in breast cancer, and the downregulation of these claudins appearing to be linked to a more aggressive phenotype and with poor clinical outcome [1–8]. In contrast, CLDN3, CLDN4 and CLDN5 appear to be overexpressed in breast cancers [9–13]. However, determining whether claudins play a tumor-suppressive or oncogenic role is dependent on both the individual claudin and the type of cancer. Perhaps the most well-studied claudin in breast cancer is CLDN4, which has been found to be highly aberrant and has a role to define a subgroup of breast cancer [2,14–17]. As well as an indicator for disease progression and prognosis in various cancers, some claudins have also been shown to be indicators for cancer cell response to anticancer drugs. For example, CLDN4 overexpression, an indicator of poor clinical outcome in patients with ovarian cancer, is an indicator of resistance to cisplatin in this cancer type [18,19], and a similar role for CLDN6 is seen in cervical adenocarcinoma [20]. In breast cancer cells such as MCF-7, CLDN6 expression confers cellular resistance to drugs such as 5-fluouracil (5-FU) and adriamycin [21]. From a clinical point of view, high CLDN2 in colorectal cancer appears to be linked to poor outcome in those receiving 5-FU treatment [22] and CLDN1 and CLDN7 appear to form a claudin signature to predict the clinical response to chemotherapies in colorectal cancer [23]. The CLDN2 link may be via the paracellular passage of drugs in lung cancer cell models [24,25]. In breast cancer and particularly in triple-negative breast cancer (TNBC), CLDN1 is a good indicator for cancer cell response to paclitaxel, doxorubicin and 5-FU [26]. However, in colorectal cancer, CLDN1 often becomes upregulated following chemotherapies, and overexpression of CLDN1 confers cells' resistance to oxaliplatin [27].

Compared with other members, CLDN9 is one of less well-studied claudins. CLDN9 protein is encoded by the *CLDN9* gene located on chromosome 16p13.3 and is 217 amino acids in size (22.8 kDa). It was first reported as a gene product similar to a gene in rats called Rat Ventral Protein.1 (RVP.1) [28,29]. Mouse studies revealed that this gene is required for the preservation of sensory cells in ears and the gene deficiency is associated with deafness [30,31].

CLDN9 is known to be a key binding protein for pathogens. It is a coreceptor to hepatitis C virus [32,33], and is also a receptor for the *Clostridium perfringens* enterotoxin (CpE) in the gut [34]. In gastric cancer, strong CLDN9 protein staining in the tissues of diffuse type is associated with high mortality [35], a finding seemingly replicable in a cell line of gastric cancer [36]. CLDN9 is generally well expressed in the pituitary gland and pituitary oncocytomas and it was seen to be highly linked to the invasive subtype of the tumors [37,38]. In cervical cancers, CLDN9, amongst other claudins, is seen to associate with lymphatic invasion [39] and is one of the few genes in endometrial cancer to predict a survival of the patients [40,41]. CLDN9 was found to be linked with metastasis of in vivo lung cancer models and with cell migration and invasiveness in vitro [42].

Studies on CLDN9 in breast cancer is rather rare. It has been reported to be low or negative along with CDLN6, CLDN12 and CLDN13 in breast tissues [43]. We have previously studied the role of claudins and tight junction molecules including CLDN19, CLDN20, and ZOs in breast cancer [6,7,44]. In the present study, we identified CLDN9 as potentially an important factor in assessing the clinical outcome of the patients and also patient response to drug treatment, and that this connection appears to be hormone receptor-dependent. We further developed cell models expressing differential CLDN9 expression levels and different hormonal receptor status, to confirm that in Her-2-positive breast cancer cells (SKBR3), knockdown of CLDN9 rendered the cells more sensitive to chemodrugs when Her-2 inhibitor is present. Collectively, CLDN9 is a prognostic indicator for breast cancer and a predictor for patient therapeutic response in Her-2-negative breast cancers.

2. Materials and Methods

2.1. Cell Lines

The following human breast cancer cell lines, MCF7, MDA-MB-231 and SKBR3, were purchased from ATCC (American Type of Cell culture) (purchased via LGC Standards, Teddinton, England, UK) and cultured in Dubecco's Modified Eagle Medium (DMEM) with 10% foetal calf serum (FCS) (Sigma-Aldrich, Dorset, UK) and 1 × antimicrobial solutions (Sigma-Aldrich, Dorset, England, UK).

2.2. Mammary Tissue Cohort

Breast cancer tissue and background normal tissues (n = 127, and n = 33) were collected immediately after surgical removal of breast cancer at the University Hospital of Wales, as reported previously [45]. None of the patients received chemotherapy prior to surgery. Written informed consent was required and obtained from patients, with a follow-up study with a median follow-up period of 120 months conducted after the surgery. Pathological, clinical and follow-up information were obtained from clinical records and used for subgroup analyses. Of the all the patients, nine patients died of reasons unrelated to breast cancer and were excluded from the subgroup analysis. Sample tissue was sectioned using a cryostat (Leica CM1950) (Leica Biosystems Ltd., Newcastle, England, UK). The samples were collected under ethical approval (Bro Taf Health Authority; ethics approval numbers 01/4303 and 01/4046). Part of the frozen tissue sections were used for routine histological evaluation while the remaining sections were blended and homogenized before being subject to Tri Reagent RNA (Meck Sigma Aldrich, Dorset, UK) extraction for further genetic analysis.

2.3. Key Research Materials

A mouse monoclonal antibody to GAPDH(SC-32233), a goat polyclonal antibody to CLDN9(SC-17672) and a mouse monoclonal antibody to CLDN9(SC-398836) were, respectively, purchased from Santa-Cruz Biotechnologies Inc. (Santa Cruz, CA, USA). A mouse monoclonal antibody to ZO-1(33-9100) was, respectively, purchased from Thermo Fisher Scientific Inc. (Thermo Fisher Scientific, Loughborough, UK). siRNA targeting human CLDN9 was obtained from Santa-Cruz Biotechnologies Inc. Chemo-drugs, including gemcitabine (GEM), docetaxel (DOC), cisplatin (CIS), methotrexate (MTX) and docetaxel were purchased from Sigma-Aldrich (Dorset, UK). A broad-spectrum permanent Her-2 inhibitor, neratinib, was obtained from PUMA Biotechnologies Inc. (Los Angeles, CA, USA). These drugs were dissolved in DMSO, further diluted with DMEM to a desired concentration and stored at −20 °C until use. Other chemicals were purchased from Merck unless otherwise stated.

2.4. Creation of CLDN9 Knockdown Cell Models

Breast cancer cell lines MCF7, MDA MB 231 and SKBR3 were used to create sub-lines with CLDN9 knockdown. Transient knockdown using siRNA (SC-43050) from Santa Cruz Biotechnology Inc., Dallas, TX, USA, was used on the cells. The sequence of siRNA for CLDN9 were as follows: Sense: GAGCAUUUGUAACUGGAAAtt. Antisense: UUUCCAGUUACAAAUGCUCtt. Transfection of cancer cells was carried out using transfection kits (SC-36868 and SC-29528) purchased from Santa Cruz Biotechnology Inc., by following the manufacturer's instructions. The effect of knockdown was verified using PCR, qPCR and Western blotting analyses.

2.5. Analysis of Gene Transcript by PCR and QPCR

Quantitative and qualitative gene transcript analyses were undertaken using real-time quantitative RT-PCR and QPCR by employing the Amplifluor Molecular Beacon system. Reactions were prepared in a MicroAmp fast Optical 96-well plate (Greiner Bio-One Ltd., Gloucestershire, UK) using primers specific to the molecule of interest. In addition to unknown samples, reactions were prepared for a known standard that was run alongside the unknown samples. Once all samples and unknowns were added to the plate, the plate was sealed with optical seals (PrimerDesign, Southampton, UK) and the sample was subjected to a StepOne Plus qPCR system (Thermo Fisher Scientific, Waltham, MA, USA). Relative copy numbers of the samples were calculated as part of the systematic analysis, in accordance with the standard curve, and were subsequently exported to Excel 2019 (Microsoft Inc., Redmond, WT, USA) for further analysis. Qualitative PCR product was separated using 1% agarose and image obtained from a UV imager. Sequences used in the study were as follows: QPCR for *CLDN9*: 5′GTGCCCTCTGTGTCATTG′3 and 5′ACTGAACCTGACCGTACATCCACACACGTGGTACACT′3, ZO-1: 5′CCACATACAGA TACGAGTCCTC′3 and 5′ACTGAACCTGACCGTACAGTAACTGCGTGAATATTGCT′3; ZO-2: 5′CAAAAGAGGATTTGGAATTG′3 and 5′ACTGAACCTGACCGTACAGAGCACA TCAGAAATGACAA′3; ZO-3: 5′CTGACATGGAGGAGCTGA′3 and 5′ACTGAACC TGACCGTACAGCTTAGCTTCCCTTCTGACT′3), GAPDH (5′AAGGTCATCCATGACAA CTT′3 and 5′ACTGAACCTGACCGTACAGCCATCCACAGTCTTCTG′3) and CK19 (5′AGC CACTACTACACGACCAT′3 and 5′ACTGAACCTGACCGTACATCGATCTGCAGGACAA TC′3).

2.6. Western Blotting

Proteins were extracted from cultured cells with RIPA buffer and quantified using a BioRad protein quantitation kit (Bio-Rad Laboratories, Hertfordshire, UK). The samples were treated with 2 × Laemmle sample buffer, boiled for 5 min at 100 °C and then loaded to 12% SDS PAGE gel for electrophoresis. The protein transfer from the gel onto the PVDF membrane that had been preactivated with methanol was subsequently accomplished using a semidry transfer technique. A 10% milk powder solution was utilized for membrane blocking. The blots were incubated with the respective primary antibody to CLDN9 and GAPDH, and then exposed to the secondary antibody that was HRP-conjugated before being visualized with EZ-ECL solution (Geneflow Ltd., Litchfield, UK).

2.7. Cellular Response to Chemotherapy Drugs

Breast cancer cells, with or without CLDN9 knockdown, were seeded into 96-well plates and treated with serially diluted drugs before they were incubated in the incubator. The concentrations of the drugs were, respectively, chosen based on their known IC_{50} and previous studies. After 72 h, the cells were fixed with 4% formalin, stained with 0.5% crystal violet and extracted with 10% acetic acid after washing. The absorbance was measured at 595 nm using a spectrophotometer to detect their respective cell densities. The percentage drug toxicity was calculated as follows:

$$\text{Percentage drug toxicity} = [(\text{Absorbance in untreated well} - \text{Absorbance in drug treated well}) / \text{Absorbance in untreated well}] \times 100$$

The scatterplots of percentage toxicity versus drug concentration were plotted, with the best fit curve used to calculate the respective IC_{50} value.

2.8. Immunohistochemical Staining of CLDN9 Protein

CLDN9 staining was carried out using a breast cancer tissue microarray BR1503f (US Biomax, Inc., Derwood, MD, USA), which had 75 cases of breast cancer tissues (in total, 128 samples were utilized for the analysis). After dewaxing and rehydration, the tissue microarray was subjected to antigen retrieval followed by thorough washing in PBS. After blocking nonspecific binding with 10% horse serum, the primary anti-CLDN9 antibody was added and incubated overnight at 4 °C (final concentration: 2 ug/mL). Following extensive washing in PBS, the secondary antibody and tertiary reagents were added, with biotinylated secondary antibody first followed by avidin–biotin amplification with a commercial kit (Vectastain Elite Universal ABC kit, Vector Laboratories Ltd., Peterborough, UK). After further washing with PBS, the slides were further incubated with 3,3′-diaminobenzidine (DAB) solution to allow for chromogenic detection. Finally, the tissue sections were counterstained with hematoxylin, washed thoroughly in tap water and then subjected to dehydration through a graded series of ethanol, prior to being cleared in xylene and mounted in DPX mounting solution (Sigma-Aldrich, Dorset, UK). Staining was visualized using a Leica DM1000 LED microscope (Leica Biosystems Ltd., Newcastle, England, UK). Negative controls were prepared by omitting the primary antibody and only using the secondary universal antibody from the Vectastain Elite ABC kit. The staining pattern and intensity were evaluated by two independent researchers, as previously reported [46].

2.9. *Immunofluorescence (IFC)*

Pretreatment was carried out with an 8-well chamber slide with the medium in the incubator overnight. 10×10^4 cells were seeded into each well of the chamber slide. After 36 h of incubation, the medium was discarded, and the cells were fixed with 4% formalin. After washing with PBS, the cells were permeabilized with 0.1% Triton × 100 (diluted with PBS) for 5 min. Nonspecific binding was blocked with 8% horse serum (dilute with PBS) for 2 h, and the slide then incubated with primary antibody at a concentration of 1:100 overnight at 4 °C. The slides were subsequently incubated with secondary antibodies tagged with either fluorescein isothiocyanate (FITC, 1:500) or tetramethylrhodamine isothiocyanate (TRITC, 1:500) (Sigma-Aldrich, Dorset, UK), together with 6-Diamidino-2-phenylindole (DAPI, 1:1000) (Merck Millipore, Watford, UK). The slides were washed and mounted with FluoSave (Calbiochem, Nottingham, England, UK) in preparation for photographing. Images were captured using an Olympus microscope and photographed with a Hamamatsu digital camera.

2.10. *Patients' Response to Chemotherapies and Evaluation*

We used a comprehensive public database that contains breast cancer patients with their therapeutic options recorded [47]. The database took the approach of ROC (receiver operating characteristic curve), allowing for classification of patients' sensitivity to a therapy. Here, the AUC (area under the curve) values and the statistical value for sensitivity to treatment were recorded. Additionally, the levels of the respective gene expression of the gene of interest were also displayed together with their statistical power (using a Mann–Whitney U test).

2.11. *Statistical Methods*

Statistical analyses were carried out using SPSS (version 27.0). Groupwise comparisons were conducted using a Kruskal–Wallis test and ANOVA where applicable. Pairwise comparisons were performed using a Mann–Whitney U test, as indicated in the text. Kaplan–Meier method and log rank test were used to run survival analysis. Univariate and multivariate analyses were conducted using Cox regression model. Classification analysis was achieved using the receiver operating characteristic (ROC) method.

3. Results

3.1. *Transcript Levels of CLDN9 in Breast Cancer Tissues*

Breast cancer tissues expressed significantly elevated levels of CLDN9 transcript ($p = 0.035$) (Table 1). Patients who died of breast cancer and who developed breast cancer-related incidence had a raised level of the CLDN9 transcript, although these were marginally statistically significant. It was interesting to note that levels of CLDN9 transcript were significantly correlated with transcripts of ZO-1 and ZO-3, but not ZO-2 (Table 2).

Table 1. CLDN9 in mammary tissues and breast cancer tissues.

Category	Subgroup	*n*	CLDN9 (Median (Q1–Q3)	*p* Value *
Tissue type	Normal	33	15 (4–764)	0.035
	Tumour	127	77 (4–4342)	
Grade	1	24	1500 (158–21,800)	
	2	43	41 (3–5418)	0.025
	3	58	31 (4–1442)	0.0036
TNM staging	1	70	125 (6–6170)	
	2	40	57 (3–1467)	0.18
	3	7	226 (32–32,600)	0.59
	4	4	89 (20–5529)	0.76
Clinical outcome	Disease free	90	80 (7–6505)	
	Died of BrCa	16	729 (11–16,375)	0.071
	All BrCa Incidence	28	125 (4–1467)	0.05
ER status	Negative	75	32 (4–700)	0.09
	Positive	38	924 (4–18,520)	
Her2	Her2(−)	57	81 (3–5165)	0.75
	Her2(+)	55	75 (7–1970)	

* Mann–Whitney U test.

Table 2. Correlation levels of the CLDN9 transcripts with that of ZO-1, ZO-2 and ZO-3 (Spearman ranked method).

	Spearman's Correlation with CLDN9	
ZO-1	Correlation Coefficient	0.297 **
	Significance (2-tailed)	0.001
ZO-2	Correlation Coefficient	−0.084
	Significance (2-tailed)	0.374
ZO-3	Correlation Coefficient	0.252 *
	Significance (2-tailed)	0.011

* indicating significance $p < 0.05$; ** indicating significance $p < 0.001$.

3.2. CLDN9 Protein Expression in Mammary Tissues

We evaluated the protein distribution in representative samples of mammary tissues and breast cancer tissues (Figure 1 and Table 3). As shown in Figure 1, normal mammary tissues indicated the presence of CLDN9 protein in areas representing tight junctions (Figure 1, arrows in green). In tumor tissues, the staining appeared to be more diffuse and seem in cytoplasmic regions rather than in the junctional regions (Figure 1, open red arrows), a pattern similar to that found with the ZO family proteins, as previous reported [48]. The analysis of the CLDN9 staining (Table 3) also indicated the pattern changing between normal tissues and tumor tissues.

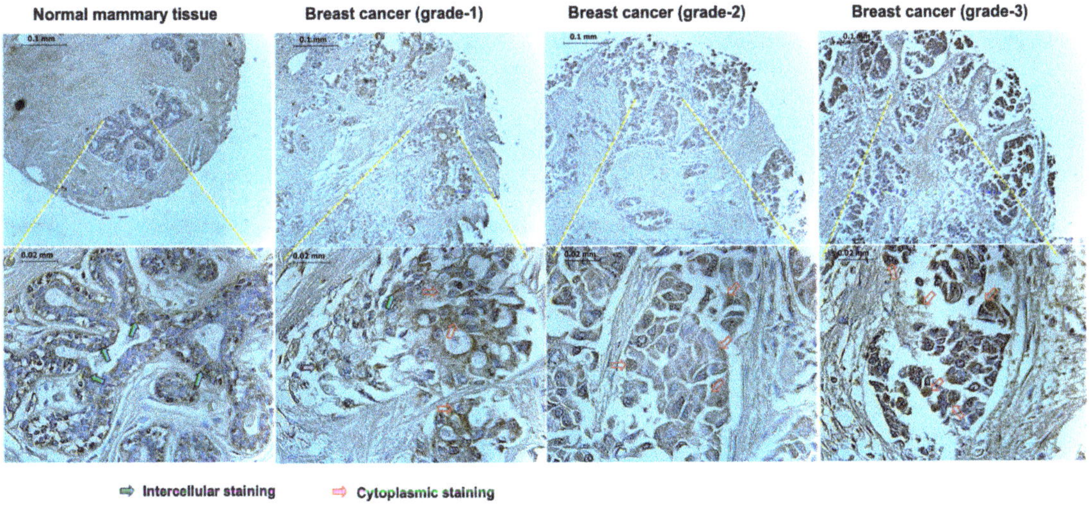

Figure 1. Immunohistochemical detection of the CLDN9 protein in normal mammary tissue (left panel) and breast cancer tissues of different grade (right three panels). In normal tissues, CLDN9 was seen in the residual mammary epithelial cells at the apical regions (green arrows) and to a degree in cytoplasmic regions (open red arrows). In breast cancer, however, the staining was seen in a diffused pattern and was primarily in the cytoplasmic region of the cells (open red arrows).

Table 3. Analysis of the CLDN9 staining in the breast cancer TMA (BR1503f).

	Intensity						Statistical Significance	
	Negative to Weak (0–1)	Moderate to Strong (2–3)	Membrane		Nucleus			
			Positive	Negative	Positive	Negative	Chi Value	*p* Value
Normal ($n = 3$)	0	3	3	0	0	3		
Tumor ($n = 128$)	74	54	19	109	23	105	19.85	0.0013
Grade1 ($n = 4$)	2	2	1	3	2	2		
Grade2 ($n = 57$)	29	28	8	49	13	44	1.849	0.8696
Grade3 ($n = 28$)	16	12	2	26	2	26	7.264	0.2018
T1 ($n = 4$)	4	0	0	4	0	4		
T2 ($n = 69$)	35	34	11	58	11	58	5.191	0.393
T3 ($n = 25$)	15	10	5	20	7	18	4.885	0.4301
T4 ($n = 15$)	9	6	1	14	2	13	3.216	0.6667
HER-2- ($n = 81$)	52	29	12	69	13	68		
HER-2+ ($n = 3$)	1	2	0	3	0	3	2.272	0.8104
HER2++ ($n = 9$)	3	6	2	7	0	9	5.273	0.3834

Table 3. *Cont.*

	Intensity						Statistical Significance	
	Negative to Weak (0–1)	Moderate to Strong (2–3)	Membrane		Nucleus			
			Positive	Negative	Positive	Negative	Chi Value	*p* Value
HER2+++ (*n* = 29)	15	14	5	24	7	22	2.431	0.7869
ER- (*n* = 59)	33	26	10	49	13	46		
ER+ (*n* = 18)	10	8	3	15	1	17	2.519	0.7736
ER++ (*n* = 20)	12	8	4	16	1	19	3.168	0.674
ER+++ (*n* = 25)	16	9	2	23	5	20	1.662	0.8937

3.3. CLDN9 and Patient Clinical Outcome

Patients with high levels of CLDN9 had a shorter, yet statistically nonsignificant overall survival (OS) than those with low levels ($p = 0.054$) (Figure 2). A weaker link was seen with relapse-free survival (RFS) ($p = 0.157$). Owing to the nature that ZO proteins are critical anchorage proteins for CLDN9 in the cells and that CLDN9 was significantly correlated with ZO-1 and ZO-3, we further integrated the expression pattern of CLDN9, ZO-1 and ZO-3 and analyzed against patient outcome. As can be seen in Figure 2, the expression signature of integrated CLDN9, ZO-1 and ZO-3 had a marked value in significantly predicting both OS ($p = 0.013$, hazard ratio (HR) = 0.1472) and RFS ($p = 0.024$, HR = 1.1148). Integrated expression also showed a high significant independent value in multivariate analysis (Table 4). The CLDN9/ZO expression does provide an excellent prediction value for OS ($p = 0.004$) and for RFS ($p = 0.010$).

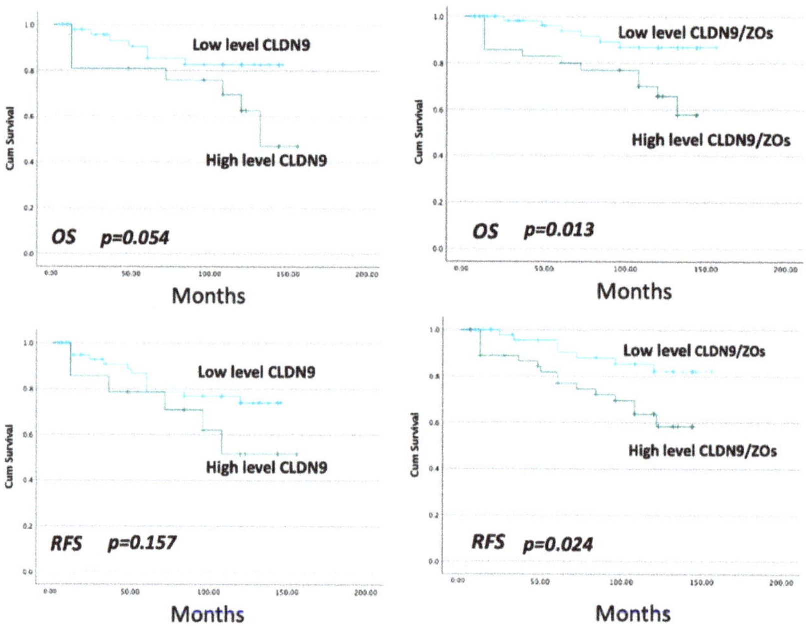

Figure 2. CLDN9 (**left panel**) and integrated expression of CLDN9/ZO-1/ZO-3 (**right panel**) in evaluating the overall survival (OS) and relapse-free survival (RFS) using Kaplan–Meier survival analysis.

Table 4. Multivariate analysis of the CLDN9/ZO expression signature against clinical outcome.

Factors	OS		RFS	
	Hazard Ratio	*p* Value *	Hazard Ratio	*p* Value *
CLDN9/ZO signature	2.033	0.004	1.239	0.010
NPI **	3.028	0.089	2.068	0.045
Grade	1.287	0.432	1.275	0.530
TNM staging	1.034	0.902	1.412	0.033
ER status	2.022	0.266	3.896	0.008
Her-2 status	3.016	0.083	7.697	0.008

* Cox regression method. ** NPI: Nottingham Prognostic Index.

3.4. CLDN9 and Hormone Status Outcome

CLDN9 was found to be expressed at different levels depending on receptor status (Figure 3). This inspired us to further investigate if the expression signature had value in the subgroups with different receptor status. As shown in Figure 3, analysis in subgroups did not markedly improve the prediction in overall survival in subgroups with different hormone receptor status, suggesting that this signature may operate well irrespective of the receptor status of breast cancer.

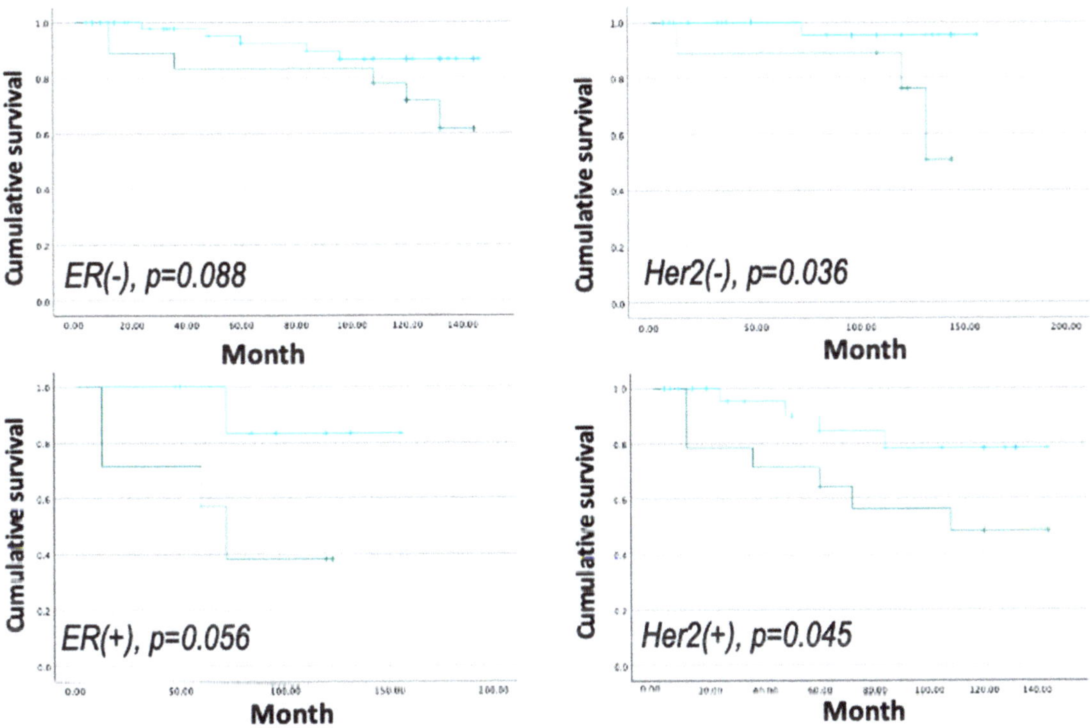

Figure 3. CLDN9/ZO expression signature and the overall survival (OS) in subgroups of patients with different receptor status.

3.5. CLDN9 and Patient Response to Drug Treatment

From previous reports that claudins are important markers in predicting patient response to therapies, we conducted an analysis on the available TCGA dataset (accessed via ROCplot.com) and explored the possible link between CLDN9 and patient response to drugs.

3.5.1. High Expression of CLDN9 and Patient Resistance to Chemotherapies

CLDN9 transcript levels have a significant impact on breast cancers and patient responses. Shown in Figure 4A,B are analyses based on the ROC model in the pathological response (A) and 5-year RFS response (B). It was clear that by both assessment methods, CLDN9 is highly significant in distinguishing patients' responses. When the levels of CLDN9 were analyzed in the responders (sensitive) and non-responders (resisted) to chemotherapies, the non-responders (resisted) had significantly higher levels of CLDN9 transcript than the responders (Figure 4C).

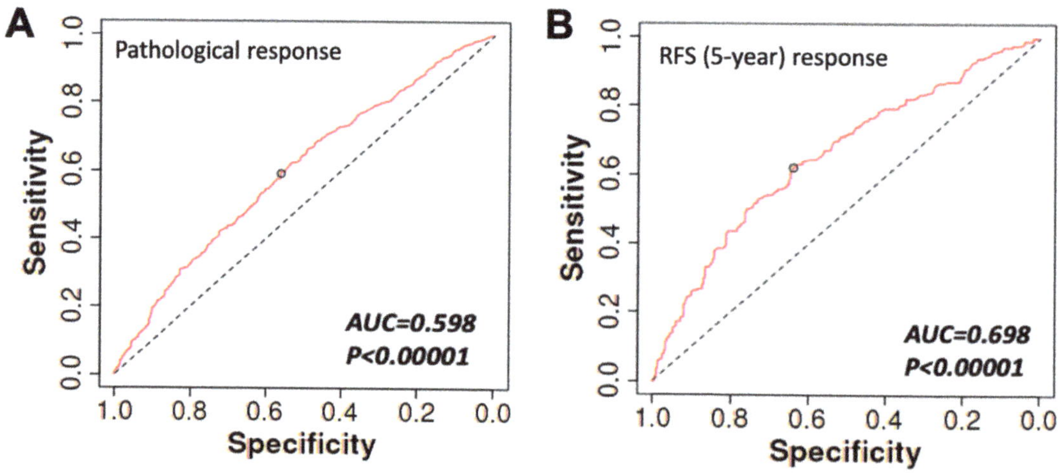

C	Non-responders			Responders			p*
	n=	median	Min-Max	*n*=	Median	Min-Max	
Pathological response	1100	166	3 - 1805	532	92	2 - 1362	<0.00001
5-year RFS response	220	219	10 - 750	256	138	5 - 1352	<0.00001

Figure 4. High levels of CLDN9 transcript are linked to patient drug resistance. (**A**,**B**): ROC analysis of patient pathological (**A**) and 5-year relapse-free survival (RFS) (**B**) to chemotherapies. There is a significant connection between CLDN9 transcript and patients' responses. (**C**) CLDN9 transcript levels in the respective group. Patients who resisted (non-responders) chemotherapies had significantly higher levels of CLDN9 compared with those who responded to treatment in both pathological assessment and RFS assessment. Data from the TCGA dataset (ROCplot.com, accessed on 23 June 2023). * by Mann-Whitney U test.

3.5.2. Expression of CLDN9 and Patient Resistance to Chemotherapies in Relation to Hormone Receptor Status and Molecular Subtypes

When receptor status was considered, it was found that patient response to chemotherapies in ER positive and ER negative tumors showed the same trend that patients who resisted chemotherapies had significantly high levels of CLDN9 transcript compared to the sensitive group ($p < 0.001$ in both groups) (Figure 5 and Table 5). The same link between CLDN9 expression and drug resistance was seen in Her-2-negative tumors ($p = 0.0000000027$). However, it was noteworthy that in Her-2-positive tumors, there was no significant difference between responders and non-responders, leading us to investigate further using in vitro methods (Figure 5 and Table 5).

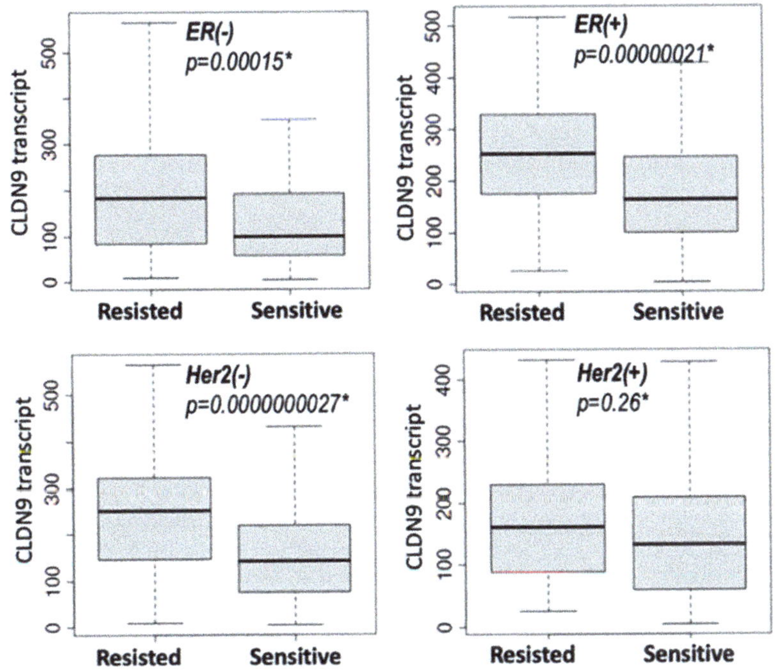

Figure 5. CLDN9 transcript expression and patient response to chemotherapies in relation to hormone receptor status (5-year RFS survival) (data from the TCGA dataset ROCplot.com, accessed 23 June 2023). High levels of CLDN9 transcript in ER-positive and negative both linked to resistance (**top panel**), as well as Her-2 negative tumors (**bottom left**). However, Her-2 positive tumors had little connection with drug responses (all drugs). * Mann–Whitney U test. Shown are median and the interquartile range of the CLDN9 transcript.

Table 5. CLDN9 expression in breast tumors with different receptor statuses and with different molecular subgroups. Data derived from ROCplot (accessed on 23 June 2023).

5-Year RFS Response		Non-Responders			Responders			p *
		n	Median	Min–Max	n	Median	Min–Max	
ER status	ER(−)	111	182	10–564	115	99	6–922	0.000034
	ER(+)	109	252	26–750	141	163	6–1315	5.9×10^{-9}
Her2 status	Her2 (−)	173	250	10–750	183	141	6–1362	7.1×10^{-11}
	Her2 (+)	47	162	25–432	73	133	5–622	0.12
ER(+)/Her2(−)		88	262	26–750	103	163	12–1352	0.000000009
ER(+)/Her2(+)		21	200	75–432	38	168	5–429	0.14
ER(−)/Her2(+)		26	140	25–343	35	122	7–622	0.13
TNBC		84	218	10–564	80	96	6–439	0.000075
Luminal-A		20	202	59–304	58	126	15–501	0.0068
Luminal-B		90	262	26–750	83	199	5–1352	0.00005

* By Mann-Whitney U test.

We subsequently explored the relationship between CLDN9 expression in molecular subtypes. As shown in Table 5, high levels in tumors where Her-2 was negative are linked to resistance. This includes ER(+)/Her-2(−) tumors (p = 0.000000009), Luminal-A (p = 0.0068) and Luminal-B (p = 0.00005) tumors where both are Her-2-negative. What is more interesting is that in triple-negative breast cancers that are both ER- and Her-2 negative, this connection was also highly significant (p = 0.000075). Conversely, in the two subgroups that were Her-2 positive, namely ER(+)/Her-2(+) and ER(−)/Her-2(+), the relationship was not significant (p = 0.14 and p = 0.13, respectively) (Table 5). Collectively, the clinical information indicates that the presence of Her-2 in breast cancer decreases the responsiveness of the cells to chemotherapies.

3.5.3. CLDN9 Is Not Connected to Endocrine or Anti-Her-2 Therapies

It is very interesting to note that levels of CLDN9 do not appear to be linked to patients' response to endocrine therapies (AUC = 0.503, p = 0.45, by 5-year RFS), or anti-Her-2 therapies (AUC = 0.513, p = 0.44, by 5-ear RFS).

3.6. Creation of CLDN9 Knockdown Breast Cancer Cell Model

To corroborate the clinical findings, we established cell models of breast cancer cells with CLDN9 knockdown, based on analyzing the baseline expression levels of CLDN9 in three breast cancer cell lines. The breast cancer cell lines MDA-MB-231 (HER-2−/ER−/PR−), MCF7 (HER-2−/ER+/PR+) and SKBR3 (HER-2+/ER−/PR−), representing the main molecular subtypes of breast cancers as portraited earlier, were employed to generate the CLDN9 knockdown cell models. As shown in Figure 6, the CLDN9 expression in each cell line exhibited a significant reduction following transfections with anti-CLDN9 siRNA, as evidenced with PCR, qPCR and Western blotting analyses.

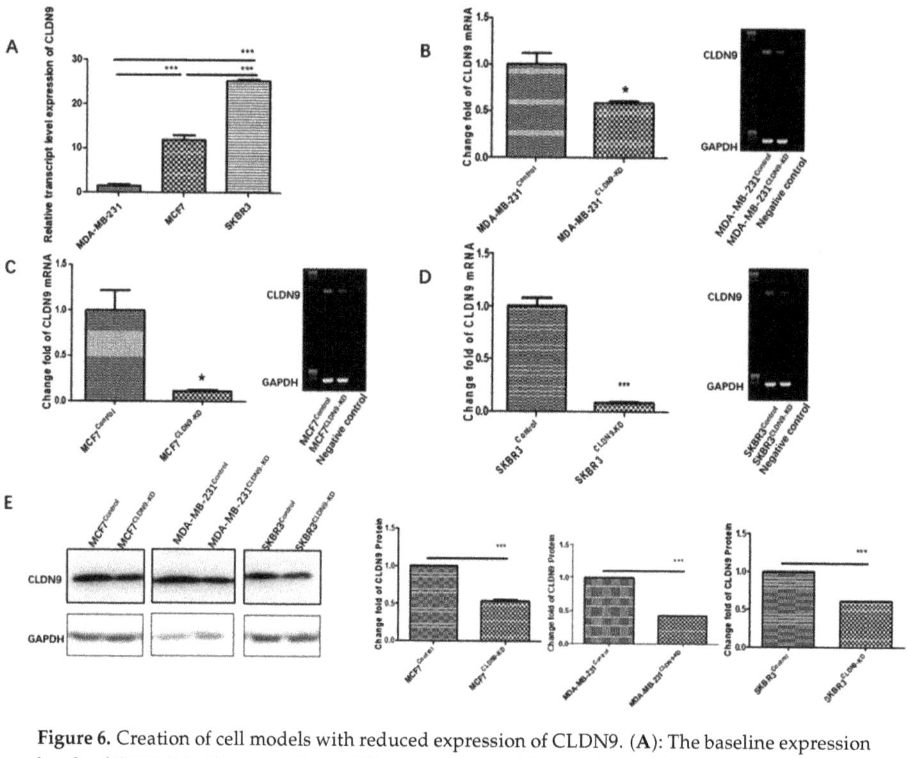

Figure 6. Creation of cell models with reduced expression of CLDN9. (**A**): The baseline expression levels of CLDN9 in the respective cell lines as shown with quantitative PCR (**B–D**): The CLDN9

transcript expression in in the respective cell line controls and CLDN9 knockdown cells as confirmed using PCR (right panel) and quantitative PCR (left panel). (**E**): Knockdown of CLDN9 protein in the respective cell lines as shown using protein blotting. * $p < 0.05$, *** $p < 0.001$ vs. control cells.

3.7. CLDN9 Expression and Cell's Response to Chemodrugs

The aim of this part of the study was to ascertain the correlation between CLDN9 expression levels and chemoresistance across various subtypes of breast cancer cells. We evaluated the IC_{50} values of methotrexate, docetaxel, cisplatin and gemcitabine in the three representative cell lines, encompassing both normal CLDN9 expression levels and those with CLDN9 knockdown. Figure 7 illustrates examples of drug responses within normal CLDN9-expressing cells and CLDN9 knockdown in the breast cancer cell models. Utilizing the cell models established in this study, we validated the responsiveness of these cell lines to chemotherapy agents. As shown in Table 6, knockdown CLDN9 in the Her-2-negative MCF7 and MDA-MB-231 cells sensitized the cells in their response to gemcitabine, docetaxel, methotrexate and, to some degree, cisplatin. However, in the Her-2(+) SKBR3 cells, knockdown CLDN9 only had marginal effects on the cells' response to these drugs (Table 6).

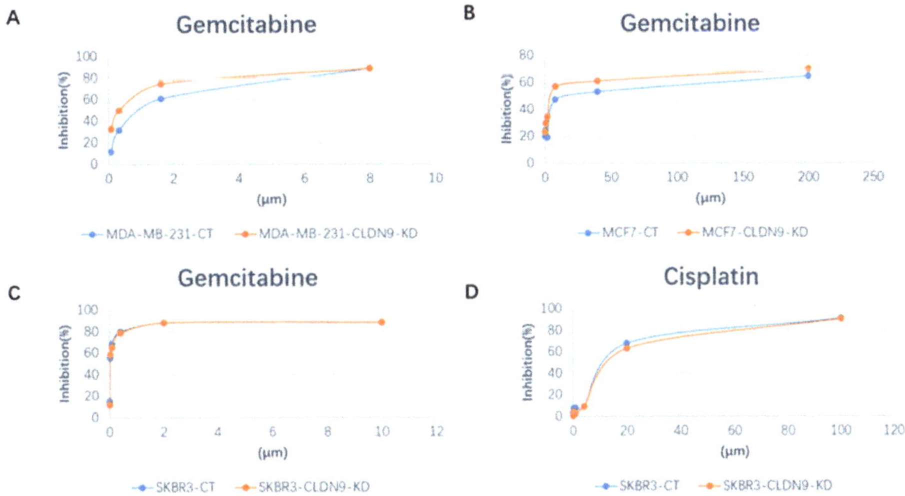

Figure 7. Implication of CLDN9 on breast cells' response to chemotherapeutic agents over broad concentration ranges. (**A**) MDA-MB-231 cells (gemcitabine 8 μM); (**B**) MCF7 cells (gemcitabine 200 μM); (**C**,**D**): SKBR3 cells (gemcitabine 10 μM and cisplatin 100 μM). CT: control; KD: knockdown

Table 6. IC_{50} of chemotherapy drugs in breast cancer cell lines with modified CLDN9 expression.

	Gemcitabine (μM)	Docetaxel (μM)	Cisplatin (μM)	Methotrexate (μM)
MDA-MB-$231^{Control}$	0.820	54.612	12.520	107.927
MDA-MB-$231^{CLDN9\text{-}KD}$	0.287	31.134	4.723	2.7834
$MCF7^{Control}$	27.261	6.575	13.462	2179.402
$MCF7^{CLDN9\text{-}KD}$	7.334	17.876	12.414	1510.961
$SKBR3^{Control}$	0.028	0.100	10.162	1012.070
$SKBR3^{CLDN9\text{-}KD}$	0.031	0.086	12.493	706.855

As indicated in Table 5, in patients bearing Her-2 breast cancer, CLDN9 appeared to have lost its value in predicting patient response to chemotherapies. In order to test if this connection can be replicated in vitro, we employed a Her-2-positive cell, SKBR3, and applied a permanent Her-2 inhibitor, neratinib, with or without chemotherapeutic drugs. As shown in Figure 8, both neratinib and chemotherapy drugs (except cisplatin) exert cytotoxicity on SKBR3 control cells. However, with CLDN9 knockdown in SKBR3 cells, cells became more sensitive to neratinib and chemodrugs. Additionally, the combination of neratinib and chemodrugs appeared to render more toxicity than using either the Her-2 inhibitor alone or the chemodrug alone. This effect was not seen with MCF7 (Figure 9) nor with MDA-MB-231 (Figure 10) cells, both of which are Her-2-negative in that the control cells and CLDN9kd cells responded similarly to neratinib, chemodrugs and their combinations, as seen in MCF-7 (Figure 9) and MDA MB-231 (Figure 10).

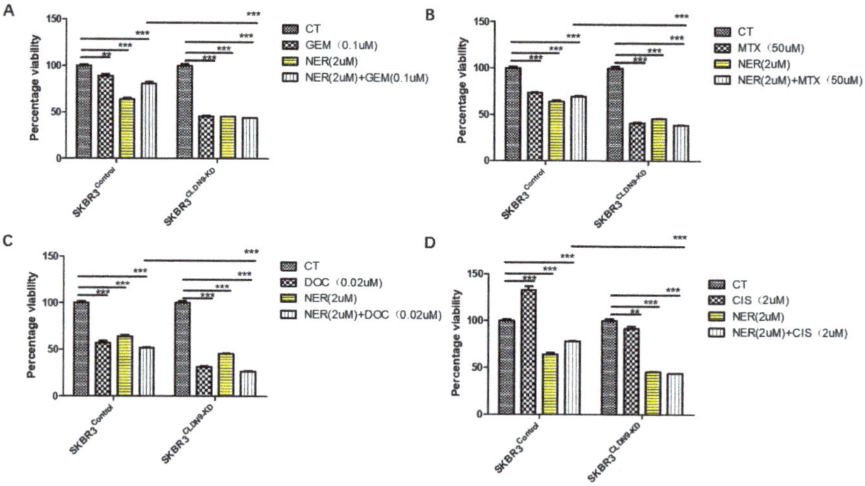

Figure 8. Inhibition of Her-2 and SKBR3 cell's response to chemotherapeutic drugs in relation with CLDN9 expression. SKBR3 wild-type cells were tested on its wild type (Her-2-positive and CLDN9-positive) and its submodel, CLDN9 knockdown (CLDN9-KD). Neratinib (NER) had a significant inhibitory effects on the growth of both models. However, following blocking of Her-2 with neratinib, the CLDN9-KD SKBR3 cells became more sensitive to Gemcitabine (**A**), Methotrexate (**B**), Docetaxel (**C**), Cisplatin (**D**) (NER plus the respective drugs in SKBR3/CLDN9-KD cells). ** $p < 0.01$, *** $p < 0.001$ vs. control cells.

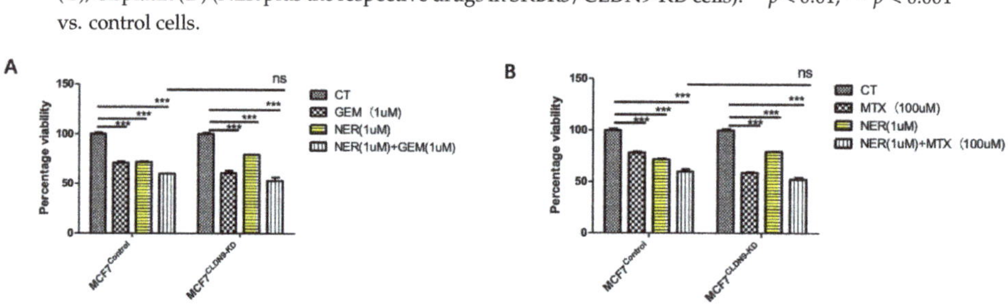

Figure 9. *Cont.*

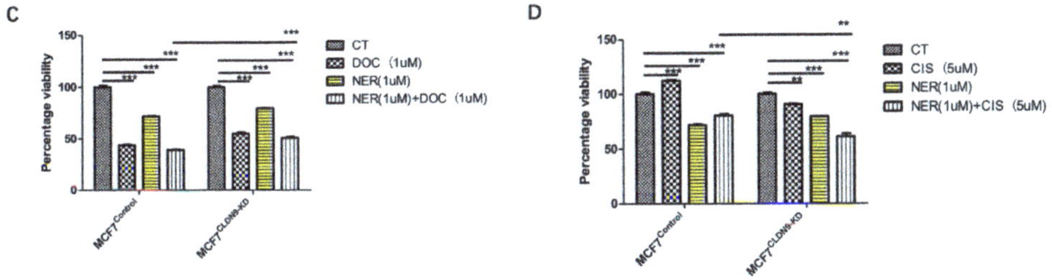

Figure 9. Inhibition of Her-2 and MCF7 cells' response to chemotherapeutic drugs in relation with CLDN9 expression. MCF7 cells were tested on its wild type (Her-2 negative and CLDN9 positive) and its submodel, CLDN9 knockdown (CLDN9-KD). Neratinib (NER) had some inhibitory effects on the growth of both models. However, blocking Her-2 with neratinib did not influence cells response to Gemcitabine (**A**), Methotrexate (**B**), Docetaxel (**C**), Cisplatin (**D**) in both cell models. ns > 0.05, ** $p < 0.01$, *** $p < 0.001$ vs. control cells.

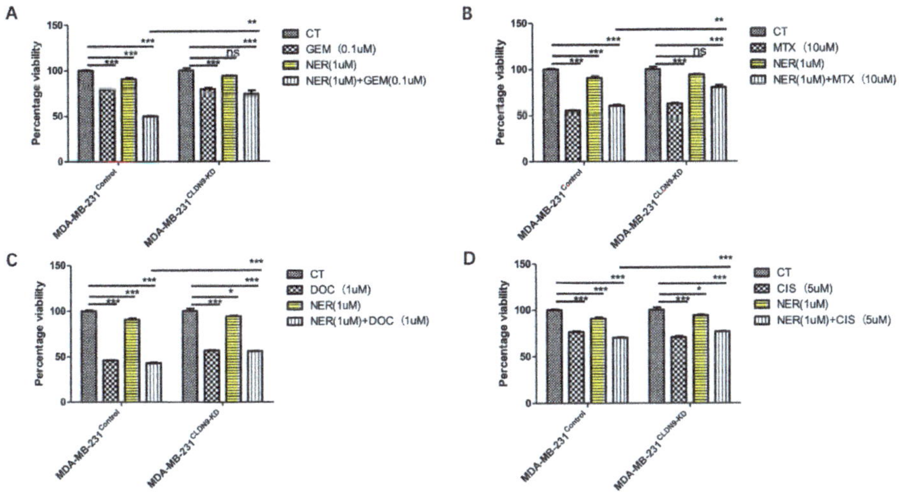

Figure 10. Inhibition of Her-2 and MDA-MB-231 cells' response to chemotherapeutic drugs in relation with CLDN9 expression. MDA-MB-231 cells were tested on its wild type (Her-2-negative and CLDN9-positive) and its submodel, CLDN9 knockdown (CLDN9-KD). Neratinib (NER) had some inhibitory effects on the growth of both models. However, blocking Her-2 with neratinib did not influence cell responses to Gemcitabine (**A**), Methotrexate (**B**), Docetaxel (**C**), Cisplatin (**D**) in both cell models. ns > 0.05, * $p < 0.05$, ** $p < 0.01$, *** $p < 0.001$ vs. control cells.

3.8. The Relationship between CLDN9 and ZO-1 and ZO-3, Respectively

To further ascertain the intrinsic relationship between CLDN9 and ZO-1, we utilized immunofluorescence staining to explore expression and localization of the ZO proteins in CLDN9 knockdowns, MCF-7, MDA-MB-231 and SKBR3 cell models. We found that CLDN9 protein was colocalized with ZO-1 proteins in the control cells (Figure 11, top panel representative images from MCF-7 cells). Following CLDN9 knockdown, most cells showed reduced levels of CLDN9 protein staining. In the CLDN9 knockdown cells, ZO-1 either remained in the tight junction areas (Figure 11 middle panel) or relocated to regions beyond tight junctions (Figure 11, middle and bottom panels) These data preliminarily demonstrated a correlation and colocalization between CLDN9 and ZO-1 protein in control cells that had expression of CLDN9 protein. When CLDN9 is lost or reduced, ZO-1 localization to tight junctions is weakened, owing to its interaction with other tight junction proteins.

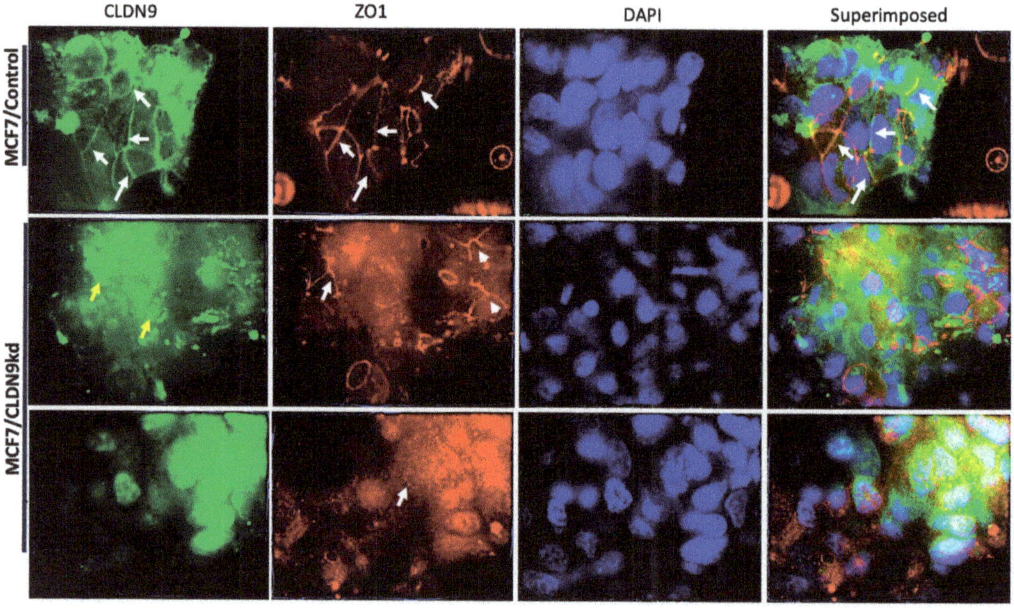

Figure 11. The correlation between CLDN9 and ZO protein in terms of colocalization and expression levels in the MCF-7 cell models. CLDN9 (green) and ZO-1 (red), as shown using IFC in wild-type (top panel) and CLDN9 knockdown (bottom two panels). In control wild type cells, CLDN9 (green) and ZO-1 (red) were seen in the tight junction areas (indicated by white arrows) and had a colocalization pattern (top right imposed image). Following CLDN9 knockdown, there were changes in the localization of ZO-1 in that it partly remained in the tight junctions owing to its interaction with other tight junction proteins (middle panel). In contrast, after losing a majority of the CLDN9 proteins, the remaining CLDN-9 was localized in regions unrelated to tight junction (yellow arrows in the middle panel and bottom panel).

4. Discussion

In this study, we report, for the first time, that claudin-9 (CLDN9) is expressed in mammary tissues and that the increased expression in breast cancer, particularly when co-expressed with its anchorage protein ZO1 and ZO3, forms a significant predictor for the survival of patients. This study further demonstrates that the expression pattern of CLDN9 has an important impact on patient response to chemotherapies, in that high levels of CLDN9 indicated chemoresistance, except in Her-2-positive tumors in which CLDN9's predictive power to drug response is lost. Using in vitro cell models, we further demonstrated that knockdown of CLDN9 in Her-2-negative breast cancer cells gave rise to cells that were more sensitive to chemotherapy drugs, a finding not seen in Her-2-positive SKBR3 cells. Additionally, blocking Her-2 with a permanent Her-2 inhibitor, neratinib, and CLDN9 knockdown in SKBR3 cells sensitized the cells to drugs, a finding not seen in Her-2-negative MCF7 and MDA-MB-231 cells.

CLDN9 has been known to be a transmembrane protein located in the tight junction area of many cell types. However, its role in cancers has been little studied and certainly not in breast cancer. In a continued effort to establish the role of tight junction and tight junctional proteins in breast cancer, we determined CLDN9 expression in breast cancer. First, we confirmed that both CLDN9 transcript and the CLDN9 protein are seen in human mammary tissues and breast cancer tissues. At transcript levels, breast tissues express significantly high levels and there is a clear indication that high levels of CLDN9 are associated with poor clinical outcome of the patients; however, this connection is not seen with Her-2-positive breast cancers. At the tight junction, CLDN9 is known to be anchored to the cytoskeleton by interacting with the subcoat proteins, Zonula Occludens (ZOs), and together they may play roles in orchestration [49]. Here, we also show that expression of CLDN9 is significantly correlated with ZO-1 and ZO-3 and the integrated co-expression of CLDN-9, ZO-1 and ZO-3 markedly improve the prognostic value in breast cancer. This therefore establishes CLDN9 in combination with ZO-1 and ZO-3 as a significant prognostic indicator for patient survival in breast cancer. Here, our limited data show that CLDN9 can be detected at locations of tight junctions particularly in normal mammary tissues. However, the pattern appears to change in breast cancer tissues, in that more cytoplasmic staining appears to be present. CLDN9 and ZO protein colocalization was partly confirmed in our cell models, in which CLDN9 and ZO-1 proteins were colocalized to the tight junctions in controls cells. However, following CLDN9 knockdown, the ZO-1 localization to the tight junction was partly affected, owing to the fact that it is an anchorage protein for rather large number of tight junction transmembrane proteins. Another intriguing observation was the expression pattern of the CLDN9 mRNA and proteins in relation to tumor grade. There appears to be high levels of *CLDN9* gene transcript in grade 1 tumors, compared with grade 2 and grade 3 tumors (Table 1). Yet, when assessed at protein levels using the immunohistochemical method (on a different cohort), there does not appear to be a significant difference between different grades (Table 3) and instead the difference seems to reside in the cellular location of the proteins, namely membranous versus cytoplasmic. This inconsistency between transcript and protein may be due to the variance in the transcript and relative smaller size of the cohort. Indeed, there does not appear to be a significant difference between different grades in a large TCGA transcript database, suggesting that CLDN9 gene transcript levels may not significantly associated with tumor grade. A larger cohort would be desirable to validate in future studies. Collectively, this needs to be further confirmed, and the membrane and cytoplasmic CLDN9 proteins need further investigation with regard to their possibly different roles.

The other highly interesting finding is the highly significant connection between CLDN9 and drug response in breast cancer. High levels of CLDN9 transcripts clearly indicate resistance to chemotherapies in all breast cancer subtypes, including triple-negative breast cancers (TNBC), other than Her-2-positive breast cancers. Where Her-2-positive is detected, there is no difference in the levels of CLDN9 between those who resisted and those who responded to chemotherapies. This is very interesting and suggests that the presence of Her-2 receptor kinase may offer a yet unidentified mechanism by which it overcomes CLDN9-mediated drug resistance. This possibility is partially confirmed in our experiment in that when Her-2 is inhibited by way of therapeutic Her-2 inhibitor and CLDN9 is reduced by genetic knockdown, we were able to sensitize the otherwise resistant Her-2-positive breast cancer to chemodrugs. Additionally, the Her-2 inhibitor and CLDN9 inhibition may have additive value in sensitizing cancer cells. This finding thus has clinical implications. In Her-2-negative breast cancers including TNBC breast cancer, high CLDN9 expression generally indicates a likelihood of patient resistance to chemotherapy, which may be reflected in the poor clinical outcome of the patients. However, in patients with Her-2-positive breast cancers, if there are concurrently low levels of CLDN9, a combination of Her-2 inhibitor and chemotherapies may offer benefits to the patient; a clinical study would be highly valuable to confirm this.

The cellular mechanism by which CLDN9 confers drug resistance remains to be elucidated. CLDN9 is not alone in this regard. Previous studies have shown that CLDN2, CLDN6 and CLDN7 are involved in patient response to chemotherapies in cancers including breast cancer and colorectal cancer, yet with no clear mechanism identified. CLDN proteins are well known components of tight junctions, the structure and function of which controls apical and paracellular permeability to molecules including certain drugs. One could argue that high levels of CLDNs in endothelial cells may result in tightly controlled paracellular permeability, resistance drugs to penetrate the endothelium, and preventing them reaching cancer cells from the circulation. This may be an important mechanism in clinical settings. Whilst this is a possibility, it is not entirely supported by the present and indeed others' findings, as all the findings were confirmed directly on cancer cells minus the presence of endothelial cells. There are likely other mechanisms unrelated to tight junction that may operate here, including the possibility of the junctional role for CLDN9 in cancer cells themselves. Tight junctions in cancer may also govern paracellular permeability and hence intratumor drug penetration; thus, CLDN9 proteins at the tight junctions in cancer cells may participate this restriction of intratumor availability in drugs. However, it is worth noting that most small compound drugs, such as the ones tested in the present study, require specific drug transporters on cell membrane, for example, SLC28A1 and SLC29A1 for gemcitabine. This again suggests other possible mechanism(s) here, including possibility of interplay between claudins and membrane drug transporters in cancer cells such as the solute carrier family (SLC). Indeed, it has been shown that CLDN12 can be affected in coordination with SLC9A3, and, in particular, in brain endothelial cells [50–52]. There have been indications that certain SLC members, namely SLC22A5, may be regulated by a common subcoat protein with claudins, such as ZO1 [53]. Thus, the shared intracellular regulatory pathway between CLDN proteins and SLC protein may also influence the response. The high correlation between CLDN9 and ZO1/ZO3 and the integrated pattern between CLDN9 and the two ZOs may provide some indirect evidence here. The brief finding that CLDN9 is also seen in the cytoplasmic regions of breast cancer may contribute to this suggestion. However, many tight junction proteins when "switched off" relocate to the cytoplasm and away from the membrane area. Presently, it is unclear what role the cytoplasmic CLDN9 protein performs in cell functions; this will be very interesting to explore in the future.

5. Conclusions

CLDN9 is expressed in mammary tissues and in breast cancer tissues. High levels of CLDN9 in breast cancer present as a potentially significant prognostic indicator for patients with breast cancer who are Her-2-negative. CLDN9 expression is also a pivotal factor in assessing patient responses to chemotherapies. Her-2 is a negating factor for the treatment response prediction value of CLDN9 and negating Her-2 and CLDN9 may enhance breast cancer cell response to chemotherapeutic drugs.

Author Contributions: W.G.J. and T.A.M. conceived the study; X.Z., F.R., E.K., Q.D., J.Z., X.L., T.A.M. and W.G.J. carried experimental work; X.Z., T.A.M. and W.G.J. conducted statistical analyses; E.D. and W.G.J. performed resource sourcing; all participated in manuscript preparation. All authors have read and agreed to the published version of the manuscript.

Funding: Xinguo Zhuang is supported by the Cardiff University–Xiamen University Strategic Partnership Fund. This work was also supported by the Xiamen Medical and Health Guidance Project (3502Z20224ZD1014), the Natural Scientific Foundation of Xiamen (No. 3502Z20227340), the Fujian Natural Science Foundation of China (No. 2022J011372) and the National Natural Science Foundation of China (No. 81871305).

Institutional Review Board Statement: This study was approved by the local ethics committee, the Bro Taf Health Authority; ethics approval numbers 01/4303 and 01/4046.

Informed Consent Statement: Informed consent was obtained from all subjects in this study.

Acknowledgments: The authors wish to acknowledge Cardiff University China Medical Scholarship.

Conflicts of Interest: The authors declare no conflict of interest.

References

1. Blanchard, A.A.; Skliris, G.P.; Watson, P.H.; Murphy, L.C.; Penner, C.; Tomes, L.; Young, T.L.; Leygue, E.; Myal, Y. Claudins 1, 3, and 4 protein expression in ER negative breast cancer correlates with markers of the basal phenotype. *Virchows Arch.* **2009**, *454*, 647–656. [CrossRef] [PubMed]
2. Kominsky, S.L.; Argani, P.; Korz, D.; Evron, E.; Raman, V.; Garrett, E.; Rein, A.; Sauter, G.; Kallioniemi, O.P.; Sukumar, S. Loss of the tight junction protein claudin-7 correlates with histological grade in both ductal carcinoma in situ and invasive ductal carcinoma of the breast. *Oncogene* **2003**, *22*, 2021–2033. [CrossRef] [PubMed]
3. Lee, J.W.; Lee, S.J.; Seo, J.; Song, S.Y.; Ahn, G.; Park, C.S.; Lee, J.H.; Kim, B.G.; Bae, D.S. Increased expressions of claudin-1 and claudin-7 during the progression of cervical neoplasia. *Gynecol. Oncol.* **2005**, *97*, 53–59. [CrossRef] [PubMed]
4. Morohashi, S.; Kusumi, T.; Sato, F.; Odagiri, H.; Chiba, H.; Yoshihara, S.; Hakamada, K.; Sasaki, M.; Kijima, H. Decreased expression of claudin-1 correlates with recurrence status in breast cancer. *Int. J. Mol. Med.* **2007**, *20*, 139–143. [CrossRef] [PubMed]
5. Kim, T.H.; Huh, J.H.; Lee, S.; Kang, H.; Kim, G.I.; An, H.J. Down-regulation of claudin-2 in breast carcinomas is associated with advanced disease. *Histopathology* **2008**, *53*, 48–55. [CrossRef]
6. Martin, T.A.; Harrison, G.M.; Watkins, G.; Jiang, W.G. Claudin-16 reduces the aggressive behavior of human breast cancer cells. *J. Cell. Biochem.* **2008**, *105*, 41–52. [CrossRef] [PubMed]
7. Martin, T.A.; Lane, J.; Ozupek, H.; Jiang, W.G. Claudin-20 promotes an aggressive phenotype in human breast cancer cells. *Tissue Barriers* **2013**, *1*, e26518. [CrossRef]
8. Kimbung, S.; Kovacs, A.; Bendahl, P.O.; Malmstrom, P.; Ferno, M.; Hatschek, T.; Hedenfalk, I. Claudin-2 is an independent negative prognostic factor in breast cancer and specifically predicts early liver recurrences. *Mol. Oncol.* **2014**, *8*, 119–128. [CrossRef]
9. Abd-Elazeem, M.A.; Abd-Elazeem, M.A. Claudin 4 expression in triple-negative breast cancer: Correlation with androgen receptors and Ki-67 expression. *Ann. Diagn. Pathol.* **2015**, *19*, 37–42. [CrossRef]
10. Tabaries, S.; Dong, Z.; Annis, M.G.; Omeroglu, A.; Pepin, F.; Ouellet, V.; Russo, C.; Hassanain, M.; Metrakos, P.; Diaz, Z.; et al. Claudin-2 is selectively enriched in and promotes the formation of breast cancer liver metastases through engagement of integrin complexes. *Oncogene* **2011**, *30*, 1318–1328. [CrossRef]
11. Ma, X.; Miao, H.; Jing, B.; Pan, Q.; Zhang, H.; Chen, Y.; Zhang, D.; Liang, Z.; Wen, Z.; Li, M. Claudin-4 controls the proliferation, apoptosis, migration and in vivo growth of MCF-7 breast cancer cells. *Oncol. Rep.* **2015**, *34*, 681–690. [CrossRef] [PubMed]
12. Sugimoto, H.; Nagahara, M.; Bae, Y.; Nakagawa, T.; Ishikawa, T.; Sato, T.; Uetake, H.; Eishi, Y.; Sugihara, K. Clinicopathologic relevance of claudin 5 expression in breast cancer. *Am. J. Clin. Pathol.* **2015**, *143*, 540–546. [CrossRef]
13. Jaaskelainen, A.; Soini, Y.; Jukkola-Vuorinen, A.; Auvinen, P.; Haapasaari, K.M.; Karihtala, P. High-level cytoplasmic claudin 3 expression is an independent predictor of poor survival in triple-negative breast cancer. *BMC Cancer* **2018**, *18*, 223. [CrossRef] [PubMed]
14. Kulka, J.; Szasz, A.M.; Nemeth, Z.; Madaras, L.; Schaff, Z.; Molnar, I.A.; Tokes, A.M. Expression of tight junction protein claudin-4 in basal-like breast carcinomas. *Pathol. Oncol. Res.* **2009**, *15*, 59–64. [CrossRef] [PubMed]
15. Lanigan, F.; McKiernan, E.; Brennan, D.J.; Hegarty, S.; Millikan, R.C.; McBryan, J.; Jirstrom, K.; Landberg, G.; Martin, F.; Duffy, M.J.; et al. Increased claudin-4 expression is associated with poor prognosis and high tumour grade in breast cancer. *Int. J. Cancer* **2009**, *124*, 2088–2097. [CrossRef] [PubMed]
16. Sauer, T.; Pedersen, M.K.; Ebeltoft, K.; Naess, O. Reduced expression of Claudin-7 in fine needle aspirates from breast carcinomas correlate with grading and metastatic disease. *Cytopathology* **2005**, *16*, 193–198. [CrossRef] [PubMed]
17. Szasz, A.M.; Tokes, A.M.; Micsinai, M.; Krenacs, T.; Jakab, C.; Lukacs, L.; Nemeth, Z.; Baranyai, Z.; Dede, K.; Madaras, L.; et al. Prognostic significance of claudin expression changes in breast cancer with regional lymph node metastasis. *Clin. Exp. Metastasis* **2011**, *28*, 55–63. [CrossRef]
18. Yoshida, H.; Sumi, T.; Zhi, X.; Yasui, T.; Honda, K.I.; Ishiko, O. Claudin-4: A Potential Therapeutic Target in Chemotherapy-resistant Ovarian Cancer. *Anticancer Res.* **2011**, *31*, 1271–1277.
19. Hicks, D.A.; Galimanis, C.E.; Webb, P.G.; Spillman, M.A.; Behbakht, K.; Neville, M.C.; Baumgartner, H.K. Claudin-4 activity in ovarian tumor cell apoptosis resistance and migration. *BMC Cancer* **2016**, *16*, 788. [CrossRef]
20. Ito, Y.; Takasawa, A.; Takasawa, K.; Murakami, T.; Akimoto, T.; Kyuno, D.; Kawata, Y.; Shano, K.; Kirisawa, K.; Ota, M.; et al. Aberrant expression of claudin-6 contributes to malignant potentials and drug resistance of cervical adenocarcinoma. *Cancer Sci.* **2022**, *113*, 1519–1530. [CrossRef]
21. Yang, M.L.; Li, Y.R.; Shen, X.F.; Ruan, Y.; Lu, Y.; Jin, X.S.; Song, P.Y.; Guo, Y.T.; Zhang, X.L.; Qu, H.N.; et al. CLDN6 promotes chemoresistance through GSTP1 in human breast cancer. *J. Exp. Clin. Cancer Res.* **2017**, *36*, 157. [CrossRef] [PubMed]
22. Paquet-Fifield, S.; Koh, S.L.; Cheng, L.; Beyit, L.M.; Shembrey, C.; Molck, C.; Behrenbruch, C.; Papin, M.; Gironella, M.; Guelfi, S.; et al. Tight Junction Protein Claudin-2 Promotes Self-Renewal of Human Colorectal Cancer Stem-like Cells. *Cancer Res.* **2018**, *78*, 2925–2938. [CrossRef] [PubMed]
23. Gowrikumar, S.; Primeaux, M.; Pravoverov, K.; Wu, C.; Szeglin, B.C.; Sauve, C.E.G.; Thapa, I.; Bastola, D.; Chen, X.S.; Smith, J.J.; et al. A Claudin-Based Molecular Signature Identifies High-Risk, Chemoresistant Colorectal Cancer Patients. *Cells* **2021**, *10*, 2211. [CrossRef] [PubMed]

24. Eguchi, H.; Matsunaga, H.; Onuma, S.; Yoshino, Y.; Matsunaga, T.; Ikari, A. Down-Regulation of Claudin-2 Expression by Cyanidin-3-Glucoside Enhances Sensitivity to Anticancer Drugs in the Spheroid of Human Lung Adenocarcinoma A549 Cells. *Int. J. Mol. Sci.* **2021**, *22*, 499. [CrossRef] [PubMed]
25. Eguchi, H.; Matsunaga, T.; Endo, S.; Ichihara, K.; Ikari, A. Kaempferide Enhances Chemosensitivity of Human Lung Adenocarcinoma A549 Cells Mediated by the Decrease in Phosphorylation of Akt and Claudin-2 Expression. *Nutrients* **2020**, *12*, 1190. [CrossRef] [PubMed]
26. Lemesle, M.; Geoffroy, M.; Alpy, F.; Tomasetto, C.L.; Kuntz, S.; Grillier-Vuissoz, I. CLDN1 Sensitizes Triple-Negative Breast Cancer Cells to Chemotherapy. *Cancers* **2022**, *14*, 5026. [CrossRef] [PubMed]
27. Cherradi, S.; Garambois, V.; Marines, J.; Andrade, A.F.; Fauvre, A.; Morand, O.; Fargal, M.; Mancouri, F.; Ayrolles-Torro, A.; Vezzo-Vie, N.; et al. Improving the response to oxaliplatin by targeting chemotherapy-induced CLDN1 in resistant metastatic colorectal cancer cells. *Cell Biosci* **2023**, *13*, 72. [CrossRef]
28. Briehl, M.M.; Miesfeld, R.L. Isolation and Characterization of Transcripts Induced by Androgen Withdrawal and Apoptotic Cell-Death in the Rat Ventral Prostate. *Mol. Endocrinol.* **1991**, *5*, 1381–1388. [CrossRef]
29. Peacock, R.E.; Keen, T.J.; Inglehearn, C.F. Analysis of a human gene homologous to rat ventral prostate.1 protein. *Genomics* **1997**, *46*, 443–449. [CrossRef]
30. Nakano, Y.; Kim, S.H.; Kim, H.M.; Sanneman, J.D.; Zhang, Y.Z.; Smith, R.J.H.; Marcus, D.C.; Wangemann, P.; Nessler, R.A.; Banfi, B. A Claudin-9-Based Ion Permeability Barrier Is Essential for Hearing. *PLoS Genet.* **2009**, *5*, e1000610. [CrossRef]
31. Ramzan, M.; Philippe, C.; Belyantseva, I.A.; Nakano, Y.; Fenollar-Ferrer, C.; Tona, R.; Yousaf, R.; Basheer, R.; Imtiaz, A.; Faridi, R.; et al. Variants of human CLDN9 cause mild to profound hearing loss. *Hum. Mutat.* **2021**, *42*, 1321–1335. [CrossRef] [PubMed]
32. Zheng, A.H.; Yuan, F.; Li, Y.Q.; Zhu, F.F.; Hou, P.P.; Li, J.Q.; Song, X.J.; Ding, M.X.; Deng, H.K. Claudin-6 and claudin-9 function as additional coreceptors for hepatitis C virus. *J. Virol.* **2007**, *81*, 12465–12471. [CrossRef] [PubMed]
33. Meertens, L.; Bertaux, C.; Cukierman, L.; Cormier, E.; Lavillette, D.; Cosset, F.L.; Dragic, T. The tight junction proteins claudin-1,-6, and-9 are entry cofactors for hepatitis C virus. *J. Virol.* **2008**, *82*, 3555–3560. [CrossRef] [PubMed]
34. Vecchio, A.J.; Stroud, R.M. Claudin-9 structures reveal mechanism for toxin-induced gut barrier breakdown. *Proc. Natl. Acad. Sci. USA* **2019**, *116*, 17817–17824. [CrossRef] [PubMed]
35. Rendon-Huerta, E.; Teresa, F.; Teresa, G.M.; Xochitl, G.S.; Georgina, A.F.; Veronica, Z.Z.; Montano, L.F. Distribution and expression pattern of claudins 6, 7, and 9 in diffuse- and intestinal-type gastric adenocarcinomas. *J. Gastrointest. Cancer* **2010**, *41*, 52–59. [CrossRef] [PubMed]
36. Zavala-Zendejas, V.E.; Torres-Martinez, A.C.; Salas-Morales, B.; Fortoul, T.I.; Montano, L.F.; Rendon-Huerta, E.P. Claudin-6, 7, or 9 Overexpression in the Human Gastric Adenocarcinoma Cell Line AGS Increases Its Invasiveness, Migration, and Proliferation Rate. *Cancer Investig.* **2011**, *29*, 1–11. [CrossRef] [PubMed]
37. Hong, L.C.; Wu, Y.G.; Feng, J.; Yu, S.Y.; Li, C.Z.; Wu, Y.T.; Li, Z.Y.; Cao, L.; Wang, F.; Zhang, Y.Z. Overexpression of the cell adhesion molecule claudin-9 is associated with invasion in pituitary oncocytomas. *Hum. Pathol.* **2014**, *45*, 2423–2429. [CrossRef]
38. Higashi, A.Y.; Higashi, T.; Furuse, K.; Ozeki, K.; Furuse, M.; Chiba, H. Claudin-9 constitutes tight junctions of folliculo-stellate cells in the anterior pituitary gland. *Sci. Rep.* **2021**, *11*, 21642. [CrossRef]
39. Zhu, J.Y.; Wang, R.G.; Cao, H.Y.; Zhang, H.P.; Xu, S.Y.; Wang, A.Y.; Liu, B.; Wang, Y.T.; Wang, R.C. Expression of claudin-5,-7,-8 and-9 in cervical carcinoma tissues and adjacent non-neoplastic tissues. *Int. J. Clin. Exp. Pathol.* **2015**, *8*, 9479–9486.
40. Wang, Z.H.; Zhang, Y.Z.; Wang, Y.S.; Ma, X.X. Identification of novel cell glycolysis related gene signature predicting survival in patients with endometrial cancer. *Cancer Cell Int.* **2019**, *19*, 296. [CrossRef]
41. Endo, Y.; Sugimoto, K.; Kobayashi, M.; Kobayashi, Y.; Kojima, M.; Furukawa, S.; Soeda, S.; Watanabe, T.; Higashi, A.Y.; Higashi, T.; et al. Claudin-9 is a novel prognostic biomarker for endometrial cancer. *Int. J. Oncol.* **2022**, *61*, 135. [CrossRef] [PubMed]
42. Sharma, R.K.; Chheda, Z.S.; Das Purkayastha, B.P.; Gomez-Gutierrez, J.G.; Jala, V.R.; Haribabu, B. A spontaneous metastasis model reveals the significance of claudin-9 overexpression in lung cancer metastasis. *Clin. Exp. Metastasis* **2016**, *33*, 263–275. [CrossRef]
43. Jia, H.Y.; Chai, X.; Li, S.J.; Wu, D.; Fan, Z.M. Identification of claudin-2,-6,-11 and-14 as prognostic markers in human breast carcinoma. *Int. J. Clin. Exp. Pathol.* **2019**, *12*, 2195–2204. [PubMed]
44. Martin, T.A.; Lane, J.; Harrison, G.M.; Jiang, W.G. The Expression of the Nectin Complex in Human Breast Cancer and the Role of Nectin-3 in the Control of Tight Junctions during Metastasis. *PLoS ONE* **2013**, *8*, e82696. [CrossRef]
45. Jiang, W.G.; Watkins, G.; Douglas-Jones, A.; Holmgren, L.; Mansel, R.E. Angiomotin and angiomotin like proteins, their expression and correlation with angiogenesis and clinical outcome in human breast cancer. *BMC Cancer* **2006**, *6*, 16. [CrossRef]
46. Xin, L.; Liu, C.; Liu, Y.; Mansel, R.E.; Ruge, F.; Davies, E.; Jiang, W.G.; Martin, T.A. SIKs suppress tumor function and regulate drug resistance in breast cancer. *Am. J. Cancer Res.* **2021**, *11*, 3537–3557.
47. Fekete, J.T.; Gyorffy, B. ROCplot.org: Validating predictive biomarkers of chemotherapy/hormonal therapy/anti-HER2 therapy using transcriptomic data of 3,104 breast cancer patients. *Int. J. Cancer* **2019**, *145*, 3140–3151. [CrossRef] [PubMed]
48. Martin, T.A.; Watkins, G.; Mansel, R.E.; Jiang, W.G. Loss of tight junction plaque molecules in breast cancer tissues is associated with a poor prognosis in patients with breast cancer. *Eur. J. Cancer* **2004**, *40*, 2717–2725. [CrossRef]
49. Martin, T.A.; Jiang, W.G. Loss of tight junction barrier function and its role in cancer metastasis. *Biochim. Biophys. Acta* **2009**, *1788*, 872–891. [CrossRef]

50. Akashi, T.; Noguchi, S.; Takahashi, Y.; Nishimura, T.; Tomi, M. L-type Amino Acid Transporter 1 (SLC7A5)-Mediated Transport of Pregabalin at the Rat Blood-Spinal Cord Barrier and its Sensitivity to Plasma Branched-Chain Amino Acids. *J. Pharm. Sci.* **2023**, *112*, 1137–1144. [CrossRef]
51. Meurer, M.; Hocherl, K. Deregulated renal magnesium transport during lipopolysaccharide-induced acute kidney injury in mice. *Pflug. Arch.-Eur. J. Physiol.* **2019**, *471*, 619–631. [CrossRef] [PubMed]
52. Goldeman, C.; Andersen, M.; Al-Robai, A.; Buchholtz, T.; Svane, N.; Ozgur, B.; Holst, B.; Shusta, E.; Hall, V.J.; Saaby, L.; et al. Human induced pluripotent stem cells (BIONi010-C) generate tight cell monolayers with blood-brain barrier traits and functional expression of large neutral amino acid transporter 1 (SLC7A5). *Eur. J. Pharm. Sci.* **2021**, *156*, 105577. [CrossRef] [PubMed]
53. Jurkiewicz, D.; Michalec, K.; Skowronek, K.; Nalecz, K.A. Tight junction protein ZO-1 controls organic cation/carnitine transporter OCTN2 (SLC22A5) in a protein kinase C-dependent way. *Biochim. Biophys. Acta-Mol. Cell Res.* **2017**, *1864*, 797–805. [CrossRef] [PubMed]

biomedicines

MDPI

Article

A CAF-Based Two-Cell Hybrid Co-Culture Model to Test Drug Resistance in Endometrial Cancers

Raed Sulaiman [1,†], Pradip De [2,3,4,†], Jennifer C. Aske [2], Xiaoqian Lin [2], Adam Dale [2], Kris Gaster [5], Luis Rojas Espaillat [6], David Starks [6] and Nandini Dey [2,3,*]

1 Department of Pathology, Avera Research Institute, Sioux Falls, SD 57105, USA
2 Translational Oncology Laboratory, Avera Research Institute, Sioux Falls, SD 57105, USA; jennifer.aske@avera.org (J.C.A.); adam.dale@avera.org (A.D.)
3 Department of Internal Medicine, University of South Dakota SSOM, USD, Sioux Falls, SD 57105, USA
4 Viecure, Greenwood Village, CO 80111, USA
5 Assistant VP Outpatient Cancer Clinics, Avera Cancer Institute, Sioux Falls, SD 57105, USA
6 Department of Gynecologic Oncology, Avera Research Institute, Sioux Falls, SD 57105, USA
* Correspondence: nandini.dey@avera.org
† These authors contributed equally to this work.

Abstract: The management of advanced or recurrent endometrial cancers presents a challenge due to the development of resistance to treatments. The knowledge regarding the role of the tumor microenvironment (TME) in determining the disease's progression and treatment outcome has evolved in recent years. As a TME component, cancer-associated fibroblasts (CAFs) are essential in developing drug-induced resistance in various solid tumors, including endometrial cancers. Hence, an unmet need exists to test the role of endometrial CAF in overcoming the roadblock of resistance in endometrial cancers. We present a novel tumor–TME two-cell ex vivo model to test CAF's role in resisting the anti-tumor drug, paclitaxel. Endometrial CAFs, both NCAFs (tumor-adjacent normal-tissue-derived CAFs) and TCAFs (tumor-tissue-derived CAFs) were validated by their expression markers. Both TCAFs and NCAFs expressed positive markers of CAF, including SMA, FAP, and S100A4, in varying degrees depending on the patients, while they consistently lacked the negative marker of CAF, EpCAM, as tested via flow cytometry and ICC. CAFs expressed TE-7 and immune marker, PD-L1, via ICC. CAFs better resisted the growth inhibitory effect of paclitaxel on endometrial tumor cells in 2D and 3D formats compared to the resistance of the tumoricidal effect of paclitaxel in the absence of CAFs. TCAF resisted the growth inhibitory effect of paclitaxel on endometrial AN3CA and RL-95-2 cells in an HyCC 3D format. Since NCAF similarly resisted the growth inhibitor action of paclitaxel, we tested NCAF and TCAF from the same patient to demonstrate the protective action of NCAF and TCAF in resisting the tumoricidal effect of paclitaxel in AN3CA in both 2D and 3D matrigel formats. Using this hybrid co-culture CAF and tumor cells, we established a patient-specific, laboratory-friendly, cost-effective, and time-sensitive model system to test drug resistance. The model will help test the role of CAFs in developing drug resistance and contribute to understanding tumor cell-CAF dialogue in gynecological cancers and beyond.

Keywords: ex vivo resistance platform; tumor–TME cellular model; CAF-mediated resistance; laboratory friendly

1. Introduction

In advanced staged endometrial cancers, the critical challenge, such as in most solid tumors, remains the management of the disease due to the development of resistance to therapy and relapse despite surgery, hormone therapy, and chemotherapy. Resistance to ongoing therapy is attributed to intrinsic refractoriness and resistant tumor cells' emergence during treatment within a short time frame [1]. More than 25% of patients with endometrial cancers are diagnosed at a stage > 1, with an invasive primary tumor, and subsequently, the conditions progresses as a metastatic disease [2]. The most commonly practiced treatment strategies for endometrial cancers involve surgery, followed by a combination of chemotherapy (paclitaxel, carboplatin/cisplatin, and doxorubicin/liposomal doxorubicin), radiation therapy alone or combined with hormonal therapy (American Cancer Society guidelines). According to the NCCN guidelines, the FDA-approved targeted therapies include multi-TKI, lenvatinib, and anti-angiogenic bevacizumab, with the recent inclusion of immunotherapy, pembrolizumab with lenvatinib, or dostarlimab for high TMB, MSI-high dMMR tumors.

The tumor microenvironment (TME) has emerged as an essential critical factor in tumor progression in solid tumors, including gynecological tumors [3]. Once inducted by tumor cells, CAFs engage with both tumor cells and all other components of the TME towards the progression of the disease and the worst outcome. As the most abundant stromal cells within TME, CAFs have a specific role in developing drug resistance in various solid tumors [4,5].

A tumor–CAF liaison exploits the host's stroma in favor of tumor progression in which CAF–tumor cell cross-talk in concert with the rest of the components of the TME as abettors of resistance to treatment, which is the bête noire of therapy [6]. We briefly overviewed the origin, activation, markers, and overall functions of CAF, with a particular reference to how different functions of CAF in an established tumor are functionally connected to the development of resistance to cancer therapy in solid tumors acting as the roadblock to therapy [7]. The current literature established the undeniable role of CAF in conversation with tumor cells and the rest of the components of TME in mediating the development of treatment resistance and a poorer outcome caused by the disease [8]. Recently, we have studied the role of CAFs in the development of resistance to the anti-angiogenic drug in ovarian cancers [9].

CAF-initiated corrupted signaling exists as part of a bi-directional conversation of CAFs with tumor cells [10]. We reviewed one example of mechanistic interactions between tumor cells and CAFs via the Wnt pathway, which acts as *Un Colpevole Comune* (a common culprit) in these cancers [11]. Given this fact, there is an urgent need to gain knowledge about CAF–tumor cell conversation, which will empower us to improve the outcome of the disease through CAF-inclusive therapy. To engage in CAF–tumor dialogue, we need to know about CAF–tumor conversation in a patient-specific manner. Hence, there is an unmet need to design a model for testing the role of endometrial CAFs in overcoming roadblocks of resistance to therapy to manage the disease. Here, we present a novel tumor–TME two-cell patient-specific ex vivo model to test CAFs' role in resisting anti-tumor drugs in a patient-specific manner. Using this hybrid model of co-culture (HyCC) of endometrial CAFs and tumors, we established a patient-specific, laboratory-friendly, cost-effective, and time-sensitive testing system, *N-of-1*, to test the role of CAFs in mediating the development of resistance to paclitaxel.

2. Materials and Methods

2.1. Patient Consent and Tissue Collection

The institutional and/or licensing committee approved all experimental protocols. Informed consent(s) (IRB approved: Protocol Number Study: 2017.053-100399_ExVivo001) was obtained from 53 patients. Resected, unfixed tumor tissue(s) and tumor-adjacent normal tissue(s) were collected from the pathology department. Tissues were collected during surgery in designated collection media as per the guidelines and relevant regulations and provided by the pathologist, depending upon the availability of the tissue on a case-by-case basis. Samples were collected in DMEM/F 12 + Glutamax 500 mL (base) supplemented with HyClone Penicillin-Streptomycin 100 × 100 mL (1%).

2.2. Cell Lines and Reagents

Human uterine fibroblasts (HUF; Primary Uterine Fibroblasts, Cat #PCS-460-010) and endometrial cells (RL-95-2 and AN3CA) were procured from ATCC and were cultured according to standard cell culture procedures. Antibodies for immunocytochemistry (ICC) were bought from Cell Marque, NOVUS, Abcam, Agilent-Dako, and Cell Signaling. All cells were used within 7–8 passages and tested negative for mycoplasma. The antibodies for WB were procured from Cell Signaling, USA.

2.3. Expression of CAF Markers via Flow Cytometry

Cells were trypsin released and rinsed in FACS buffer (1% FBS in phenol red-free RPMI). The cell number was adjusted to 1 × 10^6 cells per sample and resuspended in FACS buffer along with corresponding cell surface antibodies (FAP R&D systems, PD-L1 Miltenyi) and incubated at 4 °C for 20 min. Cells were rinsed twice with FACS buffer, then fixed for 30 min, rinsed, and resuspended in permeabilization buffer (Fix/Perm kit Miltenyi), and the corresponding intracellular antibodies (SAM and S100A4) were added for 30 min. Cells were again rinsed and resuspended in FACS buffer and analyzed on an Accuri C6 cytometer. FCS Express (DeNovo software) was used for analysis (Figure 1). Flow cytometry was performed using SMA-FITC, FAP-PE, S100A4-PerCP, and EpCAM-APC. Cells were stained for 15 min with cell surface antibodies or corresponding isotype control antibodies (Miltenyi). TCAFs and NCAFs from all patients with established CAFs at every passage of the primary culture were tested for the expression of markers to confirm specificity.

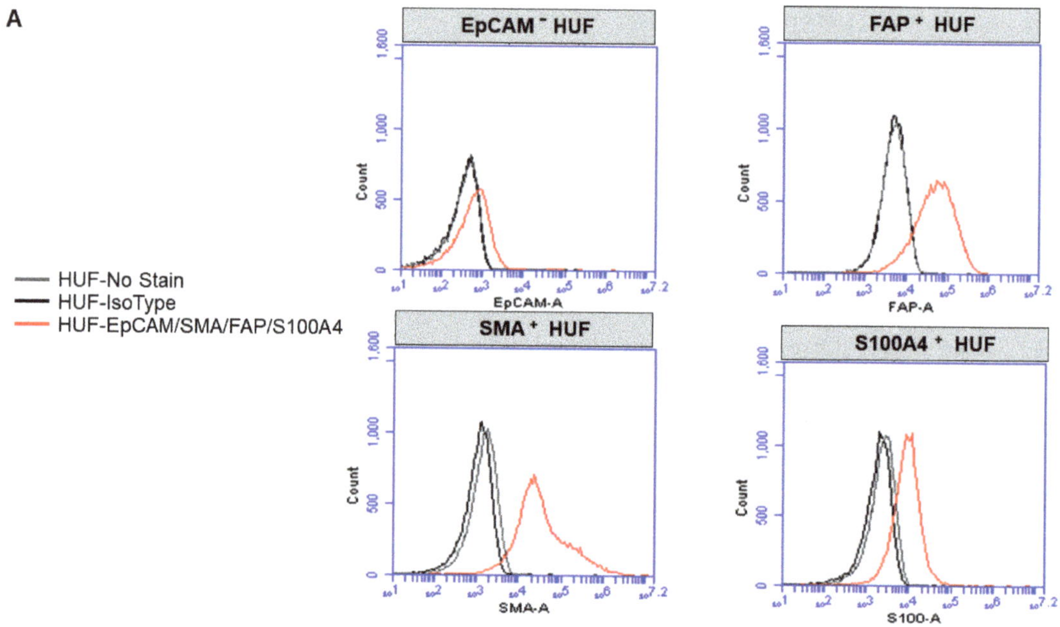

Figure 1. *Cont.*

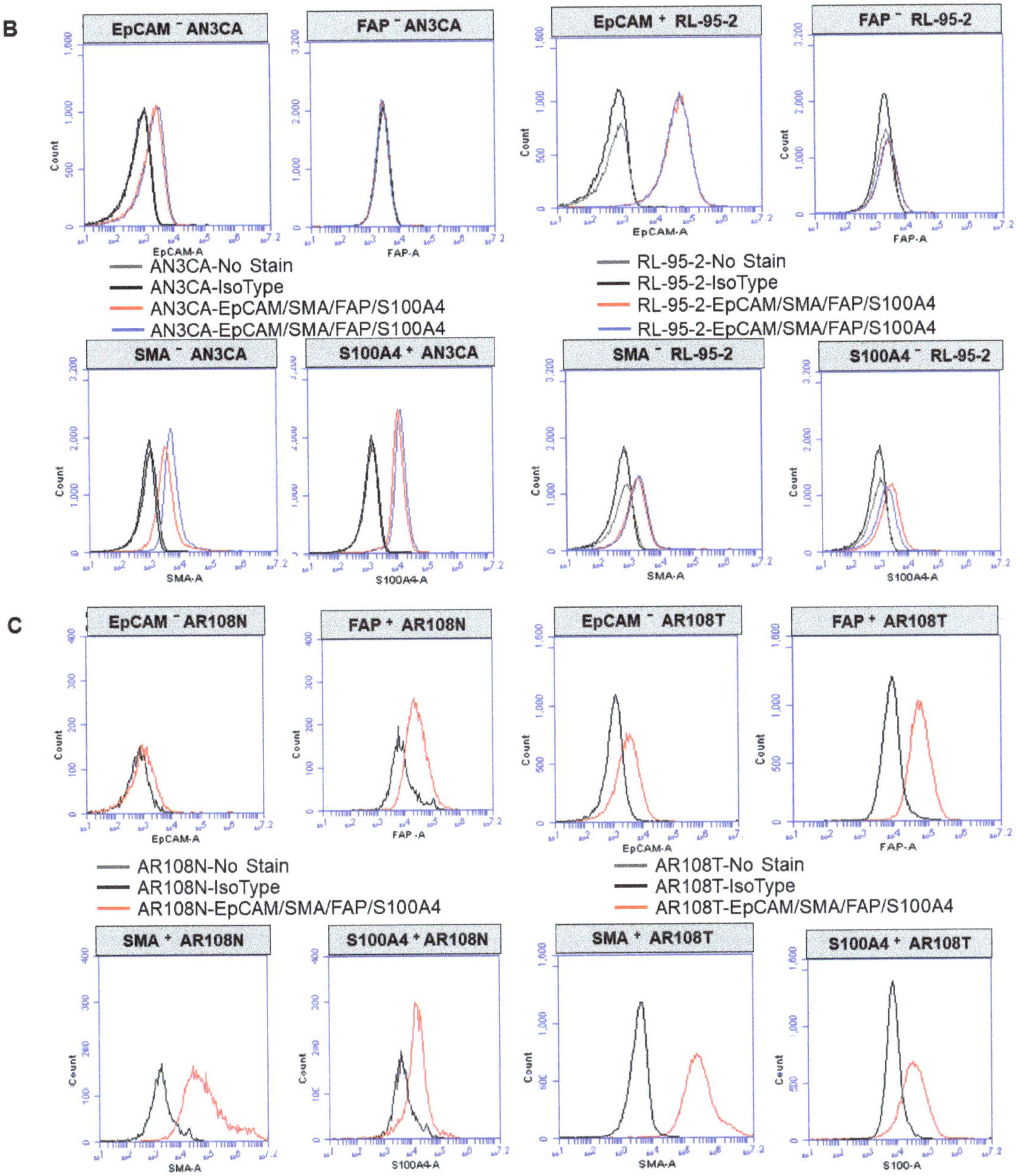

Figure 1. *Cont.*

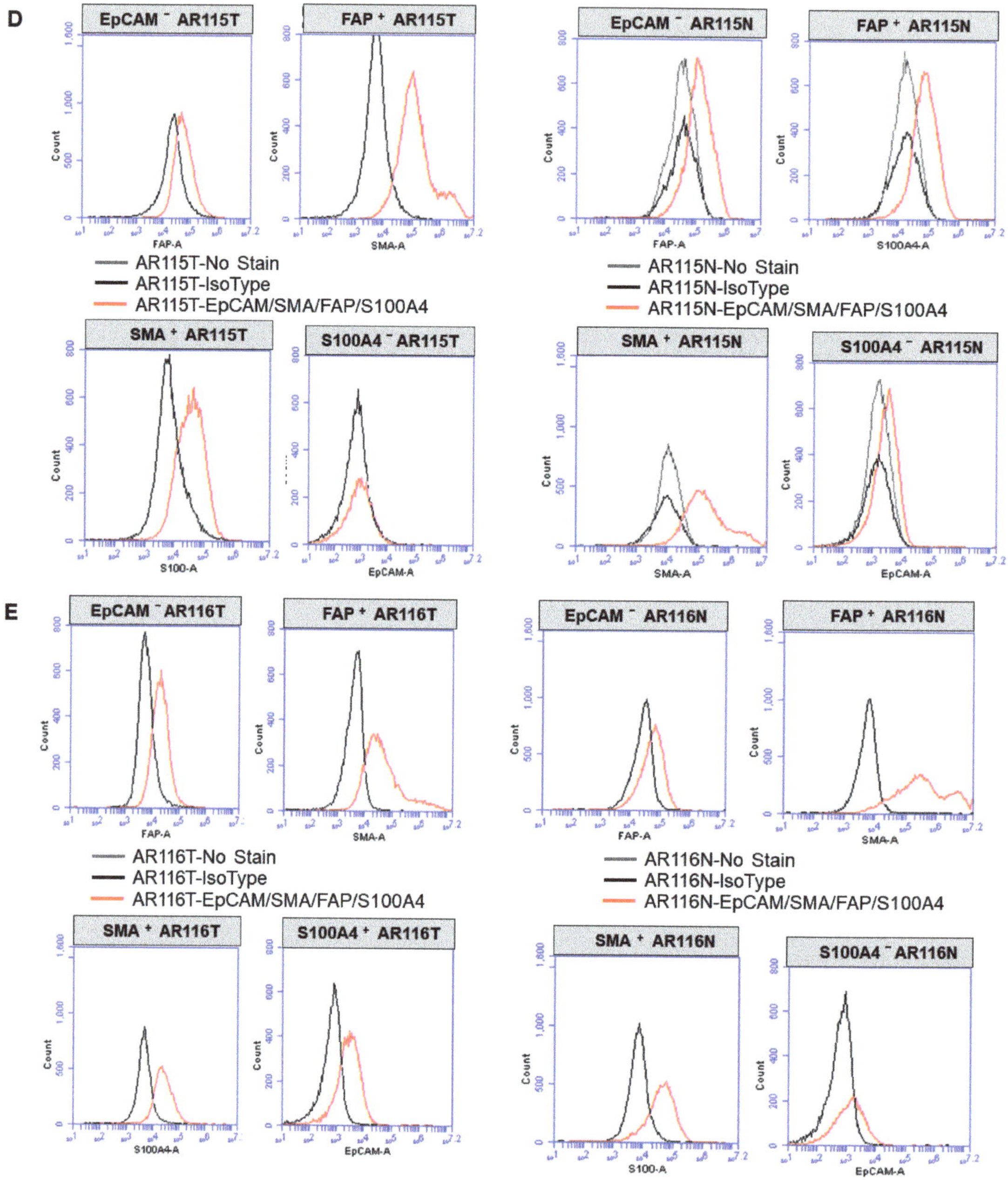

Figure 1. Characterization of enzyme digestion independently cultured CAF via flow cytometry: Flow cytometric (**A–E**) evaluation of marker proteins for CAF was conducted. Expression of positive and negative controls of CAF are tested in internal controls of HUF (**A**) and two endometrial cell lines, AN3CA and RL-95-2 (**B**). Both NCAFs and TCAF were characterized by flow cytometric expression of EpCAM, FAP, SMA, and S100A4 in three representative patients with endometrial cancers (**C–E**). The grey line represents no stain, the black line represents iso-type controls, and the red line represents the protein expression of choice.

2.4. Characterization of CAF via ICC

We characterized the CAFs by testing the subcellular expression of EpCAM, CK 8,18, SMAalpha, S100A4, TE-7, and PD-L1 (Figure 2). ICC was performed with relevant positive controls and negative controls and was evaluated by a pathologist. We used (1) endometrial tumor cells, RL-95-2 and AN3CA, as the positive control for epithelial cells; (2) HUF cells as the positive control for fibroblast cells; (3) endothelial cells, HUVEC as the negative control for EpCAM, CK 8,18, SMAalpha, S100A4, and TE-7 and as the positive control for CD31; (4) NCI-H441 cells as the positive control for PD-L1; (5) MCF7 cells were used as the negative control for PD-L1. The list of antibodies used is presented elsewhere [12].

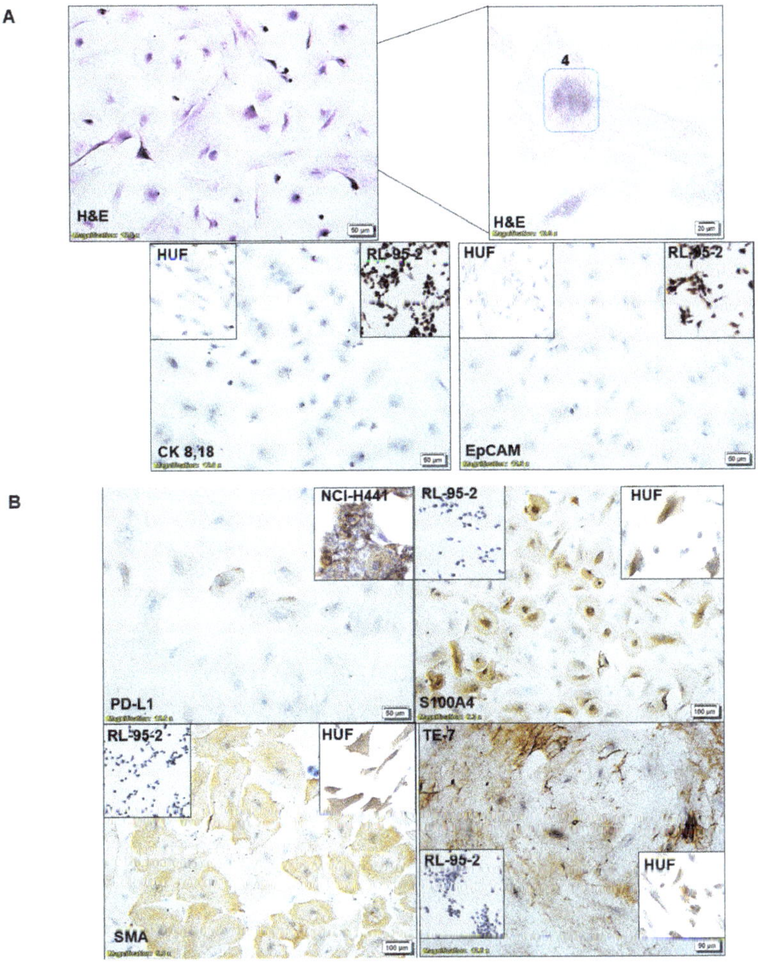

Figure 2. Characterization of enzyme digestion independently cultured CAF via ICC: Multi nuclear CAF (H&E; number in the inset refers to the number of nuclei) were negative for the expression of epithelial markers, EpCAM, and CK 8,18 (**A**). The number "4" in sub-figure A denoted the number of nuclei in the CAF. The expression of the proteins in RL-95-2 was used as the positive control, and expression in HUF was used as the negative control (insets). Expression of positive CAF markers, SMA, S100A4, TE-7, and immune marker PD-L1 are presented (**B**). The expression of the proteins in RL-95-2 was used as negative controls, and expressions in HUF and NCI-H441 were used as positive controls (insets).

2.5. *Establishment of Patient-Specific Ex Vivo Tumor–TME Two-Cell Model of Hybrid Co-Culture (HyCC)*

We designed a novel matrigel-based two-cell HyCC model comprising patient-derived primary pre-characterized CAFs and tumor cells. Figure 3 presents a diagrammatic representation of the detailed plan of the HyCC in our study. We used DiO and DiI, two family members of lipophilic fluorescent stains used to label cell membranes and other hydrophobic structures, as long-term tracer dyes. DiO stain is a green, fluorescent, lipophilic carbocyanine dye that is widely used as a lipophilic tracer. DiI is a lipophilic orange-red fluorescent dye that diffusely stains the entire cell, and it is spectrally similar to tetramethylrhodamine. The HyCC consisted of a preparatory phase and an experiment phase. The preparatory phase of the HyCC was composed of (1) the validation of positive and negative CAF markers via flow cytometry, ICC, and qRT-PCR (depending on the number of CAFs available in the same passage from which the HyCC set up); (2) the staining of verified TCAFs and NCAFs and endometrial tumor cells, AN3CA and RL-95-2 by DiO and DiI stain, respectively; (3) the testing of fluorescence signals in cells via flowcytometry and immuno-fluorescence; (4) the comparison of 2D and 3D growth patterns of unstained and stained cells; (5) the comparative testing of the effect of paclitaxel on 2D and 3D growth patterns between unstained and stained cells. The experiment phase was composed of (1) the overnight plating of DiO-CAFs on plates to obtain an 80–95% confluent monolayer of CAF; (2) the plating of DiI-stained tumor cells on DiO-CAFs with and without paclitaxel in 2D and 3D formats. Media were changed every 48 h. Double-fluorescence signals from the live cells in cultures were recorded, along with bright field photo-micrographs at different time points using dry objectives in Olympus BX43 Microscope using cellSens 1.18 LIFE SCIENCE IMAGING SOFTWARE (OLYMPUS CORPORATION). Semi-quantification was performed based on the fluorescence intensities of DiI tumor cells on DiO-TCAFs/NCAFs from 5–6 random microscopic fields of independent experiments (performed in quadruplicates). Statistical significance was determined by calculating Student's t-test at $p < 0.05$.

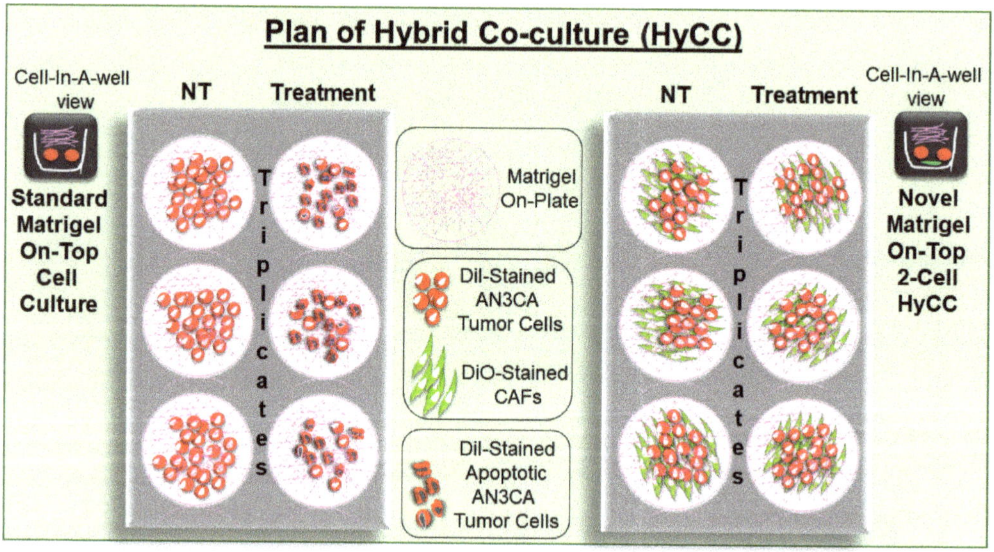

Figure 3. Schematic plan of the Hybrid Co-Culture (HyCC): The novel matrigel 2-cell HyCC of CAFs stained with DiO- and DiI-stained AN3CA cells was presented as compared to the standard matrigel cell culture for 3D clonogenic growth of cancer cells. The model was used to test the effect of the cytotoxic drug paclitaxel on tumor cells.

2.6. Testing Resistance of CAFs to Paclitaxel Using Hybrid Co-Culture of Tumor–TME Two-Cell Model

First, we tested the growth of AN3CA and RL-95-2 in HyCC on TCAF in 2D and 3D formats. Then, we tested the effect of CAF on resisting the tumoricidal effect of paclitaxel in (1) AN3CA in HyCC on TCAF in the 3D format, (2) RL-95-2 in HyCC on TCAF in the 3D format, (3) AN3CA in HyCC on NCAF in the 3D format, (4) RL-95-2 in HyCC on NCAF in the 3D format and (5) AN3CA in HyCC on both NCAF and TCAF from the same patient's samples in both 2D and 3D formats. Media were changed every 48 h. Double-fluorescence signals from the live cells in cultures were recorded, along with bright field photo-micrographs at different time points using dry objectives in Olympus BX43 Microscope using cellSens 1.18 LIFE SCIENCE IMAGING SOFTWARE (OLYMPUS CORPORATION). Semi-quantification was performed based on the fluorescence intensities of DiI tumor cells on DiO-TCAFs/NCAFs from 5–6 random microscopic fields of independent experiments (performed in quadruplicates). Significance was determined by calculating Student's t-test at $p < 0.05$.

3. Results

3.1. Ex Vivo Primary Culture- and Marker-Based Verification of CAF

The institutional and/or licensing committee approved all experimental protocols. Informed consent(s) (IRB approved Protocol Number Study: 2017.053-100399_ExVivo001) was obtained from 53 consecutive patients with endometrial tumors at any stage/grade of the disease undergoing surgery with or without a pre-treatment/history of any previous carcinoma. Primary cultures of CAFs were created from the resected samples of tumors and tumor-adjacent normal samples from patients with endometrial cancers, as per the guidelines and relevant regulations provided by the pathologist who performed grossing.

Validation of TCAF and NCAF in the ex vivo primary culture of the patient's samples was performed. Characterization of the cultured TCAF (from the resected samples of tumor samples) and NCAF (from the resected tumor-adjacent normal samples) via flow cytometry (Figure 1A–E) and ICC (Figure 2A,B) was performed. Both TCAFs and NCAFs expressed positive markers of CAF, including SMA, FAP, and S100A4, to a varying degree, depending on the patients, but they were consistently negative for negative markers of CAF, EpCAM (a positive marker of epithelial tumor cells). The expression of S100A4 was qualitative in a patient-specific manner, with it being absent in some patients, as represented in Figure 1. The expression of positive and negative controls of CAF were tested in internal controls, HUF (A), and two endometrial cell lines, AN3CA and RL-95-2 (B).

CAFs have the morphology characteristic of fibroblasts with a single nucleus, although we sometimes observed multinucleated CAFs, as represented in Figure 2A's upper panel. CAFs were negative for the ICC expression of epithelial markers, EpCAM and CK 8,18 (Figure 2A's lower panel). The ICC expression of EpCAM and CK 8,18 in RL-95-2 was used as the positive control, and the ICC expression in HUF was used as the negative control (insets). The CAFs ICC expressed positive fibroblast markers, SMA, S100A4, TE-7, and immune marker, PD-L1 (Figure 2B). The ICC expression of SMA, S100A4, TE-7, and PD-L1 proteins in RL-95-2 was used as the negative controls, and expressions in HUF (for SMA, S100A4, and TE-7) and NCI-H441 (for PD-L1) were used as the positive controls (insets).

3.2. Patient-Specific Ex Vivo Tumor–TME Two-Cell Model of Hybrid Co-Culture (HyCC)

We first tested the growth pattern of endometrial tumor cells in 2D and 3D formats of HyCC on TCAFs, as diagrammatically presented in Figure 3. The photomicrographs in Figure 4A,B show that the growth of AN3CA and RL-95-2 was significantly more extensive on day 7 compared to day 0, which was semi-quantified by the fluorescence intensity of the DiI-stained area in both 2D and 3D formats. The DiO-CAFs' fluorescence intensities, on the other hand, in the 3D format, were either increased or not significantly changed after 7 days. In contrast, the fluorescence intensity of DiO-CAF was found to be reduced in the 2D format. Figure 3 compares the standard model of the matrigel 3D clonogenic growth of cancer cells with the HyCC presented in the "Cell-In-A-Well View", which was designed to test the functional role of CAFs on the effect of the cytotoxic drug paclitaxel on tumor cells.

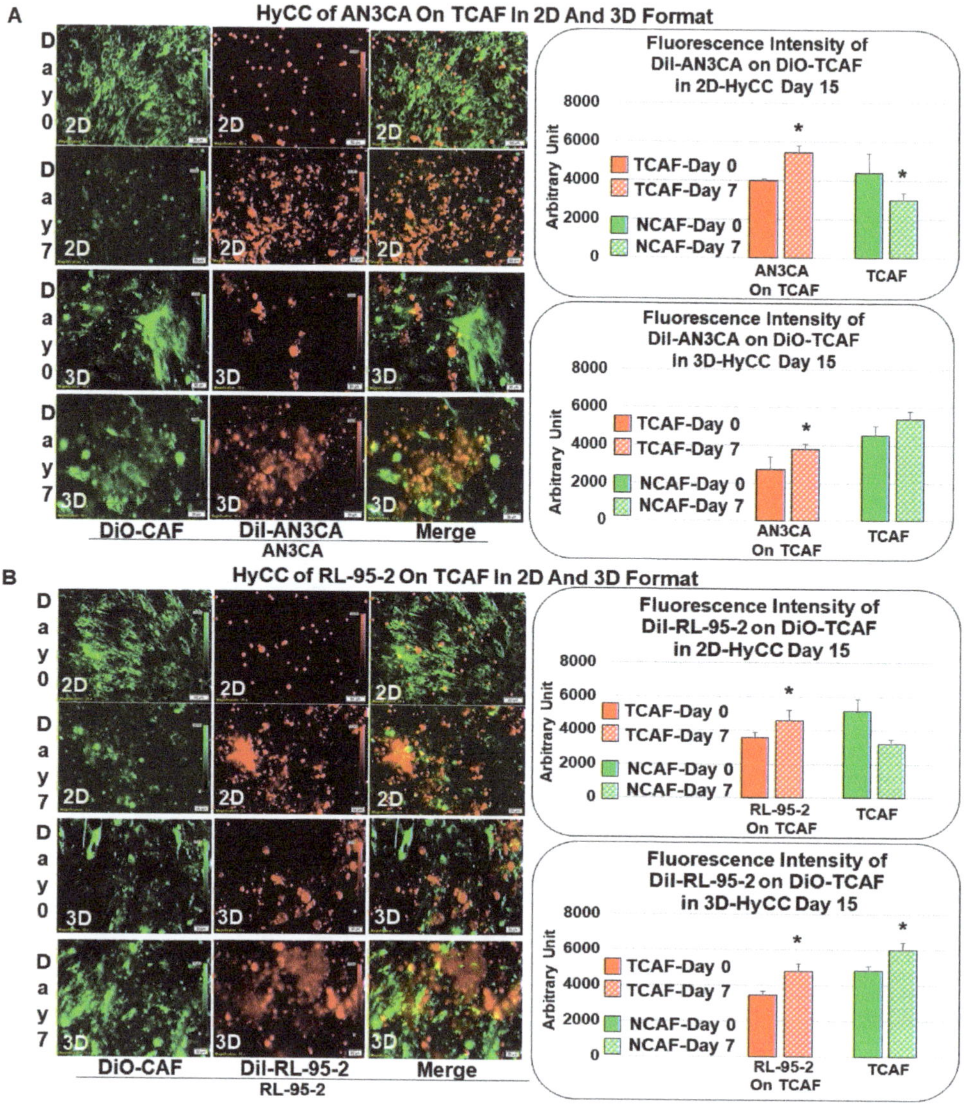

Figure 4. Growth of endometrial cells AN3CA and RL-95-2 for TCAF on TCAF in HyCC in 2D and 3D formats: Dil-stained AN3CA (**A**) and RL-95-2 (**B**) were plated on DiO-stained TCAF, and their 2D and 3D matrigel growth was recorded. The media were changed every 48 h. Photomicrographs were taken on day 0 (within 24 h for the 3D format as pictures are difficult to focus at zero hours in 3D) and on day 7. The fluorescence intensity of tumor cells on TCAF was semi-quantified from 5 random microscopic fields of independent experiments (performed in quadruplicates). Solid bars represent the fluorescence intensities of cells at day zero, and filled (sphere) bars represent the fluorescence intensities of cells at day 7. * Significance was determines by calculating Student's t-test at $p < 0.05$.

3.3. Testing the Effect of CAFs in Resisting Paclitaxel Using Ex Vivo Platform-Based Co-Culture of Tumor–TME Two-Cell Model

To test the effect of CAFs on the growth inhibitory effect of paclitaxel on endometrial cells, first, we tested the HyCC growth pattern of these cells in 2D and 3D formats (Figure 4A,B). Dil-stained AN3CA (A) and RL-95-2 (B) were plated on DiO-stained TCAF, and their 2D and 3D matrigel growth was recorded. Media were changed every 48 h. Photomicrographs were taken on day 0 (within 24 h for the 3D format as pictures are difficult to focus at zero hours in 3D) and on day 7. The fluorescence intensity of tumor cells on TCAF was semi-quantified from five random microscopic fields of independent experiments (performed in quadruplicates). Both AN3CA and RL-95-2 growth was significantly increased after 7 days of culturing. However, the clonogenic 3D growth of AN3CA is characteristically different in terms of morphology from that of RL-95-2 cells.

Once we established regular growth patterns of the cells via HyCC, we tested the effect of CAFs on the growth-inhibitory action of paclitaxel. The treatment with paclitaxel decreased the growth of both AN3CA and RL-95-2 in real time (2D), as well as in the 3D format, as shown in Figure 5A. This growth inhibitory property of paclitaxel was used as the baseline to measure the role of CAF in resisting the effect of paclitaxel on tumor cells. Having ascertained the growth pattern of DiI tumor cells of DiO-CAFs in HyCC and tested the growth inhibitory effect of paclitaxel on tumor cells, we sought to test the capacity of CAFs to resist the growth inhibitory effect of paclitaxel on tumor cells. Figure 5(Bi,Bii) shows that TCAF resisted the growth inhibitory effect of paclitaxel on both tumor cells on day 7. The treatment of cells with paclitaxel in the absence of TCAF (AN3CA-only and RL-95-2-only columns of Bi and Bii, respectively) demonstrated the profound loss of growth, as shown in the photomicrographs and bar diagrams in the inset. Once we demonstrated the role of TCAFs in resisting the growth-inhibitory effect of paclitaxel, we wanted to test whether NCAFs from another patient would exhibit a similar effect on TCAFs. Figure 6(Ai, Aii) shows that NCAFs' capability to resist the growth-inhibitory effect of paclitaxel was comparable to that of TCAFs. Interestingly, we observed no change in the DiO-CAFs, both TCAFs and NCAFs, before and after paclitaxel in the 3D format and the 2D format. We observed that both TCAFs and NCAFs from different patients similarly resisted the growth-inhibitory effect of paclitaxel on endometrial tumor cells. Considering the differences in the pathological characteristics of the tumors from which TCAFs and NCAFs were derived, it is possible that the similar resisting effects (to paclitaxel) observed in both patients were inherent to the characteristics of the TME in those two patients. Hence, from the experiments, it was unclear whether NCAFs and TCAFs behaved similarly due to the underlying pathological differences in the TMEs of the two patients. Therefore, we then tested TCAFs and NCAFs derived from the same patient to demonstrate that both TCAFs and NCAFs resisted the growth-inhibitory effect of paclitaxel in AN3CA cells in both 2D and 3D formats (Figure 7(Ai,Aii)). The insets show the growth inhibitory effect of paclitaxel alone on the tumor cells in the absence of CAFs. We also observed that the DiO-CAFs' fluorescence intensities in the 3D format were either increased parallel to the growth of tumor cells or did not significantly change after 7 days of HyCC. In contrast, the DiO-CAFs; fluorescence intensities were found to be reduced in the 2D format on every occasion (Figures 4 and 5).

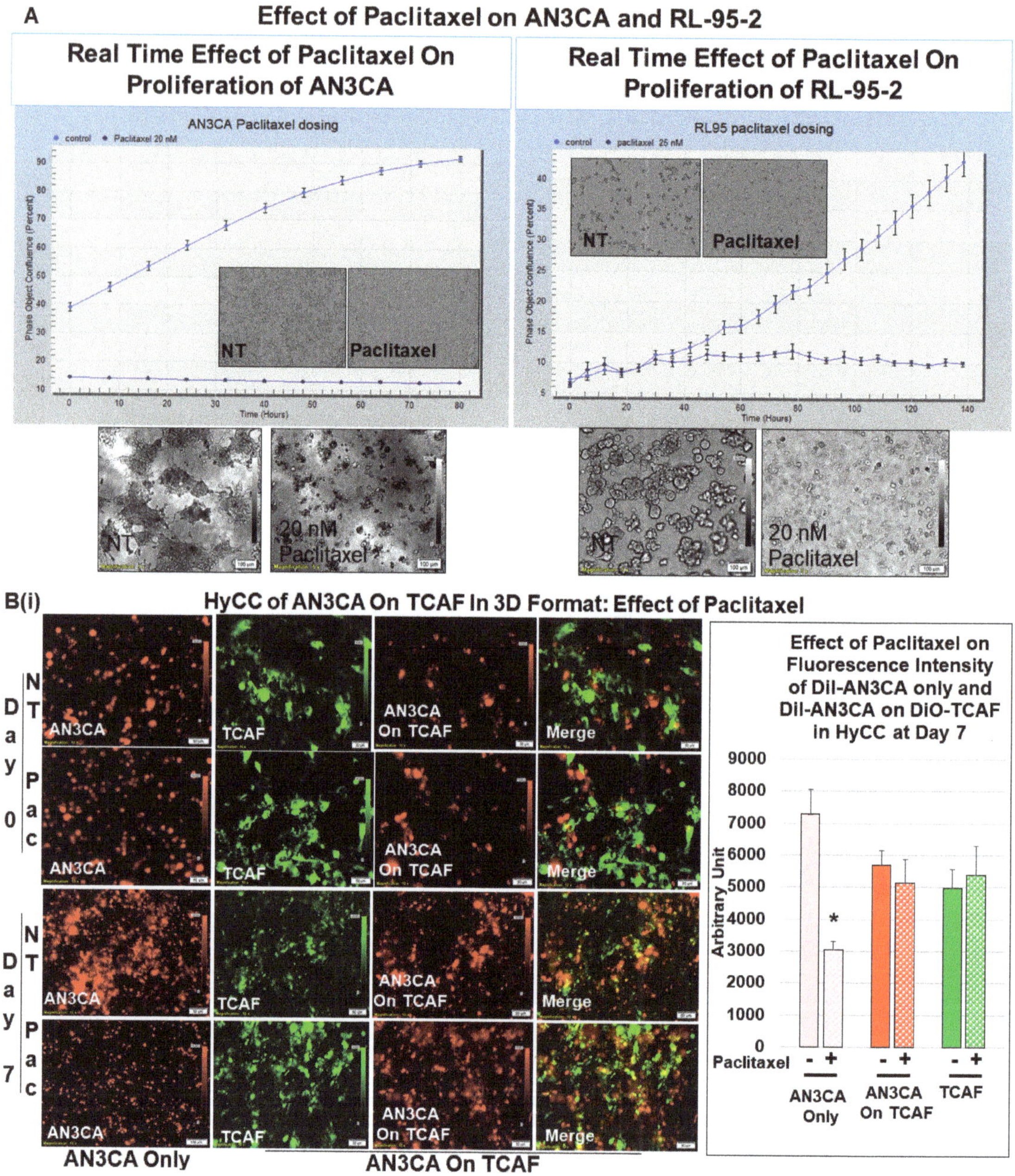

Figure 5. *Cont.*

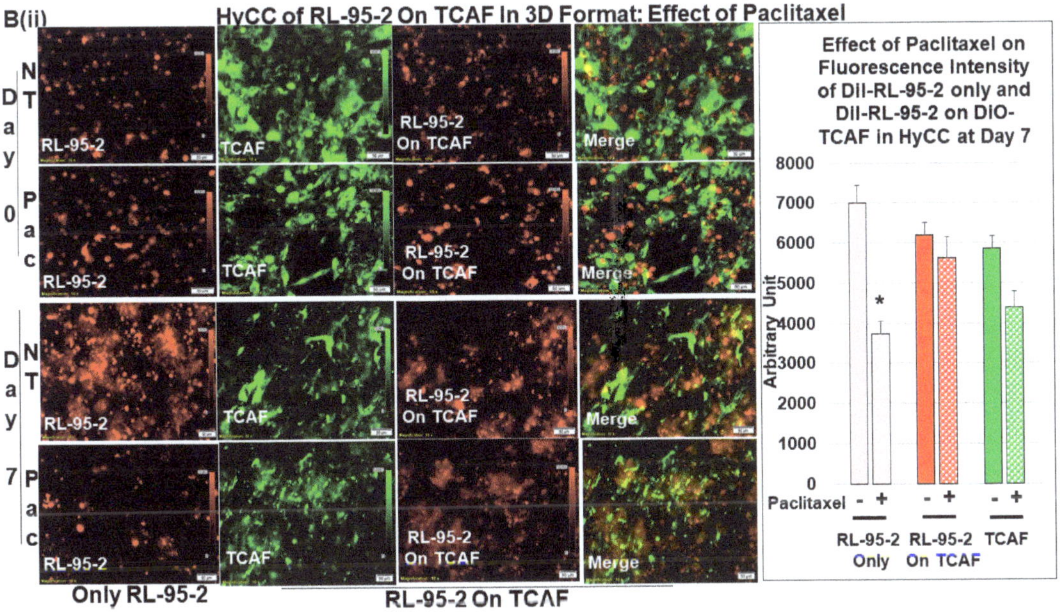

Figure 5. Effects of paclitaxel on the growth of endometrial tumor cells cultured with or without TCAF: Growth inhibitory effect of paclitaxel in real-time 2D growth and 3D on-matrigel growth of AN3CA and RL-95-2 is presented (**A**). Effect of paclitaxel on the same cells co-cultured on the TCAFs (**B**). TCAF resisted the growth inhibitory effect of paclitaxel on AN3CA (**Bi**) and RL-95-2 (**Bii**) in an HyCC 3D format. The media were changed every 48 h. The fluorescence intensity of tumor cells on TCAF was semi-quantified from 5 random microscopic fields of independent experiments (performed in quadruplicates). Solid bars represent the fluorescence intensities of cells at day zero, and filled (sphere) bars represent the fluorescence intensities of cells at day 7. * Significance was found by calculating Student's t-test at $p < 0.05$.

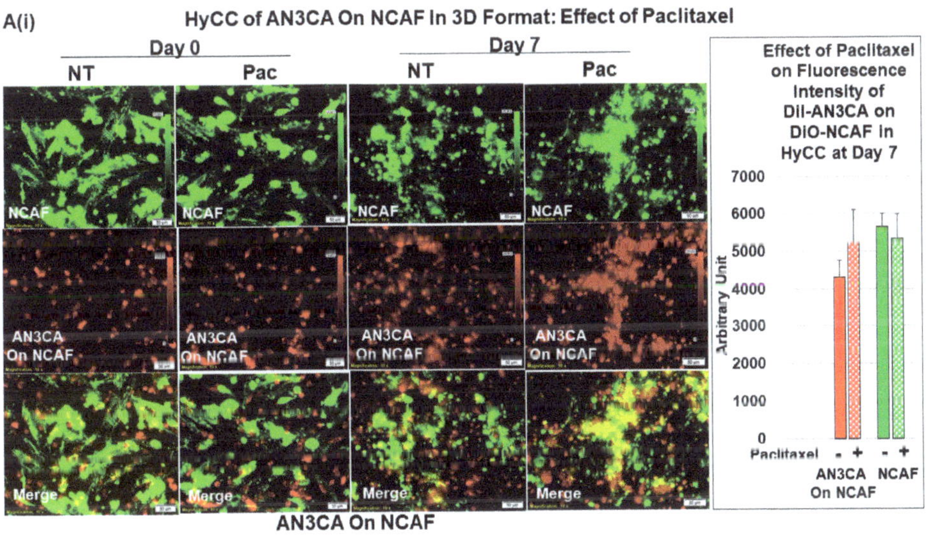

Figure 6. *Cont.*

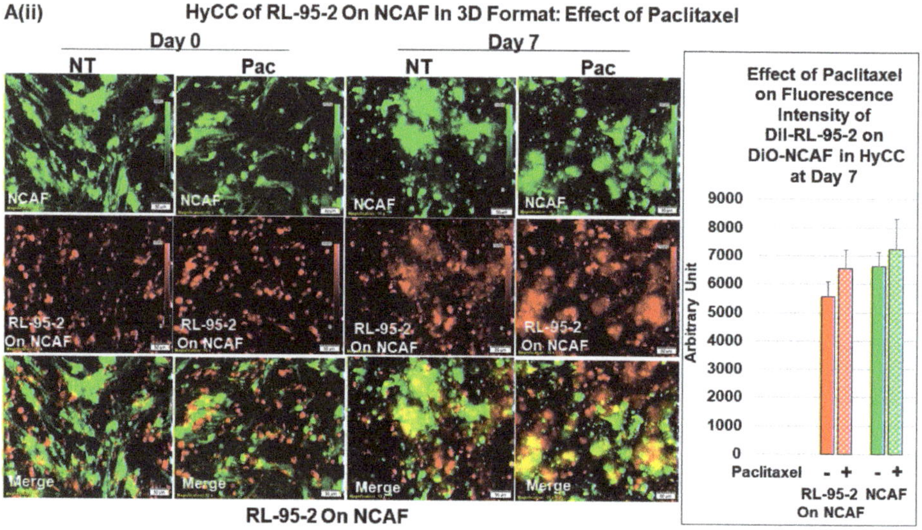

Figure 6. Effects of paclitaxel on the growth of endometrial tumor cells cultured on NCAF: Effect of paclitaxel on the endometrial cells co-cultured on the NCAFs (**A**). NCAFs resisted the growth inhibitory effect of paclitaxel on AN3CA (**Ai**) and RL-95-2 (**Aii**) in an HyCC 3D format. The media was changed every 48 h. The fluorescence intensity of tumor cells on TCAF was semi-quantified from 5 random microscopic fields of independent experiments (performed in quadruplicates). Solid bars represent the fluorescence intensities of cells at day zero, and filled (sphere) bars represent the fluorescence intensities of cells at day 7.

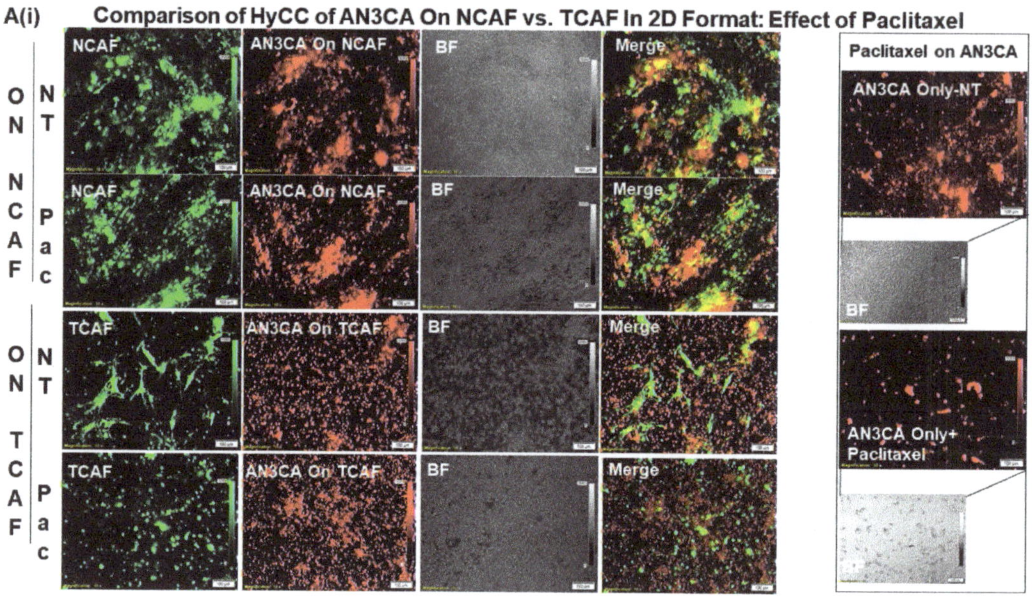

Figure 7. *Cont.*

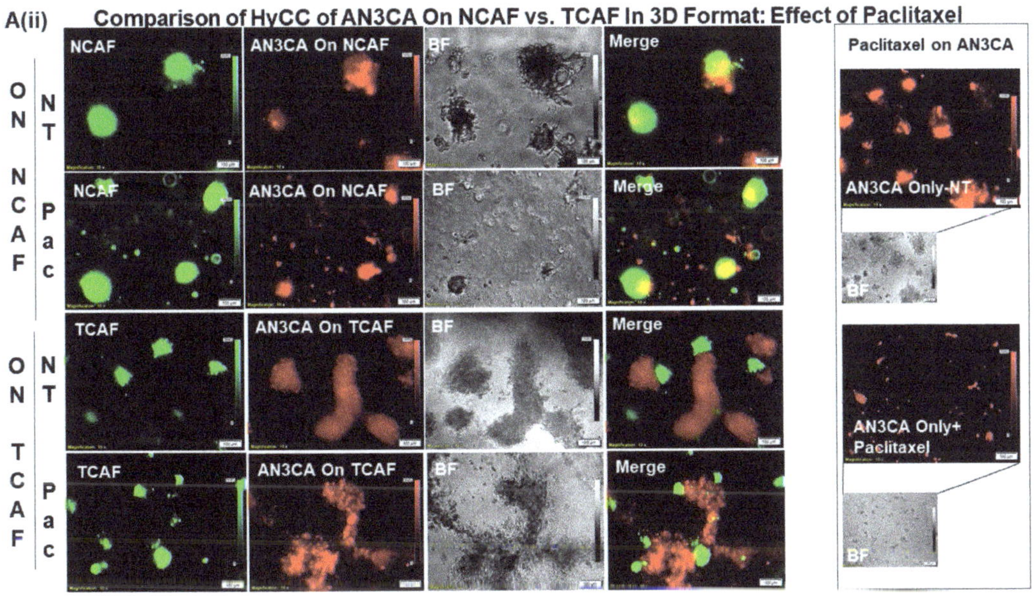

Figure 7. Effects of paclitaxel on the growth of endometrial tumor cells cultured on NCAF and TCAF of the same patient: Protective Action of NCAF-P4 and TCAF-P5 of the same patient in resisting the tumoricidal effect of paclitaxel in AN3CA in 2D (**Ai**) and 3D matrigel (**Aii**) formats are presented. Insets (**Ai,Aii**) show the growth inhibitory effect of paclitaxel in DiI-stained regular AN3CA cells, which were not plated on CAFs (without HyCC). The media were changed every 48 h. The fluorescence intensity of tumor cells on TCAF was semi-quantified from 5 random microscopic fields of independent experiments (performed in quadruplicates).

4. Discussion

Cancer progression has long been understood to be fundamentally determined and characterized by dynamic changes in signaling between the stroma and embedded tumor cells [13]. CAFs, which are often found in primary and metastatic solid tumors, including endometrial cancers, are highly versatile, plastic, and resilient cells that are essentially involved in cancer progression and treatment failure. As the most involved contributing members of TME by volume, CAFs have an undeniable role in the progression of advanced disease [4,5,14,15] and in the development of treatment resistance in several solid tumors [16–19]. In the past several years, a number of preclinical and translational studies revealed several functions of CAFs in mediating critical tumor phenotypes to cause a poor outcome for patients with the disease [20]. The functions mentioned above of CAFs are just beginning to reveal an array of direct/indirect interactions with tumor cells and the rest of the components of TME, tumor-associated immune cells, and tumor-associated endothelial cells [21–23].

From the point of view of the entirety of the tumor, the role of CAF has been mathematically determined. Benjamin Wölfl et al. mathematically framed the tumor–TME interaction as a model of an evolutionary game [24] to conceptualize and analyze biological interactions where the tumor cells' fitness is not only influenced by their own genomic traits but also by the traits of cells within the TME, including CAFs. Their model implied that the progression of cancer is an interactive evolutionary competition between these different cell types, which is an interaction that can be explained through Lotka–Volterra competition equations and their extension term, the term "Deadlock game" and the term "Leader game", in the context of the presence or absence of drugs and/or cancer-associated fibroblasts. Building on the concept, Kaznatcheev et al. [25] demonstrated that cancer-associated fibroblasts qualitatively switch the type of game from "Leader to Deadlock" being played by the in vitro population using their non-small-cell lung cancer model. In their system, they viewed an untreated tumor as being similar to DMSO + CAF, and thus, played the Leader game. A treatment with Alectinib (Alectinib + CAF) or eliminating CAFs (through a stromal-directed therapy) switched the game into a "Deadlock game." Our data conceptually support their model. Our study experimentally explains their extension terms of "Deadlock game" and "Leader game". Our model provides an experimental testing platform of the "Deadlock game" mode with a chemotherapy drug in culture, from which it switches to the 'Leader game' mode in the presence of CAF (paclitaxel-treated cells plated on CAFs in HyCC). Their report validates our ex vivo two-cell HyCC-based model of drug resistance, which can be used to test clinically viable drug combinations and outcome data in the future.

In lieu of a deterministic role of CAF influencing the disease outcome [4,26], several studies provided evidence that supports CAF-inclusive therapy for the clinical management of the disease [27]. Indeed, 37 CAF-directed/inclusive clinical trials have been currently instituted and are ongoing at the NIH (https://clinicaltrials.gov/ct2/home accessed on 12 December 2022). The majority of these clinical trials are instituted in advanced/metastatic solid tumors (*ClinicalTrials.gov Identifier: NCT05547321*), including pancreatic ductal adenocarcinoma (PDAC), hepatocellular carcinomas, and malignant tumors of the breast, colon, prostate, lung, and ovary, explaining the interest of the scientific community in the clinical relevance of CAF in these organ types, especially in PDAC (*ClinicalTrials.gov Identifier: NCT05262855*). Although CAF-directed clinical trials are just beginning to reveal the clinical relevance of CAFs, the role of CAFs in endometrial cancers, among other gynecological cancers, is surprisingly limited. We have recently reported the first study to prove the clinical relevance of CAF in endometrial cancers [12]. As more of these studies in the future will strengthen our knowledge of CAF signaling in endometrial cancers, we will need to develop a personalized approach to test the role of CAF in developing resistance to drug/combination(s). Here, we designed a new model system to test the function of patient-derived endometrial CAFs in resisting the growth-inhibitory effect of a chemotherapy drug, paclitaxel. Paclitaxel is routinely used along with platinum agents to treat advanced endometrial cancers, and the administration of paclitaxel (weekly) as a single agent to patients with recurrent or metastatic disease is the most commonly practiced approach in clinics [28,29] (https://www.cancer.org/ accessed on 12 December 2022).

We observed a discrepancy in DiO-CAFs' fluorescence intensities between the 2D and 3D formats after 7 days of HyCC. As expected, the fluorescence intensities of DiO-CAFs in 3D formats were increased with time in contrast to the fluorescence intensities of DiO-CAFs in 3D formats. The primary components of matrigel are four major basement membrane ECM proteins, including laminin and collagen IV. We used BD Matrigel™ Basement Membrane, which contains laminin, collagen IV, TGF-beta, an epidermal growth factor, an insulin-like growth factor, and a fibroblast growth factor. Since 3D culturing involved using the growth media with matrigel, it is theoretically possible that the CAFs growth pattern was additionally supported by matrigel, with special reference to TGF-beta and fibroblast growth factor. We also noted a characteristic morphological formation of CAFs, both TCAFs and NACFs, when the cultures were in 3D matrigel. Our novel model will provide a unique option to test the resistance to a drug or a combination in a primary CAF derived from the post-surgical tissues of either particular patient planning to undergo prospective adjuvant therapy or post-surgery surveillance.

The power of our model is that, though it was developed and tested using endometrial CAFs, it can be adapted to any organ type of cancer(s) for testing the development of resistance to adjuvant therapy drugs/combination(s) post-surgery. Recently, we studied the role of patient-derived primary ovarian CAFs in resisting the anti-angiogenic effect of lenvatinib on HUVEC cells [9]. Our ex vivo two-cell HyCC-based model of drug resistance provides a cost-effective and laboratory-friendly way to establish CAF, and then test the functions of fibroblasts in each patient with advanced and recurrent endometrial cancers and other solid tumors. Our enzymatic digestion-independent method of establishment of CAF needs a simple method of culturing, followed by CAF characterization via standard flow cytometry, ICC, and qRT-PCR. The HyCC model is a vital fluorescent dye-based 3D matrigel assay that can be performed in any standard laboratory, including a community-based cancer center such as the Avera Cancer Institute. Thus, the model can serve as a patient-specific real-time testing modality for the function of fibroblasts in patients with advanced and recurrent endometrial cancers. The single most significant limitation of the model is that the number of CAFs derived from the available patient tissue limits the experimental setup. Hence, the testing power of the model is tissue-limited and can only be applied to test the development of resistance to adjuvant therapy. We have tried to use 2–3 biopsy tissue cores to test the amount of the volume of the starting tissue sample, with no success. Additionally, in our study, some tumor tissue samples did not yield CAFs. It is possible that the status of CAF is a determinist factor here, as most of these tissues were obtained from patients with grade 1/stage I diseases. Additionally, as the starting material is a post-surgical resected tumor tissue sample, the model will not be applicable to neoplasms of inaccessible organ types, such as in some central nervous system tumors, unless a tumor can be obtained following surgical resection.

As the results of current CAF-directed clinical trials will be incorporated into the current literature, we will begin to become more confident about a CAF-inclusive treatment plan to manage the advanced form of the disease, especially in the context of the development of treatment resistance in solid tumors. Our model system can provide a unique opportunity for the personalized system to test the development of resistance to the same drugs/combinations received by patients as an adjuvant treatment in clinics. During the entire course of the development and management of a tumor, from the preneoplastic stage to the terminal stage of metastasis and resistance, CAFs play an undeniably complex and multifaceted role. Acknowledging the challenges of CAF biology, the contribution of CAF cannot be ignored as a mechanistic determinant of resistance and clinical outcome of a treatment modality, chemotherapy, radiotherapy, targeted therapy, and immunotherapy. The study of CAF biology is on the cusp of the next wave of discovery that will facilitate the tailored targeting of the cancer types in combination with targeting tumor cells to optimize the clinical benefit and disease management.

5. Patent

The study presented in the MS is part of a patent application (United States Patent and Trademark Office); application number 16/875,910.

Author Contributions: R.S.: The pathologist provided the evaluation of CAF based on IHC staining. P.D.: The Senior Scientist interrogated the genomic alteration of each tumor sample and matched the genomic alteration with the drug combination for the testing. P.D. helped in the analysis of data and writing the MS. J.C.A.: performed the flow cytometry work. J.C.A. supervised the ex vivo study logistics and helped with the EndNote library work. X.L.: The Research Assistant Lead standardized and performed tissue processing, ICC staining for CAF markers, H&E stains, and IHC. XL maintained the record of the histology work. A.D.: The Research Associate obtained patient consent and provided technical assistance with record keeping. A.D. helped in the preparation of figures, tables, and statistical correlations. K.G.: They provided insight into the overall logistical management of the ex vivo study and coordinated the clinical logistics of the study. L.R.E.: The surgeon provided clinical insight into endometrial and ovarian tumors and corresponding blood samples. D.S.: The surgeon provided clinical insight into endometrial and ovarian tumors and corresponding blood samples. N.D.: The Senior Scientist conceptualized and supervised the study, wrote the MS, and analyzed the data. All authors have read and agreed to the published version of the manuscript.

Funding: The study was funded entirely by Avera Cancer Institute.

Institutional Review Board Statement: Anonymized tissue samples were collected at surgery from patients with endometrial cancers following their informed (IRB approved: Protocol Number Study: 2017.053-100399_ExVivo001) consent.

Informed Consent Statement: Informed (IRB approved: Protocol Number Study: 2017.053-100399_ExVivo001) consent for giving resected tissue was obtained from 72 enrolled patients with endometrial cancers.

Data Availability Statement: Not applicable.

Acknowledgments: We acknowledge Avera Cancer Institute for funding the entire study. We acknowledge every patient and their family for their participation in the ex vivo study at the Avera Cancer Institute.

Conflicts of Interest: The authors declare no conflict of interest.

References

1. Brasseur, K.; Gevry, N.; Asselin, E. Chemoresistance and targeted therapies in ovarian and endometrial cancers. *Oncotarget* **2017**, *8*, 4008–4042. [CrossRef] [PubMed]
2. Plataniotis, G.; Castiglione, M.; Group, E.G.W. Endometrial cancer: ESMO Clinical Practice Guidelines for diagnosis, treatment and follow-up. *Ann. Oncol.* **2010**, *21* (Suppl. S5), v41–v45. [CrossRef] [PubMed]
3. Belli, C.; Trapani, D.; Viale, G.; D'Amico, P.; Duso, B.A.; Della Vigna, P.; Orsi, F.; Curigliano, G. Targeting the microenvironment in solid tumors. *Cancer Treat. Rev.* **2018**, *65*, 22–32. [CrossRef] [PubMed]
4. Chen, Y.; McAndrews, K.M.; Kalluri, R. Clinical and therapeutic relevance of cancer-associated fibroblasts. *Nat. Rev. Clin. Oncol.* **2021**, *18*, 792–804. [CrossRef]
5. Verginadis, I.I.; Avgousti, H.; Monslow, J.; Skoufos, G.; Chinga, F.; Kim, K.; Leli, N.M.; Karagounis, I.V.; Bell, B.I.; Velalopoulou, A.; et al. A stromal Integrated Stress Response activates perivascular cancer-associated fibroblasts to drive angiogenesis and tumour progression. *Nat. Cell Biol.* **2022**, *24*, 940–953. [CrossRef] [PubMed]
6. De, P.; Aske, J.; Sulaiman, R.; Dey, N. Bete Noire of Chemotherapy and Targeted Therapy: CAF-Mediated Resistance. *Cancers* **2022**, *14*, 1519. [CrossRef]
7. De, P.; Aske, J.; Dey, N. Cancer-Associated Fibroblast Functions as a Roadblock in Cancer Therapy. *Cancers* **2021**, *13*, 5246. [CrossRef]
8. Ucaryilmaz Metin, C.; Ozcan, G. Comprehensive bioinformatic analysis reveals a cancer-associated fibroblast gene signature as a poor prognostic factor and potential therapeutic target in gastric cancer. *BMC Cancer* **2022**, *22*, 692. [CrossRef]
9. Sulaiman, R.; De, P.; Aske, J.C.; Lin, X.; Dale, A.; Koirala, N.; Gaster, K.; Espaillat, L.R.; Starks, D.; Dey, N. Patient-Derived Primary Cancer-Associated Fibroblasts Mediate Resistance to Anti-Angiogenic Drug in Ovarian Cancers. *Biomedicines* **2023**, *11*, 112. [CrossRef]
10. Pradip, D.; Jennifer, A.; Nandini, D. Cancer-Associated Fibroblasts in Conversation with Tumor Cells in Endometrial Cancers: A Partner in Crime. *Int. J. Mol. Sci.* **2021**, *22*, 9121. [CrossRef]
11. De, P.; Aske, J.C.; Dale, A.; Rojas Espaillat, L.; Starks, D.; Dey, N. Addressing activation of WNT beta-catenin pathway in diverse landscape of endometrial carcinogenesis. *Am. J. Transl. Res.* **2021**, *13*, 12168–12180. [PubMed]
12. Sulaiman, R.; De, P.; Aske, J.C.; Lin, X.; Dale, A.; Gaster, K.; Dey, N. Characterization and Clinical Relevance of Endometrial CAFs: Correlation between Post-Surgery Event and Resistance to Drugs. *Int. J. Mol. Sci.* **2023**, *24*, 6449. [CrossRef]
13. Bissell, M.J.; Labarge, M.A. Context, tissue plasticity, and cancer: Are tumor stem cells also regulated by the microenvironment? *Cancer Cell* **2005**, *7*, 17–23. [CrossRef] [PubMed]
14. Li, C.; Teixeira, A.F.; Zhu, H.J.; Ten Dijke, P. Cancer associated-fibroblast-derived exosomes in cancer progression. *Mol. Cancer* **2021**, *20*, 154. [CrossRef] [PubMed]
15. Joshi, R.S.; Kanugula, S.S.; Sudhir, S.; Pereira, M.P.; Jain, S.; Aghi, M.K. The Role of Cancer-Associated Fibroblasts in Tumor Progression. *Cancers* **2021**, *13*, 1399. [CrossRef] [PubMed]
16. McAndrews, K.M.; Chen, Y.; Darpolor, J.K.; Zheng, X.; Yang, S.; Carstens, J.L.; Li, B.; Wang, H.; Miyake, T.; Correa de Sampaio, P.; et al. Identification of Functional Heterogeneity of Carcinoma-Associated Fibroblasts with Distinct IL6-Mediated Therapy Resistance in Pancreatic Cancer. *Cancer Discov.* **2022**, *12*, 1580–1597. [CrossRef]
17. Nicolas, A.M.; Pesic, M.; Engel, E.; Ziegler, P.K.; Diefenhardt, M.; Kennel, K.B.; Buettner, F.; Conche, C.; Petrocelli, V.; Elwakeel, E.; et al. Inflammatory fibroblasts mediate resistance to neoadjuvant therapy in rectal cancer. *Cancer Cell* **2022**, *40*, 168–184.e113. [CrossRef]
18. Su, S.; Chen, J.; Yao, H.; Liu, J.; Yu, S.; Lao, L.; Wang, M.; Luo, M.; Xing, Y.; Chen, F.; et al. CD10(+)GPR77(+) Cancer-Associated Fibroblasts Promote Cancer Formation and Chemoresistance by Sustaining Cancer Stemness. *Cell* **2018**, *172*, 841–856.e816. [CrossRef]
19. Kieffer, Y.; Hocine, H.R.; Gentric, G.; Pelon, F.; Bernard, C.; Bourachot, B.; Lameiras, S.; Albergante, L.; Bonneau, C.; Guyard, A.; et al. Single-Cell Analysis Reveals Fibroblast Clusters Linked to Immunotherapy Resistance in Cancer. *Cancer Discov.* **2020**, *10*, 1330–1351. [CrossRef]
20. Sahai, E.; Astsaturov, I.; Cukierman, E.; DeNardo, D.G.; Egeblad, M.; Evans, R.M.; Fearon, D.; Greten, F.R.; Hingorani, S.R.; Hunter, T.; et al. A framework for advancing our understanding of cancer-associated fibroblasts. *Nat. Rev. Cancer* **2020**, *20*, 174–186. [CrossRef]

21. Desbois, M.; Wang, Y. Cancer-associated fibroblasts: Key players in shaping the tumor immune microenvironment. *Immunol. Rev.* **2021**, *302*, 241–258. [CrossRef] [PubMed]
22. Tsoumakidou, M. The advent of immune stimulating CAFs in cancer. *Nat. Rev. Cancer* **2023**, *23*, 258–269. [CrossRef] [PubMed]
23. Unterleuthner, D.; Neuhold, P.; Schwarz, K.; Janker, L.; Neuditschko, B.; Nivarthi, H.; Crncec, I.; Kramer, N.; Unger, C.; Hengstschlager, M.; et al. Cancer-associated fibroblast-derived WNT2 increases tumor angiogenesis in colon cancer. *Angiogenesis* **2020**, *23*, 159–177. [CrossRef]
24. Wolfl, B.; Te Rietmole, H.; Salvioli, M.; Kaznatcheev, A.; Thuijsman, F.; Brown, J.S.; Burgering, B.; Stankova, K. The Contribution of Evolutionary Game Theory to Understanding and Treating Cancer. *Dyn. Games Appl.* **2022**, *12*, 313–342. [CrossRef]
25. Kaznatcheev, A.; Peacock, J.; Basanta, D.; Marusyk, A.; Scott, J.G. Fibroblasts and alectinib switch the evolutionary games played by non-small cell lung cancer. *Nat. Ecol. Evol.* **2019**, *3*, 450–456. [CrossRef] [PubMed]
26. Orimo, A.; Gupta, P.B.; Sgroi, D.C.; Arenzana-Seisdedos, F.; Delaunay, T.; Naeem, R.; Carey, V.J.; Richardson, A.L.; Weinberg, R.A. Stromal fibroblasts present in invasive human breast carcinomas promote tumor growth and angiogenesis through elevated SDF-1/CXCL12 secretion. *Cell* **2005**, *121*, 335–348. [CrossRef] [PubMed]
27. Bejarano, L.; Jordao, M.J.C.; Joyce, J.A. Therapeutic Targeting of the Tumor Microenvironment. *Cancer Discov.* **2021**, *11*, 933–959. [CrossRef]
28. Markman, M.; Fowler, J. Activity of weekly paclitaxel in patients with advanced endometrial cancer previously treated with both a platinum agent and paclitaxel. *Gynecol. Oncol.* **2004**, *92*, 180–182. [CrossRef]
29. Hoskins, P.J.; Swenerton, K.D.; Pike, J.A.; Wong, F.; Lim, P.; Acquino-Parsons, C.; Lee, N. Paclitaxel and carboplatin, alone or with irradiation, in advanced or recurrent endometrial cancer: A phase II study. *J. Clin. Oncol.* **2001**, *19*, 4048–4053. [CrossRef]

biomedicines

MDPI

Article

Protein-Functionalized Microgel for Multiple Myeloma Cells' 3D Culture

Juan Carlos Marín-Payá [1], Sandra Clara-Trujillo [1,2], Lourdes Cordón [3,4], Gloria Gallego Ferrer [1,2], Amparo Sempere [3,5] and José Luis Gómez Ribelles [1,2,*]

[1] Centre for Biomaterials and Tissue Engineering, CBIT, Universitat Politècnica de València, 46022 Valencia, Spain
[2] Biomedical Research Networking Center on Bioengineering, Biomaterials and Nanomedicine (CIBER-BBN), 28029 Valencia, Spain
[3] Centro de Investigación Biomédica en Red de Cáncer (CIBERONC), Instituto Carlos III, 20029 Madrid, Spain
[4] Hematology Research Group, Instituto de Investigación Sanitaria La Fe (IISLAFE), 46026 Valencia, Spain
[5] Haematology Department, Hospital Universitari i Politècnic La Fe, 46026 Valencia, Spain
[*] Correspondence: jlgomez@ter.upv.es

Abstract: Multiple myeloma is a hematologic neoplasm caused by an uncontrolled clonal proliferation of neoplastic plasma cells (nPCs) in the bone marrow. The development and survival of this disease is tightly related to the bone marrow environment. Proliferation and viability of nPCs depend on their interaction with the stromal cells and the extracellular matrix components, which also influences the appearance of drug resistance. Recapitulating these interactions in an in vitro culture requires 3D environments that incorporate the biomolecules of interest. In this work, we studied the proliferation and viability of three multiple myeloma cell lines in a microgel consisting of biostable microspheres with fibronectin (FN) on their surfaces. We also showed that the interaction of the RPMI8226 cell line with FN induced cell arrest in the G0/G1 cell cycle phase. RPMI8226 cells developed a significant resistance to dexamethasone, which was reduced when they were treated with dexamethasone and bortezomib in combination.

Keywords: multiple myeloma; microgel; fibronectin; dexamethasone; bortezomib

1. Introduction

Multiple myeloma (MM) accounts for approximately 10% of all hematological malignancies and 2% of all cancers [1,2]. It is characterized by the proliferation of neoplastic plasma cells (nPCs) in the bone marrow [2]. The development of a realistic 3D culture system is growing in relevance due to the important role played by the extracellular matrix (ECM) components and the stromal cells in their interactions with nPCs, promoting their survival and generating drug resistance [3–6]. Although new therapies have been developed, MM still remains an incurable disease due to the nPCs' protection against chemotherapy or receptor-targeting drugs [7–9].

Tissue engineering has contributed to the development of 3D culture systems using inert materials, which acquire a bioactive capacity when functionalized with ECM biomolecules [10,11], promoting cell-material interaction and generating a similar response to that expected in vivo. The cellular and noncellular components of the tumor niche is the tumor microenvironment (TME), described by the International Cancer Microenvironment Society as an ecosystem composed of tumor cells, resident and infiltrating non-tumor cells, and molecules present in proximity to these cells. The ECM constitutes not only a structural scaffold but also plays a key role in the spread of cancers, as cell migration into and out of the TME greatly depends on the cell adhesion to ECM components. Tumor development is highly dependent on the physiological state of the tumor-specific microenvironment that will provide pro and antitumor signals [10–12].

In the case of multiple myeloma, it is demonstrated that the 3D support allows a closer emulation of the bone marrow environment than the conventional 2D cultures [12,13]. We have recently proposed the use of microgels for this purpose. The culture medium consists of an agglomerate of biomimetic microspheres and tumor cells suspended in a liquid culture medium [13,14]. Microspheres have frequently been used as drug delivery systems in the development of cancer models, and there are also references to their use as a three-dimensional support in the culture of solid tumors [15–18].

Recent and ongoing research has shown that properly recreating cell–cell and cell–ECM interactions is critical to developing reliable tumor models in vitro. We hypothesize that in the development of an in vitro disease model for multiple myeloma, the presence of ECM components presented by biomaterial in a cell-accessible manner is essential, while at the same time, the culture medium needs to allow the motility of cells that normally are grown in suspension. In this sense, our proposed biomimetic microgel would fulfill both requirements. In the future, it will be necessary to incorporate co-culture with adherent stromal cells, for which previous literature shows different strategies [6].

The correlation between tumor cell adhesion to fibronectin and the generation of drug resistance is well established [19]. In this work, we study the biological response to an environment formed by a microgel that exposes FN on the surface of microspheres and thus at distances from cells that are on the order of cellular dimensions. As the exhibition of adhesion motifs in FN is pivotal in cell–ECM interactions, mimicking the in vivo FN conformation when developing an artificial environment becomes of primary interest. Although cell activity produces FN fibrillogenesis in vivo modeling the ECM environment, certain biomaterials have been shown to induce FN spreading and networking on their surfaces previous to cell seeding. This feature is important in MM, since plasma cells are not expected to remodel the FN present on the surface of the supporting biomaterial used for cell culture. Physical absorption of FN onto poly(ethyl acrylate) has been demonstrated to preserve FN fibrillar networks and bioactivity by promoting their fibrillogenesis [20–23].

In cell–FN interaction, adhesion occurs through β1-integrins α4β1 (VLA-4) and α5β1 (VLA-5) (16). This adhesion induces a nuclear accumulation of p27kip1 proteins via down-regulation of Jab1 [7,24], preventing the serine 10 phosphorylation of p27Kip1, as well as the p27kip1 nuclear exportation. The resulting increase in the p27kip1 protein level inhibits cyclin A- and cyclin E-dependent CDK2 kinase activity [7,24,25], generating a G0/G1 cell cycle arrest and thus an intrinsic drug resistance to dexamethasone (DEX) in the RPMI8226 MM cell line [7], known as cell adhesion-mediated drug resistance (CAM-DR) [9,25].

Patients are currently treated in clinical practice with MM corticosteroids and proteasome inhibitors. The corticosteroid DEX induces apoptosis through the activation of intrinsic apoptotic pathways [26,27], downregulation of antiapoptotic genes, and upregulation of pro-apoptotic genes [26,28]. The recently introduced reversible inhibitor of the 26S proteasome complex bortezomib (BRZ) [29] induces the cleavage of survival proteins such as Mcl1 and triggers apoptosis [30,31]. In the RPMI8226 cell line, it has been found that BRZ overcomes the FN CAM-DR effect through the downregulation of VLA-4, and in combination with DEX, reduces the CAM-DR effect [29]. The drug dosage can be studied in MM patients by comparing the effect of a low dose versus high-dose DEX [32,33]. Microgels, agglomerates of functionalized microspheres in a liquid culture medium, have also been proposed as a 3D environment for MM cell cultures [13,14].

This work aims to develop a 3D culture system based on a microgel made up of poly(ethyl acrylate) copolymers functionalized with ECM biomolecules, such as FN, and capable of mimicking both the bone marrow microenvironment by interacting with the tumor cells and the CAM-DR effect generated by the biomolecules. After validation, the system can be tested in vitro with primary cells from patients.

2. Materials and Methods

2.1. Polymerization and Microspheres Production

The microgel was made of a random ethyl acrylate (EA) and ethyl methacrylate (EMA) copolymer (Sigma-Aldrich, St. Louis, MO, USA), poly (EA-co-EMA) synthetized by a block polymerization protocol by mixing 50% EA and 50% EMA with 30% acetone *w*/*w* as a solvent. Initiation of the reaction was induced by 0.5% benzoin (Sigma-Aldrich) and a 24-h incubation of the solution in an ultraviolet (UV) irradiation chamber. After polymerization, the polymer was dissolved in acetone and reprecipitated four times with deionized water to remove the monomer residue and any remaining low molecular weight substances. The material was then dissolved in chloroform (3g/100 mL, Scharlab, S.L., Barcelona, Spain) and mixed with 5% *w*/*w* magnetic ferrite nanoparticles (EMG 1300M, Ferrotec, Santa Clara, CA, USA), facilitating the manipulation of the microspheres and their magnetic separation from cells or any aqueous medium.

The microspheres were produced by an oil-in-water emulsion technique, where the continuous phase was a 0.5% polyvinyl alcohol aqueous solution (PVA, Sigma-Aldrich), and the dispersed phase was a 3% *w*/*v* magnetic copolymer solution in chloroform. The emulsion was carried out in a volumetric flask by stirring 200 mL of the continuous phase at 2000 rpm for 15 min, adding 20 mL of the oil phase afterwards with the help of a funnel, and agitating for 10 min. Then, 150 mL of ultrapure water (UPW) was added to promote chloroform evaporation and stirred for 24 h. To complete chloroform evaporation overnight, the solution was later transferred to a beaker and stirred at 700 rpm. In order to remove the remaining PVA, the solution was transferred to 50 mL tubes by a series of four washings with UPW. After air drying the solution at room temperature, a field-emission scanning electron microscope (FESEM, ULTRA 55 model, ZEISS, Oberkochen, Germany) was used to observe the microspheres' sizes and surfaces. The dispersion of the microspheres in culture conditions after manual stirring was confirmed in a stereo zoom microscope (MZ APO model, Leica Microsystems, Wetzlar, Germany). To determine microsphere sizes, 3 different emulsions were analyzed per condition (750, 1500, and 2000 rpm). At least 100 microspheres were measured for each emulsion, coming from 3 independent FESEM pictures. Image analysis was performed using the ImageJ software (National Institutes of Health, Bethesda, MD, USA).

2.2. Sterilization and Functionalization

A sterilization protocol was carried out, before using the microspheres for cell culture assays, by UV irradiation for 30 min and then incubating overnight in PBS with 3% penicillin/streptomycin. The day after, the microspheres were washed twice with PBS, UV-irradiated for 30 min, incubated overnight with PBS with 1% penicillin/streptomycin, and washed again with PBS. Once sterilized and before cell seeding, the microspheres were coated with 20 µg/mL of human plasma FN (Sigma-Aldrich) by physical adsorption for 1 h and washed twice with RPMI 1640 culture medium without FBS in order to functionalize them.

2.3. Characterization

The microspheres were characterized by a micro-BCA assay using a Pierce BCA Protein Assay Kit (Thermo Fisher Scientific, Waltham, MA USA). Acrylate microspheres without FN were used as a baseline. This assay was carried out in quadruplicate and read on a Victor3 Plate Reader (PerkinElmer, Waltham, MA, USA). Finally, the assessment was performed by calculating an average of the samples at 570 and 550 nm.

2.4. Cell Culture

The RPMI8226 MM cell line, kindly provided by Dr. Beatriz Martin (Josep Carreras Leukaemia Research Institute), was grown in an orbital shaker at 300 rpm in RPMI 1640 medium (Gibco, Thermo Fisher) and supplemented with 15% fetal bovine serum (FBS, Gibco), 1% L-glutamine (Sigma-Aldrich), and 1% penicillin/streptomycin (Gibco, 10.000 U/mL). The U226-B1 and MM.1S cell lines, purchased from the American Type

Culture Collection (ATCC, Rockville, MD, USA), were grown as described above but supplemented with 10% FBS, 1% L-glutamine, and 1% penicillin/streptomycin the culture medium. Cell culture assays were performed under different conditions: cell suspension culture without microspheres (SUSP), cell suspension with uncoated microgel (M), and cell suspension with microgel coated with FN (MFN). Different microspheres/liquid-medium volume ratios were employed for the design of the 3D culture system in a final volume of 500 μL of microgels of 17.2% *v*/*v* microspheres (M17), 13.8% *v*/*v* (M13), 10.3% *v*/*v* (M10), and 6.8% *v*/*v* (M6). As it was a cell culture in suspension, a partial renewal of the liquid medium by adding 400 μL of fresh medium was performed every day by mixing for 15 minutes to facilitate the distribution of nutrients, precipitating cells and microgel by stopping the mixing after 1 h, and finally, carefully removing 400 μL of culture medium. Cell seeding was performed in triplicate on P24 plates. Both microgel and cell suspension were added consecutively in the proper ratios to obtain the desired microsphere volume fraction. In order to avoid the adhesion of the RPMI8226 cells to the other proteins present in the FBS (which can be adsorbed onto the microspheres' surfaces) in the FN experiments, the cells were prior seeded in a culture medium without FBS for 1 h without mixing, and 1 h later, the appropriate amount of FBS was added to the wells.

To perform flow cytometry assay and before cell staining, the cells and the magnetic acrylate microspheres were magnetically sorted out to prevent cytometer clogging.

2.5. Proliferation Assay

The influence on cell proliferation of the ratio of the volume occupied by the microspheres to that of the liquid culture medium was assessed using a Quant-iT Picogreen dsDNA Kit (Invitrogen, Thermo Fisher). For this, 5×10^4 cells were grown at days 2, 5, and 7, including three replicas per condition. A lysis-buffered solution to digest the samples was prepared for 100 mL; amounts of 0.648 g of sodium phosphate monobasic (Panreac), 0.653 g of sodium phosphate dibasic (Panreac Quimica SLU, Barcelona, Spain), and 1 mL of EDTA (Gibco, 0.5 M) were dissolved in UPW at pH 6.5. Papain at 3.875 U/mL (Sigma-Aldrich) and 1.5 mg/mL of L-cysteine were added to the lysis buffer on the day of the experiment. For each sample, including the acrylate microspheres without cells as a baseline condition, 500 μL of lysis buffer was added and stirred at 60 for 18 h. Finally, DNA was stained with the PicoGreen reagent as indicated by the manufacturer and subsequently analyzed on an opaque Optiplate96F plate (PerkinElmer) using a Victor3 plate reader (PerkinElmer) at 485/535 nm.

2.6. Cell Cycle Assay and CAM-DR Effect on RPMI 8226 Cell Line

A flow cytometry cell cycle assay was performed using the DNA-PREP kit (Beckman Coulter, San Diego, CA, USA) to study the influence of adhesion of the RPMI8226 cells to FN on cell arrest. A total of 1.5×10^5 cells were grown on days 2, 3, and 5, including three replicates per condition that were merged in a single tube and analyzed in the cytometer. Once available, the cells were washed twice with PBS at 300 g for 5 min in the centrifuge. Cells were processed with the DNA-PREP kit following the manufacturer's instructions, which include a permeabilization step and staining with propidium iodide. The samples were acquired in a Navios flow cytometer and data were analyzed on the Kaluza Version 2.1 software (Beckman Coulter).

FN-induced CAM-DR was studied in the presence of BRZ and high-dose DEX. Three conditions were studied after growing 1.5×10^5 cells to expose them to: (i) 5 nM of BRZ for 48 h [34,35], (ii) 10^3 μM of DEX [14] at 8.75 mg/mL (fortecortin, Merck KGaA, Darmstadt, Germany) for 72 h, and iii) a DEX+BRZ combination, with only DEX for the first 24 h and both drugs for the last 48 h. A viability assay was carried out using annexin-V (Miltenyi Biotec, Bergisch Gladbach, Germany) and 7-amino-actinomycin D (7-AAD, Becton Dickinson, San Jose, CA, USA). Briefly, cells were washed twice with PBS, centrifuged at 300 g for 5 min, and resuspended in 300 μL of binding buffer solution (Miltenyi Biotec). Cells were stained with 10 μL of annexin-V conjugated with fluorescein isothiocyanate and

5 µL of CD138 monoclonal antibody conjugated with BD Horizon V500 (Becton Dickinson, San Jose, CA, USA), then incubated for 20 min at room temperature in the dark. Before staining with 7-AAD, the cells were washed with 1–2 mL of binding buffer solution and resuspended in 200 µL of binding buffer solution. Finally, the cells were stained with 10 µL of 7-AAD, incubated for 15 min at room temperature in the dark, and acquired in a FACSCanto-II flow cytometer (Becton Dickinson, San Jose, CA, USA). Data analysis was undertaken using Kaluza software.

2.7. Statistical Analysis

Statistically significant differences between groups were determined by applying one-way ANOVA analysis (Bonferroni test) on the IBM SPSS Statistics Version 20.0 software at a significance of ≤ 0.05.

3. Results

3.1. Acrylate Microsphere Production and Characterization

Microgel microparticles were produced by an oil-in-water emulsion technique capable of controlling microdroplet size by varying the agitation speed. A representative FESEM image is shown in Figure 1a, while Figure 1b shows the dispersion of the microspheres in the liquid medium. The copolymer microsphere coagulation occurred due to the evaporation of the solvent throughout the continuous liquid phase. Three different tests were carried out at 750, 1,500, and 2,000 rpm agitation speeds. Figure 1 shows that as the stirring speed increased, the microsphere size decreased, obtaining an sizes of 27 ± 13 µm at 750 rpm (Figure 1c), 10 ± 4 µm at 1,500 rpm (Figure 1d), and 9 ± 4 µm at 2000 rpm (Figure 1e). The cell culture tests were performed using the conditions that provided a similar microsphere size to nPCs (9-20 µm) [36]. A 2000 rpm rotation speed was used, since the differences in the particle sizes with respect to 1500 rpm were small, but higher speeds favored chloroform evaporation, thus obtaining higher efficiency at the end of the process (data not shown).

Microspheres functionalized with FN by physical absorption were characterized by a micro-BCA assay to quantify the amount of FN present on them, for which coating was performed in triplicate. The mass fraction of FN was found to be 0.4 ± 0.07 µg per mg of microspheres.

The suspension of microspheres and cells was maintained under gentle orbital agitation. The characteristics of the environment of the cells cultured in this system were simulated by finite element analysis in a recent work [37].

3.2. Proliferation Assay

The effect on cell proliferation of the volume occupied by the microspheres was tested at the beginning with the RPMI8226 cell line (Figure 2a). PicoGreen tests were performed on days 2, 5, and 7. On day 2, the differences between the microsphere volume ratios were not significant, nor were they statistically significant, even when the mean value was higher than in the SUSP-RPMI control. However, on day 5, all the conditions with microgel present were found to have significant differences *vs* SUSP-RPMI, especially the M13-RPMI, with a difference of $2.52 \times 10^5 \pm 2.49 \times 10^4$ cells, with respect to the SUSP-RPMI condition. On day 7, the number of cells reached a value close to 1.5×10^6 cell/well (containing 500 µL of culture medium), which is the recommended limit for this cell line. The culture seemed to be saturated, and no significant differences were found between the samples. A viability assay was performed to confirm that there was no interference of the cell death with the PicoGreen analysis, obtaining around $88.9 \pm 1.9\%$ of cell viability under all conditions (Figure 2b). Based on these results, the M13 condition was used for the subsequent assays, and we determined whether this behavior was replicated in different multiple myeloma cell lines. Significant differences were found in both cell lines on days 5 and 7, with a difference on day 7 of $1.66 \times 10^5 \pm 1.6 \times 10^4$ cells of M13-U266 *vs* SUSP-U266 and $2.37 \times 10^5 \pm 3.74 \times 10^4$ cells of M13-MM.1S *vs* SUSP-MM.1S (Figure 2c). Due to the higher proliferation of RPMI8226 than U266-B1 and MM.1S, both cell lines did not reach the previous saturation level, while cell viability was around 90.2 ± 2.4% in U266 and 89.9 ± 2.3% in MM.1S (Figure 2d).

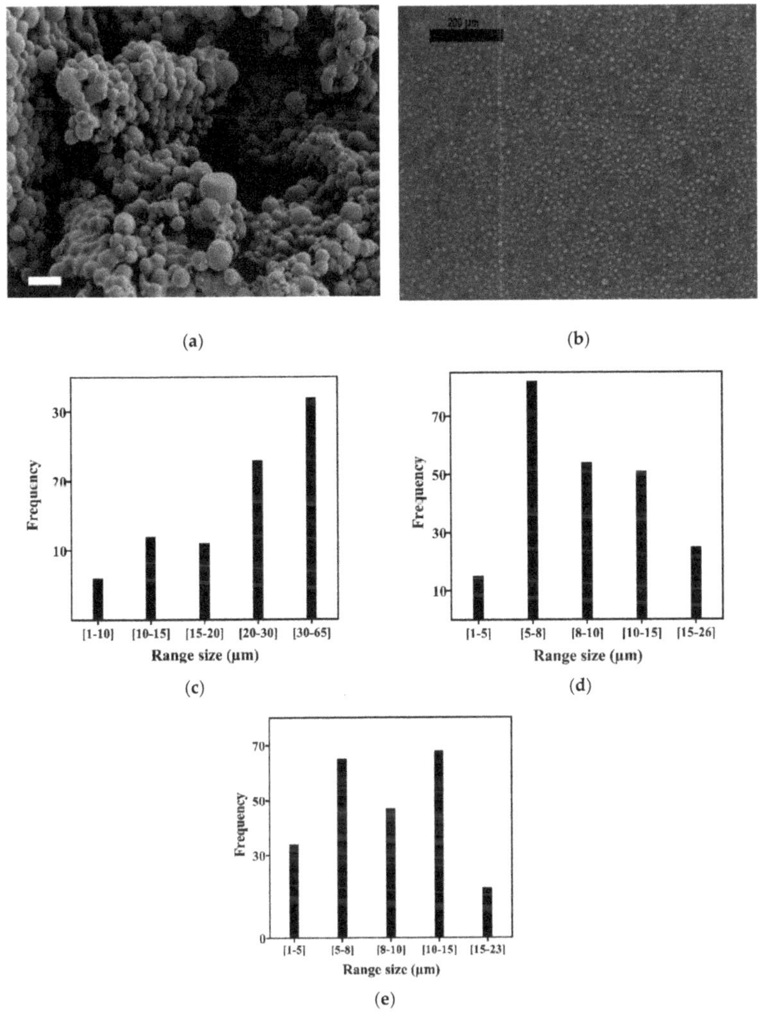

Figure 1. Physical microsphere characterization: (**a**) FESEM image of the microspheres (bar: 20 µm), (**b**) Image of microgel dispersion in a P24 well, (**c**) size distribution at 750 rpm, (**d**) 1500 rpm, and (**e**) 2000 rpm.

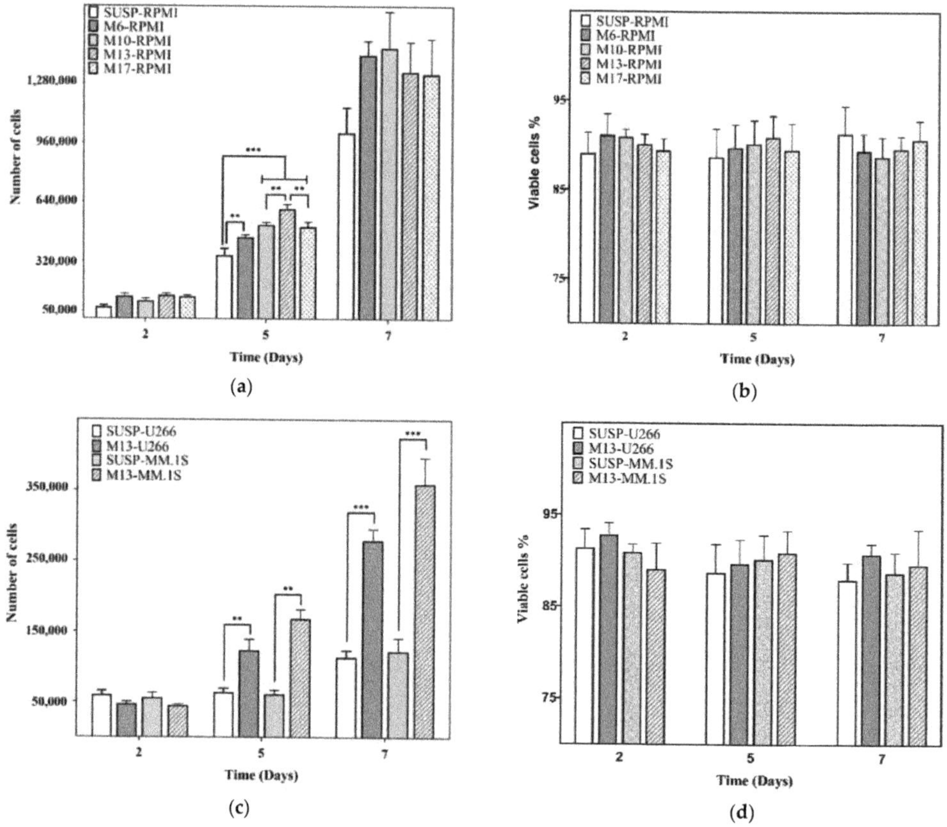

Figure 2. Evaluation of cell proliferation (**a**) and viability (**b**) on the RPMI8226 cell line at different volume ratios of uncoated microgels. Assessment of cell proliferation (**c**) and viability (**d**) on U266 and MM.1S cell lines at M13 volume ratio. SUSP: suspension culture without microspheres; M: suspension culture with different volume ratios of uncoated microspheres; M6: 6.89% v/v of microspheres; M10: 10.34% v/v of microspheres; M13: 13.8% v/v of microspheres; M17: 17.2% v/v of microspheres. Graphics depict mean ± standard deviation. Level of statistical significance: (**) p-value ≤ 0.01, (***) p-value ≤ 0.001.

3.3. Cell Cycle Analysis on the RPMI 8226 Cell Line

The adhesion of nPCs to FN was studied by a cell cycle assay on days 2, 3, and 5. As shown in Figure 3A, on day 2, the MFN-RPMI condition exhibited a higher arrest in the G0/G1 phase than the others, the differences being 18% and 23% with respect to the SUSP-RPMI and the M13-RPMI, respectively. On day 3, differences reduced to approximately 10% and 16% with respect to the SUSP-RPMI and the M13-RPMI, respectively. On day 5, there were differences of 7% and 10% regarding the SUSP-RPMI and M13-RPMI, respectively.

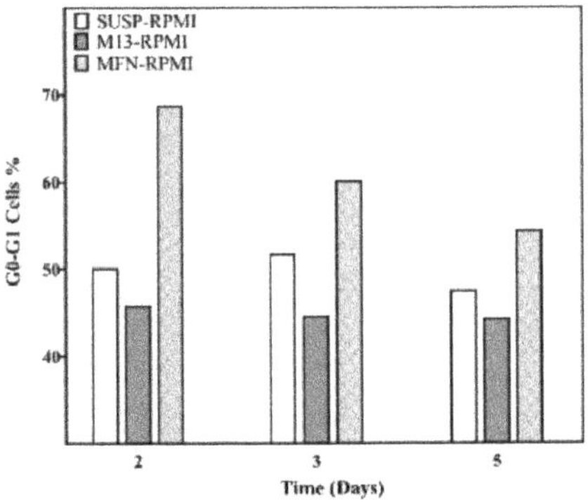

(a)

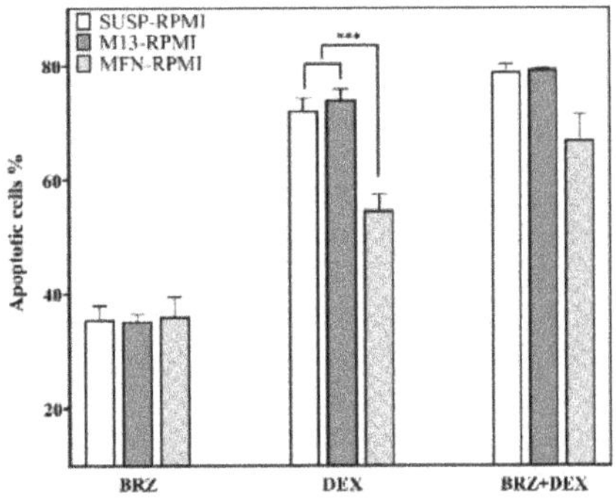

(b)

Figure 3. (**a**) Proportion of cells in the G0/G1 phase. (**b**) Results of drug resistance and the synergistic activity effect of both drugs on the RPMI8226 cell line viability. SUSP: suspension culture without microspheres. M13: suspension culture with 13.8% v/v of microspheres with uncoated microspheres. MFN: suspension culture with 13.8% v/v of microspheres coated with fibronectin. Graphic depicts mean ± standard deviation. Level of statistical significance: (***) p-value ≤ 0.001.

3.4. Drug Resistance on RPMI8226 Cell Line

To assess the CAM-DR effect of FN against high-dose DEX and BRZ, a flow cytometry viability assay was performed by preparing an untreated control for each condition that was used to normalize the viability results obtained by the effect of the drug (data not shown). After exposure to BRZ, only very similar results were observed in all the conditions, with a mortality of 35.5% in SUSP-RPMI, 35.2% in M13-RPMI, and 36% in MFN-RPMI (Figure 3b). In the only presence of DEX, the FN condition presented significant differences to SUSP-RPMI and M13-RPMI, showing a mortality of 54.6%, compared to 72% for SUSP-RPMI and 73.9% for M13-RPMI. In the presence of both drugs, there were very few differences between the SUSP-RPMI and M13-RPMI conditions, compared to those previously observed with only DEX, with a mortality of 78.8% SUSP-RPMI and 79.3% M13-RPMI. However, mortality increased by 12.2% in the FN condition with respect to that with only DEX, with a mortality of 66.8%.

4. Discussion

The development of 3D culture systems that can mimic and reproduce the tumor microenvironment and generate a similar cellular response to that produced in vivo is a common research goal at present [38,39]. However, to establish an in vitro model that provides conclusions about the effect of a certain factor on the generation of drug resistance is a real challenge. (In this work, we explored nPC adhesion to FN, but this can be easily extrapolated to interactions with other extracellular matrix components.) The environment in which the disease develops is highly complex, with several differentiated microenvironments and multiple cell–cell and cell–ECM component interactions with soluble factors [3,4]. When establishing a model aiming to isolate one of the factors that contributes to the development of the pathology, there is a risk of reaching erroneous conclusions due to the lack of other elements that affect the equation, in particular, the 3D configuration and cell mobility [38,40]. In this work, we therefore describe a model that aims to present the biomolecules to the tumor cells in a more realistic way than simply allowing the biomolecule to adsorb on the bottom of the culture well. The model described here has three fundamental characteristics: (1) it enables the mobility of tumor cells in a liquid medium, practically without restriction, as the microspheres that float along with them are equally mobile; (2) the proximity among the cells and the biomolecules under study, are at distances of the order of cell sizes with the chosen volume fraction of the microspheres;

(3) the use of a synthetic support that induces a protein conformation that helps it display its adhesion domains, allowing interaction with cells or growth factors. Its versatility in this regard is enormous, and previous literature shows multiple examples of surface functionalization that can be applied to microspheres to study specific biomolecules [41].

We developed a 3D culture environment using polyacrylate microspheres with an average size close to nPCs. This microgel configuration favors cell–cell interaction and inter-microsphere cell mobility, which in turn, allows the cells to interact with the biomolecules on the microgel surface. One of the pitfalls of this type of system with similar cell-microgel size is the generation of background interference due to the presence of the microgel during sample processing or the possibility of clogging the flow cytometer during measurement. However, due to the microspheres' magnetic properties, the cells can be separated from the material, making the platform easy to manipulate when performing cytometry cell assays.

The effect of microspheres on nPC proliferation depends on the volume occupied by the biomaterial and the liquid medium available for cell suspension. The results obtained suggest that in nPC culture conditions in the presence of a microgel, higher cell proliferation was observed in different multiple myeloma cell lines (Figure 2a,c), in good agreement with other studies [13]. This can be attributed to the configuration of the 3D culture platform, where the microspheres suspended in the liquid medium act as a scaffold. The volume excluded by the microspheres might have an increased cell–cell contact effect. On the other hand, microsphere mobility avoids the restrictions imposed when cells are cultured in scaffold pores or in a conventional hydrogel where the cells are confined, providing increased freedom for the expansion and proliferation of cells. Since the in vivo characteristics of tumoral cells include an elevated proliferation rate, we used the M13 condition, which provided higher proliferation than the other conditions (Figure 2a), to obtain an in vitro model as close as possible to the in vivo model. On day 7 of culture, we found that the different microgel conditions tended to produce similar numbers of cells (Figure 2a). This was due to the culture's high cell proliferation and volumetric limitations, in which the saturated culture presented a cell density of around 2×10^6 cells/mL, when the supplier's recommendation was not to exceed 1×10^6 cells/mL.

The presence of FN in the microspheres was confirmed by micro-BCA. RPMI8226 adhesion to FN, as well as its CAM-DR effect, had previously been studied in conventional 2D cultures, with FN adhering to the bottom of the well, resulting in nPC arrest in the G0/G1 phase of the cell cycle [7,24]. Prior to the CAM-DR assays, we therefore performed a cell cycle assay on days 2, 3, and 5 to confirm cell–FN interaction. On day 2 of culture, the MFN-RPMI condition had an increased cell percentage in the G0/G1 phase, compared to the other conditions (Figure 3a). This difference gradually decreased over time as the rest of the cells not attached to FN continued to proliferate and reduce the percentage of those arrested in the G0/G1 phase. After confirming that the culture in the microgel allowed the cells to interact with the functionalized biomolecule on the microspheres' surfaces, the microgel was tested for interference with the drug assays by evaluating the CAM-DR effect of FN against high-dose DEX and BRZ, two of the drugs used against MM in clinical practice. The effect of FN was previously reported in the RPMI8226 MM cell line. As expected, FN did not trigger any type of resistance to BRZ, as all the conditions had the same percentages of mortality (Figure 3B). However, in the case of DEX, there were significant differences between the FN and the other condition because the arrest in phase G0/G1 was found to be associated with the generation of resistance to DEX [7]. Finally, we carried out an assay by combining the actions of both drugs for 72 h, incorporating only DEX for the first 24 h and DEX + BRZ in the last 48 h. The results obtained suggest that the action of BRZ is in some way masked by the high dose of DEX, with no differences in mortality between the SUSP-RPMI and M13-RPMI conditions and the only DEX condition. However, due to the downregulation of VLA-4 produced by BRZ [29], it was found that the MFN-RPMI condition, which showed an initial resistance to DEX, overcame this CAM-DR effect and presented higher mortality than the FN condition with only DEX, confirming that the resistance was generated by the CAM-DR effect of FN.

This work was limited to verifying the viability and proliferation of the multiple myeloma cell line RPMI8226 in the microgel and the effect of fibronectin on resistance to dexamethasone and bortezomib, in order to show the capability of our 3D environment in the study of multiple myeloma. Future developments will be aimed at developing microgels exhibiting other biomolecules of the bone marrow ECM on the surfaces of the microspheres. The resistance of primary tumor plasma cells to the drugs currently used in tumor treatment will be studied.

5. Conclusions

We have shown that it is possible to consolidate the development of a 3D culture platform based on microgels. The system, easy to manipulate, could be proposed as a very useful strategy in clinical practice to test single agents or drug combinations that can produce a synergistic effect by increasing the anti-tumoral potential [30] for the treatment of hematological neoplasms. This work offers a wide range of future research perspectives, such as the development of 3D cultures with other ECM components or co-cultures with stromal cells. The validation of this platform for the culture of MM primary cells from patients could be applied to test and develop novel drugs for a better understanding of their effects and the interaction between tumoral cells and the bone marrow environment.

Author Contributions: Contribution: J.C.M.-P., A.S. and J.L.G.R. designed and directed research and wrote the manuscript; J.C.M.-P. and S.C.-T. performed research and analyzed data; S.C.-T., L.C. and G.G.F. critically reviewed the data and manuscript; A.S. and J.L.G.R. reviewed the data and the final manuscript before it was submitted. All authors have read and agreed to the published version of the manuscript.

Funding: This research was funded by the Generalitat Valenciana, grant number PROMETEO/2016/063; JLGR acknowledges that support by was funded by the Spanish State Research Agency (AEI) through the PID2019-106099RB-C41 / AEI / 10.13039/501100011033 project.

Institutional Review Board Statement: Not applicable.

Informed Consent Statement: Not applicable.

Data Availability Statement: Publicly available datasets were analyzed in this study. This data can be found at the Riunet repository of the Universitat Politècnica de València, here: [http://hdl.handle.net/10251/189950].

Acknowledgments: CIBER-BBN is an initiative funded by the VI National R&D&I Plan 2008–2011, Iniciativa Ingenio 2010, and Consolider Program. CIBER Actions were financed by the Instituto de Salud Carlos III, with assistance from the European Regional Development Fund. The kind supplying of RPMI 8226 cells by Beatriz Martin (Josep Carreras Leukaemia Research Institute) is greatly acknowledged. The Microscopy Service of the UPV (Universitat Politècnica de València) is gratefully acknowledged.

Conflicts of Interest: The authors declare no conflict of interest. The funders had no role in the design of the study; in the collection, analyses, or interpretation of data; in the writing of the manuscript, or in the decision to publish the results.

References

1. Piechotta, V.; Jakob, T.; Langer, P.; Monsef, I.; Scheid, C.; Estcourt, L.J.; Ocheni, S.; Theurich, S.; Kuhr, K.; Scheckel, B.; et al. Multiple Drug Combinations of Bortezomib, Lenalidomide, and Thalidomide for First-Line Treatment in Adults with Transplant-Ineligible Multiple Myeloma: A Network Meta-Analysis. *Cochrane Database Syst. Rev.* **2019**, *11*, CD013487. [CrossRef]
2. Firth, J. Haematology: Multiple Myeloma. *Clin. Med. J. R. Coll. Physicians Lond.* **2019**, *19*, 58–60.
3. Márk, Á.; Varga, G.; Timár, B.; Kriston, C.; Szabó, O.; Deák, L.; Matolcsy, A.; Barna, G. The Effect of Microenvironmental Factors on the Development of Myeloma Cells. *Hematol. Oncol.* **2017**, *35*, 741–745. [CrossRef] [PubMed]
4. Wang, J.; De Veirman, K.; De Beule, N.; Maes, K.; De Bruyne, E.; Van Valckenborgh, E.; Vanderkerken, K.; Menu, E. The Bone Marrow Microenvironment Enhances Multiple Myeloma Progression by Exosome-Mediated Activation of Myeloid-Derived Suppressor Cells. *Oncotarget* **2015**, *6*, 43992–44004. [CrossRef] [PubMed]

5. de la Puente, P.; Muz, B.; Gilson, R.C.; Azab, F.; Luderer, M.; King, J.; Achilefu, S.; Vij, R.; Azab, A.K. 3D Tissue-Engineered Bone Marrow as a Novel Model to Study Pathophysiology and Drug Resistance in Multiple Myeloma. *Biomaterials* **2015**, *73*, 70–84. [CrossRef]
6. Waldschmidt, J.M.; Fruttiger, S.J.; Wider, D.; Jung, J.; Thomsen, A.R.; Hartmann, T.N.; Duyster, J.; Hug, M.J.; Azab, K.A.; Jung, M.; et al. Ex Vivo Propagation in a Novel 3D High-Throughput Co-Culture System for Multiple Myeloma. *J. Cancer Res. Clin. Oncol.* **2022**, *148*, 1045–1055. [CrossRef]
7. Fei, M.; Hang, Q.; Hou, S.; He, S.; Ruan, C. Adhesion to Fibronectin Induces P27Kip1 Nuclear Accumulation through Down-Regulation of Jab1 and Contributes to Cell Adhesion-Mediated Drug Resistance (CAM-DR) in RPMI 8,226 Cells. *Mol. Cell. Biochem.* **2014**, *386*, 177–187. [CrossRef]
8. Zhou, L.; Zhang, Y.; Leng, Y.; Dai, Y.; Kmieciak, M.; Kramer, L.; Sharma, K.; Wang, Y.; Craun, W.; Grant, S. The IAP Antagonist Birinapant Potentiates Bortezomib Anti-Myeloma Activity in Vitro and in Vivo. *J. Hematol. Oncol.* **2019**, *12*, 25. [CrossRef]
9. Meads, M.B.; Hazlehurst, L.A.; Dalton, W.S. The Bone Marrow Microenvironment as a Tumor Sanctuary and Contributor to Drug Resistance. *Clin. Cancer Res.* **2008**, *14*, 2519–2526. [CrossRef]
10. Salvatore, V.; Teti, G.; Focaroli, S.; Mazzotti, M.C.; Mazzotti, A.; Falconi, M. Oncotarget 9608 Www.Impactjournals.Com/Oncotarget The Tumor Microenvironment Promotes Cancer Progression and Cell Migration. *Oncotarget* **2017**, *8*, 9608–9616. [CrossRef]
11. Wu, M.; Melody, A.S. Modeling Tumor Microenvironments in Vitro. *J. Biomech. Eng.* **2014**, *136*, 021011. [CrossRef]
12. Salo, T.; Vered, M.; Bello, I.O.; Nyberg, P.; Bitu, C.C.; Zlotogorski Hurvitz, A.; Dayan, D. Insights into the Role of Components of the Tumor Microenvironment in Oral Carcinoma Call for New Therapeutic Approaches. *Exp. Cell Res.* **2014**, *325*, 58–64. [CrossRef]
13. Clara-Trujillo, S.; Tolosa, L.; Cordón, L.; Sempere, A.; Gallego Ferrer, G.; Gómez Ribelles, J.L. Novel Microgel Culture System as Semi-Solid Three-Dimensional in Vitro Model for the Study of Multiple Myeloma Proliferation and Drug Resistance. *Biomater. Adv.* **2022**, *135*, 212749. [CrossRef]
14. Marín-Payá, J.C.; Díaz-Benito, B.; Martins, L.A.; Trujillo, S.C.; Cordón, L.; Lanceros-Méndez, S.; Ferrer, G.G.; Sempere, A.; Ribelles, J.L.G. Biomimetic 3d Environment Based on Microgels as a Model for the Generation of Drug Resistance in Multiple Myeloma. *Materials* **2021**, *14*, 7121. [CrossRef]
15. Liaw, C.Y.; Ji, S.; Guvendiren, M. Engineering 3D Hydrogels for Personalized In Vitro Human Tissue Models. *Adv. Healthc. Mater.* **2018**, *7*, 1701165. [CrossRef]
16. Nii, T.; Tabata, Y. Immunosuppressive Mesenchymal Stem Cells Aggregates Incorporating Hydrogel Microspheres Promote an in Vitro Invasion of Cancer Cells. *Regen. Ther.* **2021**, *18*, 516–522. [CrossRef]
17. Seeto, W.J.; Tian, Y.; Pradhan, S.; Minond, D.; Lipke, E.A. Droplet Microfluidics-Based Fabrication of Monodisperse Poly(Ethylene Glycol)-Fibrinogen Breast Cancer Microspheres for Automated Drug Screening Applications. *ACS Biomater. Sci. Eng.* **2022**, *8*, 3831–3841. [CrossRef]
18. Pradhan, S.; Clary, J.M.; Seliktar, D.; Lipke, E.A. A Three-Dimensional Spheroidal Cancer Model Based on PEG-Fibrinogen Hydrogel Microspheres. *Biomaterials* **2017**, *115*, 141–154. [CrossRef]
19. Hughes, V. Neighbourhood Watch. *Nature* **2011**, *480*, S48–S49. [CrossRef]
20. Campillo-Fernández, A.J.; Unger, R.E.; Peters, K.; Halstenberg, S.; Santos, M.; Salmerón Sánchez, M.; Meseguer Dueñas, J.M.; Monleón Pradas, M.; Gómez Ribelles, J.L.; Kirkpatrick, C.J. Analysis of the Biological Response of Endothelial and Fibroblast Cells Cultured on Synthetic Scaffolds with Various Hydrophilic/Hydrophobic Ratios: Influence of Fibronectin Adsorption and Conformation. *Tissue Eng. Part A* **2009**, *15*, 1331–1341. [CrossRef]
21. Ribeiro, C.; Panadero, J.A.; Sencadas, V.; Lanceros-M?ndez, S.; Tama?o, M.N.; Moratal, D.; Salmer?n-S?nchez, M.; G?mez Ribelles, J.L. Fibronectin Adsorption and Cell Response on Electroactive Poly(Vinylidene Fluoride) Films. *Biomed. Mater.* **2012**, *7*, 035004. [CrossRef] [PubMed]
22. Andreozzi, L.; Faetti, M.; Salmerón Sanchez, M.; Gómez Ribelles, J.L. Phenomenological Theory of Structural Relaxation Based on a Thermorheologically Complex Relaxation Time Distribution. *Eur. Phys. J. E* **2008**, *27*, 87–97. [CrossRef] [PubMed]
23. Salmerón-Sánchez, M.; Rico, P.; Moratal, D.; Lee, T.T.; Schwarzbauer, J.E.; García, A.J. Role of Material-Driven Fibronectin Fibrillogenesis in Cell Differentiation. *Biomaterials* **2011**, *32*, 2099–2105. [CrossRef] [PubMed]
24. Hazlehurst, L.A.; Damiano, J.S.; Buyuksal, I.; Pledger, W.J.; Dalton, W.S. Adhesion to Fibronectin via B1 Integrins Regulates P27(Kip1) Levels and Contributes to Cell Adhesion Mediated Drug Resistance (CAM-DR). *Oncogene* **2000**, *19*, 4319–4327. [CrossRef] [PubMed]
25. Vincent, T.; Mechti, N. Extracellular Matrix in Bone Marrow Can Mediate Drug Resistance in Myeloma. *Leuk. Lymphoma* **2005**, *46*, 803–811. [CrossRef]
26. Burwick, N.; Sharma, S. Glucocorticoids in Multiple Myeloma: Past, Present, and Future. *Ann. Hematol.* **2019**, *98*, 19–28. [CrossRef]
27. Chauhan, D.; Hideshima, T.; Rosen, S.; Reed, J.C.; Kharbanda, S.; Anderson, K.C. Apaf-1/Cytochrome c-Independent and Smac-Dependent Induction of Apoptosis in Multiple Myeloma (MM) Cells. *J. Biol. Chem.* **2001**, *276*, 24453–24456. [CrossRef]
28. Chauhan, D.; Auclair, D.; Robinson, E.K.; Hideshima, T.; Li, G.; Podar, K.; Gupta, D.; Richardson, P.; Schlossma, R.L.; Krett, N.; et al. Identification of Genes Regulated by Dexamethasone in Multiple Myeloma Cells Using Oligonucleotide Arrays. *Oncogene* **2002**, *21*, 1346–1358. [CrossRef]

29. Hatano, K.; Kikuchi, J.; Takatoku, M.; Shimizu, R.; Wada, T.; Ueda, M.; Nobuyoshi, M.; Oh, I.; Sato, K.; Suzuki, T.; et al. Bortezomib Overcomes Cell Adhesion-Mediated Drug Resistance through Downregulation of VLA-4 Expression in Multiple Myeloma. *Oncogene* **2009**, *28*, 231–242. [CrossRef]
30. Anderson, K. Proteasome Inhibitors in Multiple Myeloma. *Semin. Oncol.* **2009**, *36*, S20–S26. [CrossRef]
31. Podar, K.; Gouill, S.L.; Zhang, J.; Opferman, J.T.; Zorn, E.; Tai, Y.T.; Hideshima, T.; Amiot, M.; Chauhan, D.; Harousseau, J.L.; et al. A Pivotal Role for Mcl-1 in Bortezomib-Induced Apoptosis. *Oncogene* **2008**, *27*, 721–731. [CrossRef] [PubMed]
32. San Miguel, J.F.; Weisel, K.C.; Song, K.W.; Delforge, M.; Karlin, L.; Goldschmidt, H.; Moreau, P.; Banos, A.; Oriol, A.; Garderet, L.; et al. Impact of Prior Treatment and Depth of Response on Survival in MM-003, a Randomized Phase 3 Study Comparing Pomalidomide plus Low-Dose Dexamethasone versus High-Dose Dexamethasone in Relapsed/Refractory Multiple Myeloma. *Haematologica* **2015**, *100*, 1334–1339. [CrossRef] [PubMed]
33. Tabchi, S.; Nair, R.; Kunacheeva, C.; Patel, K.; Lee, H.; Thomas, S.; Amini, B.; Ahmed, S.; Mehta, R.; Bashir, Q.; et al. Retrospective Review of the Use of High-Dose Cyclophosphamide, Bortezomib, Doxorubicin, and Dexamethasone for the Treatment of Multiple Myeloma and Plasma Cell Leukemia. *Clin. Lymphoma Myeloma Leuk.* **2019**, *19*, 560–569. [CrossRef] [PubMed]
34. Cao, Y.; Qiu, G.Q.; Wu, H.Q.; Wang, Z.L.; Lin, Y.; Wu, W.; Xie, X.B.; Gu, W.Y. Decitabine Enhances Bortezomib Treatment in RPMI 8226 Multiple Myeloma Cells. *Mol. Med. Rep.* **2016**, *14*, 3469–3475. [CrossRef]
35. Ramakrishnan, V.; Mager, D.E. Network-Based Analysis of Bortezomib Pharmacodynamic Heterogeneity in Multiple Myeloma Cells. *J. Pharmacol. Exp. Ther.* **2018**, *365*, 734–751. [CrossRef]
36. Ribatti, D. The Discovery of Plasma Cells: An Historical Note. *Immunol. Lett.* **2017**, *188*, 64–67. [CrossRef]
37. Urdeitx, P.; Clara-Trujillo, S.; Gómez Ribelles, J.L.; Doweidar, M.H. Computational Modeling of Multiple Myeloma Growth and Tumor Aggregate Formation. *Comput. Methods Programs Biomed. Updat.* **2022**, *in press*. [CrossRef]
38. Kirshner, J.; Thulien, K.J.; Martin, L.D.; Marun, C.D.; Reiman, T.; Belch, A.R.; Pilarski, L.M. A Unique Three-Dimensional Model for Evaluating the Impact of Therapy on Multiple Myeloma. *Blood* **2008**, *112*, 2935–2945. [CrossRef]
39. Belloni, D.; Heltai, S.; Ponzoni, M.; Villa, A.; Vergani, B.; Pecciarini, L.; Marcatti, M.; Girlanda, S.; Tonon, G.; Ciceri, F.; et al. Modeling Multiple Myeloma-Bone Marrow Interactions and Response to Drugs in a 3d Surrogate Microenvironment. *Haematologica* **2018**, *103*, 707–716. [CrossRef]
40. Zdzisińska, B.; Roliński, J.; Piersiak, T.; Kandefer-Szerszeń, M. A Comparison of Cytokine Production in 2-Dimensional and 3-Dimensional Cultures of Bone Marrow Stromal Cells of Muliple Myeloma Patients in Response to RPMI8226 Myeloma Cells. *Folia Histochem. Cytobiol.* **2009**, *47*, 69–74. [CrossRef]
41. Spicer, C.D.; Pashuck, E.T.; Stevens, M.M. Achieving Controlled Biomolecule-Biomaterial Conjugation. *Chem. Rev.* **2018**, *118*, 7702–7743. [CrossRef]

biomedicines MDPI

Article

Intravenous Infusion of High Dose Selenite in End-Stage Cancer Patients: Analysis of Systemic Exposure to Selenite and Seleno-Metabolites

Olof Breuer [1], Ola Brodin [2,3], Ali Razaghi [2], David Brodin [4], Bente Gammelgaard [5,*] and Mikael Björnstedt [2,*]

1 Department of Laboratory Medicine, Division of Clinical Pharmacology, Karolinska Institutet, Karolinska University Hospital-Huddinge, SE-14186 Stockholm, Sweden
2 Department of Laboratory Medicine, Division of Pathology, Karolinska Institutet, Karolinska University Hospital-Huddinge, SE-14186 Stockholm, Sweden
3 Tema Cancer, Karolinska University Hospital-Solna, SE-14186, Stockholm, Sweden
4 Department of Biosciences and Nutrition, Karolinska Institutet, SE-14183 Huddinge, Sweden
5 Department of Pharmacy, Faculty of Health and Medical Sciences, University of Copenhagen, Universitetsparken 2, 2100 Copenhagen, Denmark
* Correspondence: bente.gammelgaard@sund.ku.dk (B.G.); mikael.bjornstedt@ki.se (M.B.)

Abstract: Cancer is one of the main causes of human death globally and novel chemotherapeutics are desperately required. As a simple selenium oxide, selenite is a very promising chemotherapeutic because of pronounced its dose-dependent tumor-specific cytotoxicity. We previously published a first-in-man systematic phase I clinical trial in patients with cancer (from IV to end-stage) (the SECAR trial) showing that selenite is safe and tolerable with an unexpectable high maximum tolerated dose (MTD) and short half-life. In the present study, we analyzed the selenium species in plasma samples, from the patients participating in the SECAR trial and from various time points and dose cohorts using LC-ICP-MS. In conclusion, selenite, selenosugars, and 1–2 unidentified peaks that did not correspond to any standard, herein denoted ui-selenium, were detected in the plasma. However, trimethylated selenium (trimethylselenonoium) was not detected. The unidentified ui-selenium was eluting close to the selenium-containing amino acids (selenomethionine and selenocysteine) but was not part of a protein fraction. Our data demonstrate that the major metabolite detected was selenosugar. Furthermore, the identification of selenite even long after the administration is remarkable and unexpected. The kinetic analysis did not support that dosing per the body surface area would reduce interindividual variability of the systemic exposure in terms of trough concentrations.

Keywords: selenotherapy; selenosugar; cancer; selenite; selenium

1. Introduction

Selenium is an antioxidant at nutritional levels; many publications report growth-inhibitory and cytotoxic effects of redox-active selenium compounds in from moderate to high concentrations against cancer [1–3]. Selenite, the most widely studied selenium species, is an inorganic selenium salt that is highly reactive and metabolized already in the gastric mucosa cells due to a reaction with reduced glutathione [3]. Furthermore, selenite is an excellent substrate for mammalian Thioredoxin reductase (TrxR) with Km, Vmax, and Kcat numbers close to the natural substrate Thioredoxin (Trx). Under anaerobic conditions, the reaction between selenite and TrxR stops after the consumption of three molecules of NADPH, showing that selenite is reduced to the highly reactive selenide. As soon as oxygen is admitted to the system, a linear reaction is reassumed. Selenide efficiently redox cycles with oxygen, producing reactive oxygen species (ROS) until the system is exhausted of thiols and/or NADPH [3,4]. The non-stoichiometric production of ROS and the resulting oxidative stress is a major mechanism of the growth inhibitory and antineoplastic effects of selenite, thus founding the basis for the application of selenite as a chemotherapeutic agent against cancer [3,5]. For instance, many publications show that tumor cells are more sensitive compared to normal cells [6–8]. Additionally, selenite is more efficiently assimilated and accumulated in highly drug-resistant neoplastic cells [1,4]. The higher sensitivity together with the more efficient uptake by tumor cells is explained by the resistant phenotype expressed by cancer cells of which a characteristic feature is a higher level of thiols intracellularly, increased reducing the extracellular environment, and efficient mechanisms to withhold a reducing extracellular milieu. Of particular interest are the MRP family of proteins and the Cystine/Glutamate (Xct) antiporter, both known to convey drug resistance [3,7]. We have previously shown that these factors potently increase selenite uptake using extracellular reduction mediated by cysteine that is constantly supplied in the extracellular space through the Xct-antiporter and MRPs [9].

Selenide is crucial for the specific incorporation of selenocysteine into selenoproteins. Furthermore, selenide may undergo methylation to methylselenol and dimethyldiselenide, which is volatile and causes the well-known signs of selenium overload, garlic odor in breath, and trimethylselenonium that is excreted in urine. The leading urinary excretory metabolite, however, is the selenosugar, *Se*-methylseleno-*N*-acetylgalactosamine (SeGal) [3].

Due to the high reactivity with thiols, selenite may not be detected in the plasma after oral administration. To study the pharmacology of selenite and monitor any effect, intravenous administration is required. Previously, selenite has not been identified in plasma. In contrast, we reported that selenite could be detected in plasma using LC-ICP-MS after intravenous infusions [10].

In an early clinical study on high doses per oral sodium selenite as an adjunct to radiotherapy in prostate cancer, one conclusion was that pharmacokinetic studies are important to find optimal treatment schedules [11]. In 2015, we published the first-in-man phase I clinical trial in patients with end-stage malignancies. The tolerance was unexpectedly high with a maximal tolerable dose of 10.2 mg/m^2 body surface. However, the half-life was short (18.5 h) and there were no signs of selenosis. Antineoplastic effects have been reported to peak after 48 h of exposure explaining why the observed effects were relatively sparse [12].

To study the metabolism of high-dose selenite in cancer patients, we analyzed the plasma samples from various time points and dose cohorts using LC-ICP-MS and studied the kinetic parameters. The purpose of the present study was thus to explore which metabolites appeared after a high dose of intravenous selenite infusion and how long the active drug selenite could be traced in the plasma samples. In addition, the relationships between systemic exposures and tumor growth inhibition were explored.

2. Materials and Methods

2.1. Clinical Samples

The clinical samples were obtained from the previously published SECAR study (Table 1) [12]. Three patients were included in each cohort except the highest which included 6 patients. In total, 10 cohorts were needed before the maximum tolerable dose could be defined. The starting dose was 0.5 mg/m^2 body surface and, for the first 4 cohorts, the dose was escalated by 0.5 mg/m^2 body surface. The plasma samples were extracted before the start of selenite treatment. After finishing the fourth cohort without any dose limiting toxicity an amendment to the Ethical Committee and the Medical Product Agency was approved (see Institutional Review Board Statement) allowing dose escalations of 50% between cohorts and the inclusion of other malignancies except lung carcinoma. The protocol for the first four cohorts included two periods of 10 days of infusion separated by a week of rest. However, from the fifth cohort onwards, the third and fourth treatment weeks and the treatment-free week were omitted and thus the patients received only 10 treatments. Dosing was not performed during weekends. Both groups were followed by chemotherapy of the same formulation that the patients before inclusion developed resistance to. The evaluation of the systemic exposure to the parent compound selenite and seleno-metabolites following the first 2 weeks of dosing at dose levels of from 1.0 up to

15.3 mg/m^2 is reported here. The tumor status according to RECIST was measured using a CT-scan just before and after the selenite treatment. The elapsed time between CT-scans was approximately 7 weeks in the first 13 patients and 4 weeks for the remaining patients.

Table 1. Characteristics of patients. Total dose calculated according to body surface area. ASAT (aspartate-aminotransferase) with normal value (76≤). Plasma-creatinine with normal value (100≤). **Abbreviations:** Met: metastatic disease and L: locally advanced disease.

Age	Tumor Histology	Dose/m^2 (mg/m^2)	Total Dose (mg)	Stage	ASAT (μkat/L)	P-Creatinine (μM)
62	Squamous cell carcinoma lung	1	1.7	L	0.42	61
73	Adenocarcinoma, lung	1	2.1	Met	0.46	106
76	Squamous cell carcinoma, lung	1	1.8	Met	0.29	52
58	Non-small cell lung cancer	1.5	2.5	Met	1.49	65
58	Squamous cell, lung cancer	1.5	3.1	Met	0.51	38
37	Adenocarcinoma, lung	1.5	3	Met	0.40	65
67	Small-cell lung cancer	2	4.6	Met	0.57	84
62	Large cell lung cancer	2	3.4	Met	0.30	51
75	Adenocarcinoma, lung	3	4.8	L	0.38	87
45	Adenocarcinoma, lung	3	5	Met	0.22	67
59	Adenocarcinoma, lung	3	6.1	Met	0.60	81
65	Squamous cell carcinoma, lung	4.5	9.8	L	1.10	65
60	Adenocarcinoma, lung	4.5	7.4	Met	0.81	54
65	Adenocarcinoma, lung	4.5	6.2	Met	0.32	37
46	Colon carcinoma	6.8	11.5	Met	0.66	54
61	Colon carcinoma	10.2	22.5	Met	1.21	85
62	Non-small cell lung cancer	10.2	17.6	Met	0.37	61
53	Colon carcinoma	15.3	29	Met	1.85	58
65	Malignant mesothelioma	15.3	32.6	L	0.25	70
41	Adenocarcinoma, lung	12.8	23.9	Met	0.45	58

2.2. Sampling Schedule

The dosing regimen was modified during the study such that doses of 3 mg/m^2 and higher were only administered during two consecutive 7-day cycles, each consisting of 5 consecutive days of daily selenite infusions. Hence, to evaluate the dose-systemic exposure relationship across the entire dose range of the study, three sampling occasions were selected that were considered comparable between all applied dosing regimens; baseline (before the 1st dose was administered), trough value at predose Day 5 (Week 1, 96 h after start), and trough value at predose Day 12 (Week 2, 264 h after the start).

2.3. Seleno-Metabolite Analysis

The plasma samples (200 μL) were transferred to Vivaspin centrifugal filter units (Sartorius) with semi-permeable membranes of molecular weight cut-off 3000 Da, centrifuged at 14,000× *g* for 30 min, and the filtrate was then transferred to LC vials and analyzed using LC-ICPMS. Isocratic chromatography was performed on a Gemini C18 column (250 × 2 mm), with 5 μm (Phenomenex) in a mobile phase consisting of 200 mM ammonium acetate in 5% methanol, pH (6.7) with a flow rate of 0.2 mL/min, and an injection volume of 10 μL. The LC-system (Agilent 1100 System) was hyphenated to a Sciex ELAN DRCe ICPMS system (Perkin Elmer) and the isotopes ^{77}Se, ^{80}Se, and ^{82}Se were monitored. The quantitation was performed using post-column isotope dilution analysis. An enriched ^{77}Se solution in the mobile phase was continuously introduced and the post-column with a syringe pump connected to the eluent flow of the LC and the three Se isotopes were monitored. The isotope ratios were calculated, mass flow chromatograms were obtained after the application of the isotope dilution equation to each data point, and peak areas were calculated using the OriginPro software. The linearity was established for selenite and selenosugar and the precision was better than 9% in the plasma samples. Based on a certified selenite standard, the accuracy was 99.5% and LOQ was 1.3 μg/L. All the details are previously described [10].

3. Results

3.1. Baseline Levels of Seleno-Metabolites in Plasma

Only selenite and total selenium were consistently quantifiable at baseline with a LOQ of 1.3 µg/L. The baseline plasma concentrations of selenite, selenosugar, total selenium, and ui-selenium are listed in Table 2.

Table 2. Mean trough concentrations of selenite and seleno-metabolites in plasma at baseline and 24 h after dose following four consecutive daily intravenous infusions of selenite during Week 1 and Week 2. Mean, SD, and range are tabulated across the entire dose range of the study.

Analyte	Time (week)	Mean Conc. (µg/L)	N	SD	Relative SD	Range
Selenite	Baseline	2.34	20	1.31	0.57	<LOQ-6.00
	1	35.9	20	56	1.56	2.40–209
	2	23.4	19	29.6	1.27	2.20–111
Selenosugar	Baseline	-	0	-	-	All < LOQ
	1	18.5	19	15	0.81	<LOQ-51.5
	2	20.6	19	17.8	0.86	2.0–67
Total selenium	Baseline	68.4	21	25.1	0.37	32–140
	1	488	20	399	0.82	119–1425
	2	516	19	430	0.83	125–1774
ui-selenium	Baseline	-	0	-	-	All < LOQ
	1	3.77	15	2.04	0.54	<LOQ-7.5
	2	4.05	14	2.03	0.45	<LOQ-8.2

3.2. Dose-Plasma Concentration of Seleno Metabolites

The samples at Week 1 or Week 2 contained selenite, selenosugar, and in some cases small amounts of one–two unidentified selenium compounds (ui-selenium). These unidentified compounds were eluting close to the selenium-containing amino acids (selenomethionine and selenocysteine) but were not co-eluting with any available standard. As the samples were exposed to ultrafiltration before analysis, the unknown species were not proteins.

Selenosugar and ui-selenium became quantifiable in the trough plasma samples upon infusion of selenite for 4 consecutive days (Table 2). In both Week 1 and 2, the statistically significant relationships were seen between the amount of selenite administered per occasion to each patient and the trough plasma concentrations for selenite (Pearson's correlation coefficient $r = 0.80$, $p < 0.001$, Week 2), ui-selenium ($r = 0.85$, $p < 0.001$, Week 2), total selenium ($r = 0.93$, $p < 0.001$, Week 2), and selenosugar ($r = 0.94$, $p < 0.001$, week 2) (Figure 1). Consistent with this result, the normalization of the plasma concentration of these compounds with respect to the dose decreased the variability in terms of the relative SD for selenite, total selenium, and selenosugar (Tables 2 and 3 and Figure 2), but less so for ui-selenium. In addition, there was only a minor fluctuation between Week 1 and Week 2 in the dose-normalized plasma concentrations indicating that a pharmacokinetic steady-state had been reached within 5 days after the start of treatment (Figure 2).

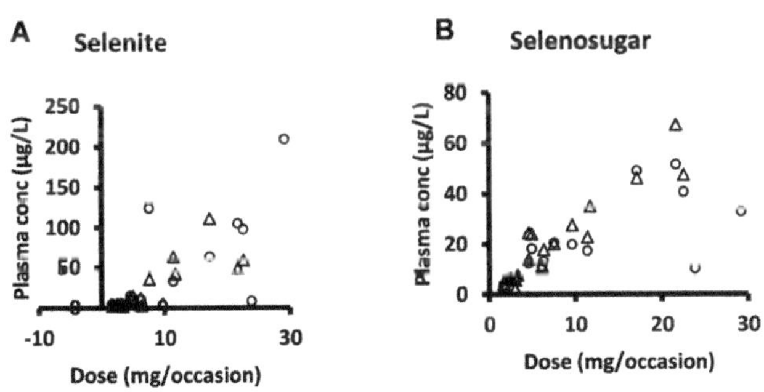

Figure 1. *Cont.*

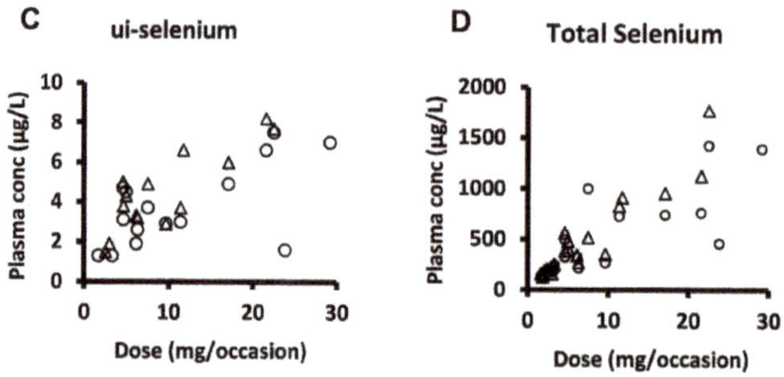

Figure 1. Plasma concentration-dose relationship for selenite and seleno-metabolites at 24 h after dose (i.e., trough value) following four consecutive daily intravenous infusions of selenite during Week 1 (circles) and Week 2 (triangles). (**A**) selenite, (**B**) selenosugar, (**C**) ui-selenium, (**D**), total selenium.

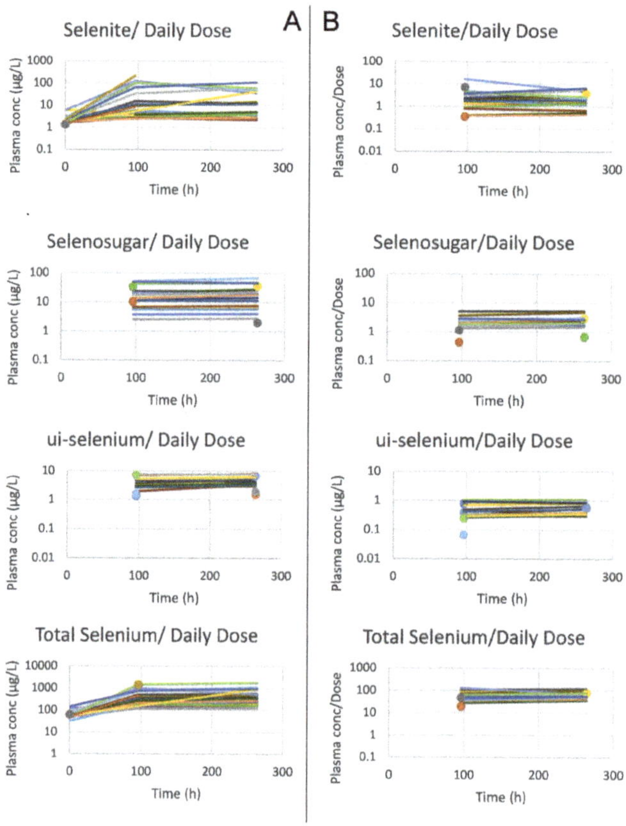

Figure 2. Mean trough concentrations per subject of selenite, selenosugar, ui-selenium and total selenium in plasma (from upper to lower panels), and before (**A**) and after (**B**) dose normalization. The samples were extracted at baseline and 24 h after dose (i.e., trough value) following four consecutive daily intravenous infusions of selenite during Week 1 and Week 2. Single observations are marked with a circle (Week 1) and triangle (Week 2).

Table 3. Dose normalized mean trough concentrations of selenite and seleno-metabolites in plasma before the 5th and 10th treatments, 24 h after respective previous infusion. Normalization was performed with respect to the amount administered on each dosing occasion per individual. Dose normalized mean, SD, and range are tabulated across the entire dose range of the study.

Analyte	Time (week)	Mean Conc/Dose (μg/L)	N	SD	Relative SD	Range
Selenite	1	3.03	20	3.58	1.18	0.35–16.4
	2	2.45	19	1.65	0.68	0.44–6.48
Selenosugar	1	2.25	19	1.00	0.44	<LOQ-5.09
	2	2.63	19	1.05	0.40	0.66–5.28
Total selenium	1	63.5	20	26.7	0.42	19.3–132
	2	71.6	19	20.3	0.28	36.7–122
ui-selenium	1	0.44	15	0.26	0.60	<LOQ-1.017
	2	0.56	14	0.23	0.41	<LOQ-1.082

3.3. Systemic Exposure of Selenite in Relation to Patient-Specific Variables

The dose levels were chosen and escalation increments were performed in relation to body surface area. However, there was no apparent relationship between the plasma concentrations normalized to the amount of selenite administered per occasion and body surface area or body weight for selenite, selenosugar, total selenium, or ui-selenium (Figure 3).

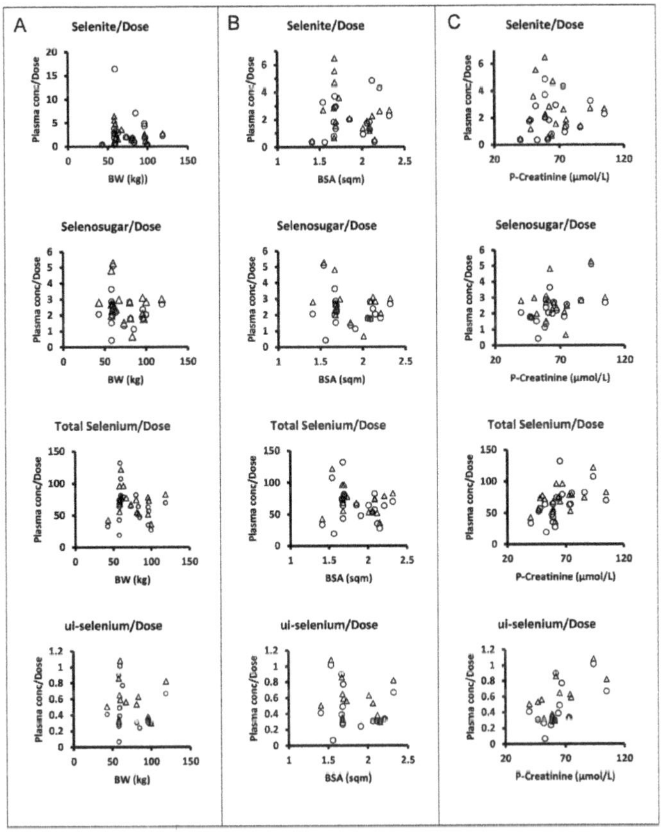

Figure 3. Dose normalized mean trough concentrations of selenite, selenosugar, total selenium, and ui-selenium in plasma (from upper to lower panels) at 24 h after dose following four consecutive daily intravenous infusions of selenite during Week 1 (circles) and Week 2 (triangles) versus individual body weight (**A**), body surface area (**B**), and plasma creatinine (**C**). Normalization was performed with respect to the amount of selenite administered per patient on each dosing occasion.

3.4. Selenite Concentration and Tumor-Growth Inhibition

A question of clinical interest is if higher plasma concentrations of selenite might increase the responses of carcinomas. An indication of a relationship between higher plasma concentration and less cancer growth and even shrinkage was found, depending on the cut-off level of growth inhibition according to RECIST evaluation before and after the selenite treatment (Figure 4). The tumor population was divided into two groups (low-RECIST and high-RECIST) with a cut-off value of 1 mm tumor growth (RECIST). The cut-off limit, 1 mm, was chosen since it might be considered to be clinically relevant.

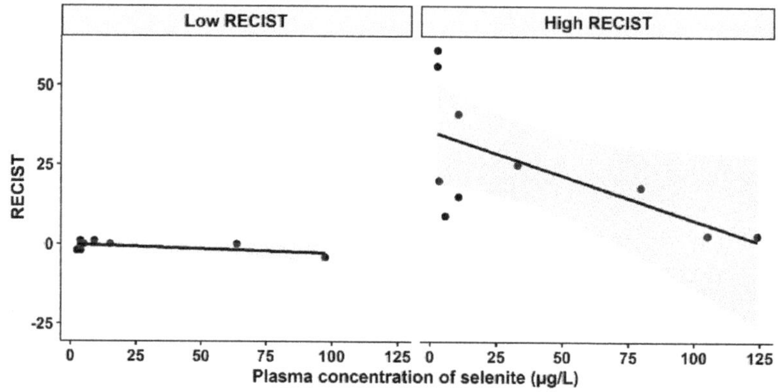

Figure 4. Relationship between RECIST and systemic exposure to selenite. Left panel: Low RECIST-group (RECIST $\leq$ 1 mm), Pearson two-sided correlation coefficient, r = −0.5467, p = 0.1609. Right panel: High RECIST group (RECIST > 1 mm), r = −0.6239, p = 0.05391. The shadowing in grey represents the confidence interval.

4. Discussion

Cancer is one of the main causes of human death globally and, despite great progress in the prognosis of some malignancies, the prognosis remains very poor for highly resistant visceral malignancies. Hence, novel chemotherapeutics are desperately required. Redox-active selenium compounds, including selenite, have remarkable tumor-specific cytotoxic properties [12]. This effect is particularly pronounced in cells that have developed resistance to cytostatic drugs. Resistant tumor cells accumulate selenium due to the facilitated uptake by transporters conferring drug resistance, including the Xct-antiporter and the MRP-superfamily of proteins [9]. Selenium has, despite this, a great body of preclinical data not yet introduced to the routine care of cancer patients due to the lack of human clinical trials. In 2015, we published a first-in-man systematic phase I clinical trial (the SECAR study) in patients with cancer (from IV to end-stage) showing that sodium selenite is safe and tolerable if 10.2 mg.m^2 MTD or less is administered [12]. The metabolic faith of selenite in humans after repeated high-dose iv infusions has so far been unknown. The available data originate at large from animal studies and not from human subjects.

After the reduction in selenite enzymatically by redox enzymes or chemically by low molecular or protein-bound thiols, the anionic highly reactive species selenide is formed [4]. Selenide is readily absorbed by cells and the extracellular reduction in selenite facilitates the selenium uptake. The reduced extracellular environment surrounding highly resistant cells, as a characteristic of the resistant phenotype, may by this mechanism explain the selective and marked tumor-specific cytotoxicity of selenite [9]. Selenide is assumed to be not only the common intermediate for all Se nutritional sources but also the checkpoint metabolite for Se utilization or excretion and, not least, methylation reactions. Se absorbed by the body within the normal nutritional Se intake range is suggested to be mostly excreted in urine as selenosugar, 1-methylseleno-N-acetyl-d-galactosamine, in mammals. Selenosugar was the main metabolite in our study, although high supranutritional levels were used.

An excessive amount of selenium is methylated to mono-di and trimethylated forms and the leading methylated urinary species is trimethylselenonium [13]. Trimethylselenonium increases with excessive selenium intake or body burden, suggesting that trimethylselenonium may serve as a urinary biomarker for both excessive selenium intake and body burden as well as a toxic dose of selenium [13]. However, most studies concerning the formation of trimethylselenonium have been performed using oral regimens. In addition, excess selenium intake may lead to the synthesis of several methylated species including volatile species, e.g., Dimethyldiselenide. The methylated volatile species will cause a distinct garlic odor in the breath. In the SECAR trial, the research nurses noted a garlic odor in the breath of a few patients. This observation indicates that volatile forms were synthesized and that one excretion pathway was from the breath in addition to urine. Despite high levels of selenite administered, trimethylselenonium was not detected in any of the cohorts investigated.

Selenoprotein P (SELENOP) is a proven biomarker of Se status [14]. Following cleavage of the signal peptide, SELENOP is mostly produced in the liver and released into the plasma. SELENOP performs two distinct tasks: it has GPX-like activity to reduce phospholipid hydroperoxide and Se transport activity to deliver Se to cells. In addition to maintaining selenoenzymes in various tissues, SELENOP is essential for Se metabolism and antioxidative defense [15]. We already hypothesized that SELENOP may reflect selenium levels in therapeutic dosages of selenite for clinical applications. For this purpose, blood samples from the same patients in the SECAR clinical trial phase I were used. The total Se was quantified using spectroscopy and SELENOP was validated using ELISA. SELENOP was increased in high-dosage selenite infusions. Thus, we concluded that circulating SELENOP is a suitable biomarker for therapeutic applications of selenite in upper intake levels [16]. Following our previous result of SELENOP, therefore, we can hypothesize that exceeding levels of selenite have been invested for SELENOP production.

Our data do not support that dosing per body surface area would reduce the interindividual variability of systemic exposure in terms of trough concentrations, since the normalization of the parent compound or its seleno-metabolites to the amount of selenite administered per dosing occasion did not show any relationship with the body surface area. However, one can discern statistically significant correlations between the plasma creatinine concentration and dose-normalized total selenium (r = 0.52, $p < 0.05$, Week 1; r = 0.53, $p < 0.05$, Week 2), ui-selenium (r = 0.61, $p < 0.05$, Week 1; r = 0.60, $p < 0.05$, Week 2), and selenosugar (r = 0.58, $p < 0.01$, Week 1; not significant Week 2). This might support that renal excretion is a contributing path to the elimination of selenium-containing compounds [17].

The lack of a strong correlation between higher plasma concentrations of selenite and cancer growth inhibition might have different explanations. The study was not designed, or dimensioned, to prove any efficacy. The carcinomas were clinically very advanced and heterogenous with both fast- and slow-growing tumors of different histology. Furthermore, this was a phase I trial in which the toxicity, safety, and kinetics were the primary and secondary endpoints, whereas the pharmacodynamic response was only included as an exploratory endpoint.

Even if the production of free oxygen radicals (ROS) is the primary mechanism for cancer cell killing, selenite also changes the redox balance of cells, which has an impact on many cellular functions such as proliferation, apoptosis, and the expression and function of membrane pumps [3]. It has also been demonstrated that high-dose selenite can inhibit the expression of cancer-stimulating genes and decrease the expression of different membrane receptors such as androgen receptors in prostate carcinoma [18] and exert immune-activation effects [1] even though the effect on soluble PDL-1 is unclear [19]. Concerning ROS production in vitro, the studies demonstrate that higher concentrations of selenite increase the effect [3] and it is reasonable to suspect that the same is also true in the clinical situation.

The concentration of selenite has been proven to be critical for effect and the dose may vary depending on the cell type [4]. Low concentrations might even stimulate tumor growth as demonstrated by several studies. Selenium is an essential part of cell culture media for serum-free conditions [4]. Most cancer cells are sensitive to from moderate to high concentrations in the span of 5–15 μM [3,4]. In fact, 15–30 μM proved to be exceptionally efficient in an ex vivo tissue culture of human pancreatic ductal adenocarcinoma with a marked specificity to tumor cell cytotoxicity and had a limited or no effect on ambient tissues [20]. The exposure time is also important and, in laboratory experiments, the maximum effect is usually observed after 48 h of exposure—a factor that might be one explanation for the lack of tumor response in the SECAR trial.

According to Figure 4, there might be two populations of tumor growth responses, whereof one consists of either slowly growing tumors or tumor growth inhibited by other means, i.e., immune activation. An example of the latter is the observation in the lower left part of Figure 4, where a patient had a tumor that neither grew nor shrank after the selenite treatment, but later regressed and after half a year had disappeared and never recurred. The tumors in the right panel of Figure 4, on the other hand, might depend on ROS production for growth inhibition. It should be noted that this is a hypothesis and our results in this study do not confirm an association between higher concentration and tumor regression.

The SECAR trial is to our knowledge the first published study where high doses of selenite were administered intravenously to advanced cancer patients. The speciation analysis showing the prolonged presence of selenite together with high tolerance (MTD 10.2 mg/m^2) and short half-life of total selenium demonstrates the potential to safely use iv administration of selenite in patients [12]. Even though this phase I study did not show any significant effect on the tumor burden, the results merit further investigations of the potential of selenium in the treatment of cancer.

5. Conclusions

The major selenium metabolite in plasma after repeated iv administrations of selenite was selenosugar. Furthermore, selenite was detected in plasma even a long time after administration. We conclude that selenite dosing per body surface area would not reduce the interindividual variability of systemic exposure in terms of trough concentrations. However, for future studies and applications of selenite, the MTD must be considered, as an adequate guide for dosing.

Author Contributions: Conceptualization, M.B. and O.B. (Ola Brodin); methodology and selenium determination, B.G.; formal analysis, all authors; writing—original draft preparation, O.B. (Ola Brodin), O.B. (Olof Breuer), A.R., M.B. and B.G.; writing—review and editing, all authors; analysis of the data and visualization, O.B. (Olof Breuer), D.B. and A.R.; funding acquisition, M.B. All authors have read and agreed to the published version of the manuscript.

Funding: This investigation was supported by grants from Cancerfonden (180429 and 201100PjF01H), Cancer och Allergifonden (206), Radiumhemmets Forskningsfonder (171023 and 201063), and Sjöbergstiftelsen to M.B. The funders had no role in the study design, data collection, analysis, the decision to publish, or the preparation of the manuscript.

Institutional Review Board Statement: The study was approved by the Ethical Committee of Stockholm and the Swedish Medical Products Agency (2006/429-31/3) latest EPN approval 2009/1983-32, registered in the EU Clinical Trial Register (Eudra CT Number: 2006-004076-13 and US NCT number NCT01959438). All patients were provided with a consent form to join the study.

Informed Consent Statement: Informed written and oral consent was obtained from all subjects involved in the study.

Acknowledgments: We would like to acknowledge Ms Lena Hernberg for handling clinical samples.

Conflicts of Interest: M.B. is listed as an inventor in a patent application for i.v. use of inorganic selenium in cancer patients and holds shares in SELEQ OY, a company involved in the development of Se-based formulations for prevention and treatment. The other authors declare no conflict of interest.

The funders had no role in the design of the study; in the collection, analyses, or interpretation of data; in the writing of the manuscript; or in the decision to publish the results.

References

1. Razaghi, A.; Poorebrahim, M.; Sarhan, D.; Bjornstedt, M. Selenium stimulates the antitumour immunity: Insights to future research. *Eur. J. Cancer* **2021**, *155*, 256–267. [CrossRef] [PubMed]
2. Razaghi, A.; Zickler, A.M.; Spallholz, J.; Kirsch, G.; Bjornstedt, M. Selenofolate inhibits the proliferation of IGROV1 cancer cells independently from folate receptor alpha. *Heliyon* **2021**, *7*, e07254. [CrossRef] [PubMed]
3. Misra, S.; Boylan, M.; Selvam, A.; Spallholz, J.E.; Bjornstedt, M. Redox-Active Selenium Compounds-From Toxicity and Cell Death to Cancer Treatment. *Nutrients* **2015**, *7*, 3536–3556. [CrossRef] [PubMed]
4. Selvam, A.K.; Szekerczes, T.; Bjornstedt, S.; Razaghi, A.; Bjornstedt, M. Methods for accurate and reproducible studies of pharmacological effects of selenium in cancer. *Methods Enzymol.* **2022**, *662*, 25–62. [CrossRef] [PubMed]
5. Fernandes, A.P.; Wallenberg, M.; Gandin, V.; Misra, S.; Tisato, F.; Marzano, C.; Rigobello, M.P.; Kumar, S.; Bjornstedt, M. Methylselenol formed by spontaneous methylation of selenide is a superior selenium substrate to the thioredoxin and glutaredoxin systems. *PLoS ONE* **2012**, *7*, e50727. [CrossRef] [PubMed]
6. Wallenberg, M.; Misra, S.; Wasik, A.M.; Marzano, C.; Bjornstedt, M.; Gandin, V.; Fernandes, A.P. Selenium induces a multi-targeted cell death process in addition to ROS formation. *J. Cell. Mol. Med.* **2014**, *18*, 671–684. [CrossRef] [PubMed]
7. Wallenberg, M.; Misra, S.; Bjornstedt, M. Selenium cytotoxicity in cancer. *Basic Clin. Pharmacol. Toxicol.* **2014**, *114*, 377–386. [CrossRef] [PubMed]
8. Nilsonne, G.; Olm, E.; Szulkin, A.; Mundt, F.; Stein, A.; Kocic, B.; Rundlof, A.K.; Fernandes, A.P.; Bjornstedt, M.; Dobra, K. Phenotype-dependent apoptosis signalling in mesothelioma cells after selenite exposure. *J. Exp. Clin. Cancer Res.* **2009**, *28*, 92. [CrossRef] [PubMed]
9. Olm, E.; Fernandes, A.P.; Hebert, C.; Rundlöf, A.-K.; Larsen, E.H.; Danielsson, O.; Björnstedt, M. Extracellular thiol-assisted selenium uptake dependent on the xc cystine transporter explains the cancer-specific cytotoxicity of selenite. *Proc. Natl. Acad. Sci USA* **2009**, *106*, 11400–11405. [CrossRef] [PubMed]
10. Flouda, K.; Dersch, J.M.; Gabel-Jensen, C.; Sturup, S.; Misra, S.; Bjornstedt, M.; Gammelgaard, B. Quantification of low molecular weight selenium metabolites in human plasma after treatment with selenite in pharmacological doses by LC-ICP-MS. *Anal. Bioanal. Chem.* **2016**, *408*, 2293–2301. [CrossRef] [PubMed]
11. Jayachandran, P.; Knox, S.J.; Garcia-Cremades, M.; Savic, R.M. Clinical Pharmacokinetics of Oral Sodium Selenite and Dosing Implications in the Treatment of Patients with Metastatic Cancer. *Drugs RD* **2021**, *21*, 169–178. [CrossRef]
12. Brodin, O.; Eksborg, S.; Wallenberg, M.; Asker-Hagelberg, C.; Larsen, E.H.; Mohlkert, D.; Lenneby-Helleday, C.; Jacobsson, H.; Linder, S.; Misra, S.; et al. Pharmacokinetics and Toxicity of Sodium Selenite in the Treatment of Patients with Carcinoma in a Phase I Clinical Trial: The SECAR Study. *Nutrients* **2015**, *7*, 4978–4994. [CrossRef] [PubMed]
13. Suzuki, K.T.; Kurasaki, K.; Okazaki, N.; Ogra, Y. Selenosugar and trimethylselenonium among urinary Se metabolites: Dose- and age-related changes. *Toxicol. Appl. Pharmacol.* **2005**, *206*, 1–8. [CrossRef] [PubMed]
14. Combs, F.; Gerald, J. Biomarkers of Selenium Status. *Nutrients* **2015**, *7*, 2209–2236. [CrossRef] [PubMed]
15. Saito, Y. Selenium Transport Mechanism via Selenoprotein P-Its Physiological Role and Related Diseases. *Front. Nutr.* **2021**, *8*, 685517. [CrossRef] [PubMed]
16. Brodin, O.; Hackler, J.; Misra, S.; Wendt, S.; Sun, Q.; Laaf, E.; Stoppe, C.; Bjornstedt, M.; Schomburg, L. Selenoprotein P as Biomarker of Selenium Status in Clinical Trials with Therapeutic Dosages of Selenite. *Nutrients* **2020**, *12*, 1067. [CrossRef] [PubMed]
17. Kobayashi, Y.; Ogra, Y.; Ishiwata, K.; Takayama, H.; Aimi, N.; Suzuki, K.T. Selenosugars are key and urinary metabolites for selenium excretion within the required to low-toxic range. *Proc. Natl. Acad. Sci. USA* **2002**, *99*, 15932–15936. [CrossRef] [PubMed]
18. Husbeck, B.; Bhattacharyya, R.S.; Feldman, D.; Knox, S.J. Inhibition of androgen receptor signaling by selenite and methylseleninic acid in prostate cancer cells: Two distinct mechanisms of action. *Mol. Cancer Ther.* **2006**, *5*, 2078–2085. [CrossRef] [PubMed]
19. Razaghi, A.; Mansouri, L.; Brodin, O.; Bjornstedt, M.; Lundahl, J. Soluble PD-L1 Expression After Intravenous Treatment of Cancer Patients With Selenite in Phase I Clinical Trial. *Front. Oncol.* **2022**, *12*, 906134. [CrossRef] [PubMed]
20. Moro, C.F.; Selvam, A.K.; Ghaderi, M.; Pimenoff, V.N.; Gerling, M.; Bozoky, B.; Elduayen, S.P.; Dillner, J.; Bjornstedt, M. Drug-induced tumor-specific cytotoxicity in a whole tissue ex vivo model of human pancreatic ductal adenocarcinoma. *Front. Oncol.* **2022**, *12*, 965182. [CrossRef] [PubMed]

MDPI AG
Grosspeteranlage 5
4052 Basel
Switzerland
Tel.: +41 61 683 77 34

Biomedicines Editorial Office
E-mail: biomedicines@mdpi.com
www.mdpi.com/journal/biomedicines

www.ingramcontent.com/pod-product-compliance
Lightning Source LLC
Chambersburg PA
CBHW061604210926
51850CB00009B/2210
9783725852499